Fluvial Hydraulics Affected by River Ice and Hydraulic Structures

Fluvial Hydraulics Affected by River Ice and Hydraulic Structures

Editor

Jueyi Sui

MDPI • Basel • Beijing • Wuhan • Barcelona • Belgrade • Manchester • Tokyo • Cluj • Tianjin

Editor
Jueyi Sui
School of Engineering
University of Northern
British Columbia
Prince George
Canada

Editorial Office
MDPI
St. Alban-Anlage 66
4052 Basel, Switzerland

This is a reprint of articles from the Special Issue published online in the open access journal *Water* (ISSN 2073-4441) (available at: www.mdpi.com/journal/water/special_issues/river_ice_hydraulics).

For citation purposes, cite each article independently as indicated on the article page online and as indicated below:

LastName, A.A.; LastName, B.B.; LastName, C.C. Article Title. *Journal Name* **Year**, *Volume Number*, Page Range.

ISBN 978-3-0365-7311-3 (Hbk)
ISBN 978-3-0365-7310-6 (PDF)

Contents

About the Editor

Jueyi Sui

Dr. Jueyi Sui is a Professor at the School of Engineering, University of Northern British Columbia, Prince George, Canada. Jueyi received his Bachelor's and Master's degrees from Hefei University of Technology, China, in 1985 and 1988, respectively. He earned his Dr.-Ing. degree in water resources engineering from Kaiserslautern University in Germany. His research interests focus on fluvial hydraulics in the presence of in-stream infrastructures and vegetation, the interaction between ice accumulation and channel bed deformation, and snow hydrology. He has published over 100 papers in peer-reviewed international journals. Currently, he serves as an associate editor for "International Journal of Sediment Research" and "Journal of Hydrology and Hydromechanics"; and acts as one of members of the editorial board for "Journal of Hydrodynamics" and "Water".

water

MDPI

Editorial

Fluvial Hydraulics Affected by River Ice and Hydraulic Structures

Jueyi Sui

School of Engineering, University of Northern British Columbia, Prince George, BC V2N 4Z9, Canada; jueyi.sui@unbc.ca

1. Introduction

Water on earth moves from one place to another by way of hydrologic processes such as precipitation, runoff, infiltration, evapotranspiration, melting, and ground-water flow. When water generated from a watershed enters a river or stream, flow conditions in the channel depend on the features of the river such as the channel slope, the properties of the bed material, and obstructions or vegetation in the channel.

Fluvial hydraulics deals with the flow of water, motion of sediment, and erosion of channel beds in rivers. Fluvial hydraulics becomes more complex as the velocity, slope, depth, and channel roughness are all subject to changes resulting from erodible beds and sediment transport due to the presence of in-stream infrastructure, vegetation in channel beds/banks, and ice cover on the water surface.

(1) *In-stream infrastructure*. The presence of in-stream structures leads to changes in the flow condition, and thus local scour around the in-stream structures. Local scour is sediment movement caused by a drastic change in flow obstructed by in-stream structures. Local scour is one of the leading causes of in-stream structure damage, and researchers have extensively investigated this issue based on flume tests, on-site measured data analysis, and numerical simulations. According to Brandimarte et al. [1], bridge damage and failure due to local scour in the vicinity of bridge piers have huge negative social and economic impacts in terms of reconstruction, the maintenance of existing structures, disruptions to traffic flow, and cost in human lives. It has been estimated that 60% of bridge failure cases in the USA are due to scour, and approximately 50 to 60 bridges annually collapse in the USA [2].

(2) *River ice*. In winter, ice cover can last up to six months in some northern regions. River ice is an important factor affecting riverbed deformation in cold regions. Under ice-covered flow conditions, the location of the maximum flow velocity is closer to the riverbed than during open-flow conditions due to the extra boundary imposed by the ice cover, which results in increased flow turbulence and bed shear stress, which typically occurs around bridge piers and abutments [3,4]. Therefore, the riverbed scour depth under ice cover is greater than in the absence of ice [5]. Under ice-covered flow conditions, local scour processes around in-stream infrastructure such as bridge piers may lead to disasters, such as the collapse of a bridge over the White River in Vermont in 1990 due to the deterioration of the bridge foundation [6] or the collapse of the Highway 16 bridge across the Bulkley River at Smithers, British Columbia, in April 1966 due to ice-induced local scour around bridge piers. Most recently, in February 2018, a bridge collapsed in Port Bruce, Ontario. Although the deformation of a riverbed under ice cover has often been a grave concern for engineers, little appears to be known about this subject [7]. Clearly, there is a critical need to study fluvial hydraulics under ice-covered flow conditions.

(3) *Vegetated channels*. In wetland environments, streams are commonly filled with either partially or fully submerged diverse aquatic plants, such as grasses, shrubs, and bushes. Both vegetated and non-vegetated zones in streams have a major impact on the hydrodynamic characteristics of flow, including velocity distributions, turbulence

Citation: Sui, J. Fluvial Hydraulics Affected by River Ice and Hydraulic Structures. *Water* **2023**, *15*, 1262. https://doi.org/10.3390/w15071262

Received: 17 March 2023
Accepted: 21 March 2023
Published: 23 March 2023

intensity, and coherent structures, as well as mass and momentum exchanges. Additionally, vegetation creates ecological habitats and plays an active role in maintaining and protecting biological diversity by providing food and shelter for fish and other aquatic creatures. There is an interaction between vegetation and bed deformation. On the one hand, vegetation influences flow structure, sediment erosion, and deposition. As a result of sediment erosion and deposition, organic materials attached to sediment particles spread throughout riverbeds/banks and affect vegetation growth and spread.

The aim of this Special Issue is to bring together research that improves our knowledge of sediment transport, local scour around in-stream infrastructure, and fluvial processes in the presence of either vegetation in a channel bed or ice cover on the water surface. It aims to include the latest advancements not only in the mechanics of sediment transport and local scour around in-stream infrastructure but also the impacts of vegetation in a channel bed on bed deformation and flow structure. Research dealing with the interaction between river ice, riverbed deformation, and in-stream infrastructure is also included.

2. Overview of This Special Issue

Before the manuscript submission deadline for this Special Issue, we received many manuscripts on fluvial hydraulics in the presence of either ice cover or in-stream infrastructure. All manuscripts have been through the normal peer-review process according to the journal requirements. In this Special Issue, twelve original contributions have been selected for publication [8–19]. Among the topics of interest for this Special Issue are:

- Turbulent flow structure, Reynolds stress, and local scour around in-stream infrastructure under ice-covered flow conditions [8–11];
- Interactions between ice jam accumulation, channel bed deformation, and in-stream infrastructure [12–14];
- The flow structure, Reynolds shear stress, and turbulence intensity of flows in the presence of vegetation in the channel bed [15–17] or under shallow flow conditions [18];
- Numerical simulation of the interaction between flow and sediment [10,19].

To equip readers with more information about this Special Issue book, a variety of innovative results from these twelve articles are summarized below:

Valela et al. [8] conducted experiments in a laboratory flume to better understand the local scour process around bridge piers under both smooth and rough ice-covered flow conditions. They found that an increase in ice cover submergence results in a greater maximum velocity and an associated increase in the near-bed velocity gradient. Rougher ice cover causes the maximum velocity close to the bed, and thus the near-bed velocity gradient is even greater. Rough ice cover generates more scour than smooth ice cover.

The work of Sang et al. [9] is about the local scour around tandem double piers under ice-covered flow conditions. They found that as the pier spacing ratio of tandem double piers increases, the scour depth around the front pier gradually decreases. When the pier spacing ratio is 5, sediment scoured around the front pier begins to deposit between the two piers. The existence of the rear pier leads to an increase in the length of the scour hole but a decrease in the depth of the scour hole around the front pier. When the pier spacing ratio is 9, the scour depth around the rear pier is the lowest. When the pier spacing ratio is more than 17, the scour around the front pier has hardly any influence on that around the rear pier.

Jafari and Sui [10] studied velocity fields and turbulence structures around spur dikes with different alignment angles under ice-covered flow conditions. They found that an increase in the upstream dike angle leads to a larger scour hole. Both the cover roughness coefficient and blockage ratio of a spur dike leads to a further increase in the turbulence kinetic energy and 3D velocity component values. The streamwise velocity contributes much to the turbulence intensity and Reynolds shear stress, leading to the development of scour holes. The lateral velocity component has the highest level of irregularities inside and outside scour holes. The Reynolds shear stress is negative inside scour holes and becomes positive toward the flow surface. It reaches its maximum slightly above scour holes. The negative values of the Reynolds stress are caused by the upward vertical momentum transport generated by a negative velocity gradient.

Zhang et al. [11] developed a theoretical model to describe the vertical distribution of longitudinal velocity, shear stress, and turbulence intensity. A two-power-law function was adopted to predict the vertical profile of velocity. It was found that the location of zero shear stress is not the same as that of the maximum velocity and is closer to the smooth boundary.

The work of Chen et al. [12] is about the waved-shape accumulation of ice jams based on fixed-bed experiments in a laboratory flume. A characteristic curve was developed to assess whether waved-shape ice accumulation occurs. An equation for calculating the ice wavelength was derived using the results of laboratory experiments. The relationship between the migration speed of ice waves and the ratio of ice discharge to the water flow rate were also studied. Furthermore, case studies were conducted with respect to ice accumulation in the St. Lawrence River, the Beauharnois Canal, and the La Grande River. The results of the case studies on ice accumulation in natural rivers also show that the relative thickness of an ice jam of 0.4 is the criterion for assessing whether an ice jam in a river should be considered an ice dam.

In the study of Hou et al. [13], it was found that the migration of ice waves affects pier scour. The interaction between ice accumulation and local scour around a pier is a very complicated process since not only does the channel bed deform, but the development of an ice jam occurs simultaneously. By conducting a series of flume experiments and by applying both continuity and momentum equations, an equation is been derived for predicting the thickness of ice waves around a pier. The thickness of the wave crest and the migration speed of the ice waves were investigated. Similar to a scour hole in a sand bed, an "ice scour hole" appeared at the bottom of the ice jam around the pier. The existence of an "ice scour hole" affects the development of ice waves.

In the study by Hu et al. [14], laboratory flume experiments were implemented to better understand the interactions between bed deformation and ice jam evolution around bridge piers. Their results show that in the presence of a local scour at bridge piers, both the ice jam thickness and water depth for flow are larger after an ice jam reaches its equilibrium state when compared with in the absence of a local scour at the pier. An equation was developed to determine the scour depth around a pier under ice-jammed flow conditions that considered related factors such as the flow Froude number, ice jam thickness, ice discharge rate, etc.

The aim of the study by Kazem et al. [15] was to experimentally investigate the characteristics of turbulence in the downstream region of a vegetation patch. The changes in turbulent structure were tracked in sequential patterns by reducing the patch size. The velocity profile, TKE, turbulent power spectra, and quadrant analysis were used to study the behavior and intensity of the turbulent structures. The results indicate that there are three different flow layers in the region downstream of the vegetation patch, including the wake layer, mixing layer, and shear layer. Flow structures, Reynolds shear stresses, TKE, and intermittent turbulent kinetic energy were investigated. As the size of the vegetation patch decreases, von Karman vortexes appear in the wake layer and form the dominant flow structures in the downstream region of a vegetation patch.

Kazem et al. [16] investigate the formation of coherent flow structures beyond vegetation patches. For a channel with a small patch, both the flow passing through the patch and the side flow around the patch have a considerable effect on the formation of flow structures beyond the patch. The von Karman vortex street splits into two parts beyond the vegetation patch: the strong part near the surface and the weak part near the bed. The middle part of the flow is completely occupied by the vertical vortex, but the horizontal vortexes cannot be detected in this region.

Nabaei et al. [17] investigated the effect of vegetation on flow structures and turbulence anisotropy around semi-elliptical abutments. When compared to channel beds without the presence of vegetation, channel beds with vegetation have a dramatically reduced primary vortex, but there is little difference in terms of the wake vortex. This causes a noticeable decrease in the Reynolds shear stress. An analysis of the Reynolds stress anisotropy indicates that the flow has a tendency to be isotropic in vegetated beds. The anisotropy profile changes from a pancake shape to a cigar shape in unvegetated channels.

This Special Issue includes interesting research about the incipient motion of bed material under hydraulically transitional flow conditions [18]. The results show that the critical Shields parameter is located under the Shields curve, showing no sediment motion. This indicates that the incipient motion of sediment particles occurs with smaller bed shear stress in a region of hydraulic transitional flow than that estimated using the Shields diagram.

The article of Ni et al. [19] deals with changes in flow and sediment in the Huai River in China. Along a river reach of this river, the incoming sediment rate from upstream of the Huai River continuously decreases. Therefore, the characteristics of channel bed deformation are affected. To investigate the interaction between flow and sediment, a one-dimensional hydrodynamic model was developed, validated using field data, and applied. It was found that the study river reach has frontal erosion and backward deposition, although the variation rate is relatively slow.

Acknowledgments: I would like to thank all the authors who contributed to this Special Issue. I would like to also thank all the reviewers for providing their valuable comments, which have greatly improved the quality of the published papers.

Conflicts of Interest: The author declares no conflict of interest.

References

1. Brandimarte, L.; Paron, P.; Di Baldassarre, G. Bridge pier scour: A review of processes, measurements and estimates. *Environ. Eng. Manag. J.* **2012**, *11*, 975–990. [CrossRef]
2. Federal Highway Administration (FHWA). *Evaluating Scour at Bridges, NH1-01-001*; FHWA: Washington, DC, USA, 1998.
3. Beltaos, S.; Miller, L.; Burrell, B.C.; Sullivan, D. Hydraulic effects of ice breakup on bridges. *Can. J. Civ. Eng.* **2007**, *34*, 539–548. [CrossRef]
4. Sui, J.; Wang, J.; He, Y.; Krol, F. Velocity profiles and incipient motion of frazil particles under ice cover. *Int. J. Sediment Res.* **2010**, *25*, 39–51. [CrossRef]
5. Sui, J.; Faruque, M.A.; Balachandar, R. Local scour caused by submerged square jets under model ice cover. *J. Hydraul. Eng.* **2009**, *135*, 316–319. [CrossRef]
6. Zabilansky, L.J. *Ice Force and Scour Instrumentation for the White River, Vermont. CRREL-SR-96-6*; Cold Regions Research & Engineering Lab: Hanover, NH, USA, 1996.
7. Sui, J.; Wang, D.; Karney, B.W. Suspended sediment concentration and deformation of riverbed in a frazil jammed reach. *Can. J. Civ. Eng.* **2000**, *27*, 1120–1129. [CrossRef]
8. Valela, C.; Sirianni, D.A.B.; Nistor, I.; Rennie, C.D.; Almansour, H. Bridge Pier Scour under Ice Cover. *Water* **2021**, *13*, 536. [CrossRef]
9. Sang, L.; Wang, J.; Cheng, T.; Hou, Z.; Sui, J. Local Scour around Tandem Double Piers under an Ice Cover. *Water* **2022**, *14*, 1168. [CrossRef]
10. Jafari, R.; Sui, J. Velocity Field and Turbulence Structure around Spur Dikes with Different Angles of Orientation under Ice Covered Flow Conditions. *Water* **2021**, *13*, 1844. [CrossRef]
11. Zhang, J.; Wang, W.; Li, Z.; Li, Q.; Zhong, Y.; Xia, Z.; Qiu, H. Analytical Models of Velocity, Reynolds Stress and Turbulence Intensity in Ice-Covered Channels. *Water* **2021**, *13*, 1107. [CrossRef]
12. Chen, P.; Sui, J.; Cao, G.; Cheng, T. Waved-Shape Accumulation of Ice Jam—Analysis and Experimental Study. *Water* **2022**, *14*, 3945. [CrossRef]
13. Hou, Z.; Wang, J.; Sui, J.; Song, F.; Li, Z. Impact of Local Scour around a Bridge Pier on Migration of Waved-Shape Accumulation of Ice Particles under an Ice Cover. *Water* **2022**, *14*, 2193. [CrossRef]
14. Hu, H.; Wang, J.; Cheng, T.; Hou, Z.; Sui, J. Channel Bed Deformation and Ice Jam Evolution around Bridge Piers. *Water* **2022**, *14*, 1766. [CrossRef]
15. Kazem, M.; Afzalimehr, H.; Sui, J. Characteristics of Turbulence in the Downstream Region of a Vegetation Patch. *Water* **2021**, *13*, 3468. [CrossRef]
16. Kazem, M.; Afzalimehr, H.; Sui, J. Formation of Coherent Flow Structures beyond Vegetation Patches in Channel. *Water* **2021**, *13*, 2812. [CrossRef]
17. Nabaei, S.F.; Afzalimehr, H.; Sui, J.; Kumar, B.; Nabaei, S.M. Investigation of the Effect of Vegetation on Flow Structures and Turbulence Anisotropy around Semi-Elliptical Abutment. *Water* **2021**, *13*, 3108. [CrossRef]

18. Shahmohammadi, R.; Afzalimehr, H.; Sui, J. Assessment of Critical Shear Stress and Threshold Velocity in Shallow Flow with Sand Particles. *Water* **2021**, *13*, 994. [CrossRef]
19. Ni, J.; Yu, B.; Wu, P. A Numerical Study of the Flow and Sediment Interaction in the Middle Reach of the Huai River. *Water* **2021**, *13*, 2041. [CrossRef]

water

MDPI

Article

Bridge Pier Scour under Ice Cover

Christopher Valela *, Dario A. B. Sirianni, Ioan Nistor, Colin D. Rennie and Husham Almansour

Department of Civil Engineering, University of Ottawa, 161 Louis Pasteur Drive, Ottawa, ON K1N 6N5, Canada; dsiri029@uottawa.ca (D.A.B.S.); inistor@uottawa.ca (I.N.); colin.rennie@uottawa.ca (C.D.R.); husham.almansour@nrc-cnrc.gc.ca (H.A.)

* Correspondence: cvale034@uottawa.ca

Abstract: Bridge pier scour is a complex process, which is influenced by many parameters, including the presence of ice cover around piers. To better understand the influence of ice on bridge pier scour, an artificial ice cover, equipped with either a smooth or a rough surface, was constructed and tested experimentally. The ice cover was positioned on the surface of the water and submerged to specified depths in order to replicate floating and fixed (pressurized) ice cover conditions, respectively. During each test, a velocity profile was collected beneath the ice cover, and after each test, a three-dimensional scan of the bed was collected to compare the resulting scour. It was discovered that the presence of an ice cover around a bridge pier increased pier scour under all conditions. Furthermore, as the ice cover was submerged deeper into the flow, the flow velocity increased, and greater scour resulted. For each level of submergence, the rough ice cover yielded increased scour depths compared to the smooth ice cover.

Keywords: local scour; ice cover; ice jam; bridge pier; sediment transport; river ice hydraulics

Citation: Valela, C.; Sirianni, D.A.B.; Nistor, I.; Rennie, C.D.; Almansour, H. Bridge Pier Scour under Ice Cover. *Water* **2021**, *13*, 536. https://doi.org/10.3390/w13040536

Academic Editor: Jueyi Sui

Received: 29 January 2021

Accepted: 17 February 2021

Published: 19 February 2021

1. Introduction

Bridges that are constructed over waterways, with piers located within the flow path, are more vulnerable to failure due to pier scour. Bridge pier scour is the erosion of bed material from the region surrounding the pier's base [1]. During high-flow events, large quantities of sediment can be removed from around and beneath the foundations of piers jeopardizing the structural integrity of the piers and consequently the bridge [2]. There are approximately 500,000 bridges in the USA National Bridge Inventory (NBI) that are built over waterways, amounting to nearly 83% of all the bridges in the NBI [2,3]. In 2011, 23,034 of such bridges in the NBI were classified as scour-critical, which, according to the USA Federal Highway Administration (FHWA), means that they are predicted to fail if a given flood event were to occur [2]. Therefore, bridge pier scour is an ongoing problem that affects many bridges in the USA, and elsewhere, ultimately endangering the safety of the public.

Bridge pier scour is caused by a three-dimensional flow separation induced by the pier. As the flow separates around the pier, three main scour-causing mechanisms arise: the horseshoe vortex upstream of the pier, flow acceleration along the sides of the pier, and wake vortices downstream of the pier (depending on the Reynolds number). When present, these components coincide and form the scour holes observed around bridge piers [1,4–7]. Detailed studies were performed by Dey et al. and Tafarojnoruz and Lauria, which examined the flow behaviour around piers by means of experimental and numerical investigations, respectively [8,9]. These studies identify the contribution of each of the aforementioned scour-causing mechanisms as the scour hole around the pier develops.

Bridge pier scour can occur under two conditions: clear water or live bed. Unlike clear-water scour, live-bed scour has the contribution of sediment from upstream, which replenishes, to some degree, the scour hole around the pier [2,6]. For this paper, only clear-water conditions will be studied.

Many parameters influence bridge pier scour. Over the years, extensive research has taken place to better understand the influence of each parameter; however, there are still some parameters for which the knowledgebase is limited. One such parameter is the presence of an ice cover around the pier. Ice on the water's surface, which acts as an upper boundary, has been found to change the flow behaviour beneath. Specifically, the velocity (u) profile under an ice cover transitions from a logarithmic shape with the maximum u near the surface to that which resembles a pipe flow with the maximum u occurring at approximately mid-depth [10]. Due to the location of the maximum u shifting downwards within the flow depth, an increased u gradient occurs near the bed, which can induce greater bed shear stresses. As a result, more scour can occur [11–13].

Ice covers can possess a wide variety of shapes and configurations influenced by various factors such as weather and location. As a result, the roughness of the underside of an ice cover can vary drastically within a given reach and season, making the ice cover roughness highly irregular [14–16]. If the ice is thermally grown only, the roughness will frequently be minimal, and the ice will have a smooth surface contacting the passing flow [12]. On the other hand, mechanical thickening processes due to ice shoving can generate a rough ice cover [17]. For example, most often, later in the season, during the breakup period, large pieces of ice can break apart into smaller ice fragments and, similar to debris, accumulate in random patterns generating a rough surface, sometimes referred to as an ice jam [11]. Due to the unsafe conditions during field measurements of an ice jam roughness, limited data are available [16]. However, Beltaos used a remote technique consisting of a floating sensor that was deployed beneath the ice jam [14]. The floating sensor was carried with the flow while recording the elevation of the underside of the ice jam for multiple kilometers. The mean elevation and the fluctuations from the mean were translated into the average thickness and the hydraulic roughness of the ice cover, respectively. For the 20 data sets collected at various locations, the absolute roughness values (taken as the 84th percentile of the absolute deviation from the mean) varied from 0.24 to 1.15 m, averaging 0.79 m, while omitting one data point with an extremely large value of 2.34 m [14]. Zare et al. utilized a bottom-mounted Acoustic Doppler Current Profiler (ADCP) over a four-month period that included spring breakup to measure u profiles continuously at a fixed location in Nelson River, Canada [10]. Hourly averaged u profiles were fit to the log law to estimate the roughness of the bottom side of the ice surface, with equivalent Manning's roughness values ranging from near zero (smooth) to as high as 0.08 (very rough). Therefore, an ice cover can possess an underside that is smooth or extremely rough.

The roughness of the underside of an ice cover, in conjunction with the bed roughness, has an influential role in shaping the corresponding u profile. If the ice and the bed's roughnesses possess similar magnitudes, the location of the maximum u will occur at approximately mid-depth. However, if the surface has a greater roughness than the bed, the location of the maximum u could occur below mid-depth. This can cause a higher u gradient next to the bed, resulting in increased shear stress and potentially greater scour [11,12,17].

Solid ice covers that span the entire width of the river, which is the focus of this study, can be either unattached or attached to the banks, such that the ice is floating or fixed, respectively. When the ice is floating, it can raise or lower depending on the stage of the river. However, when the ice is attached to the banks, it is unable to adjust to varying water levels causing pressurized conditions to occur beneath the ice. Furthermore, ice covers, especially ice jams, frequently grow in size in the vertical direction due to the mechanical thickening process [18]. When an ice cover attached to the banks grows vertically, the cross-sectional area of the river channel beneath the ice decreases, and a rise in the upstream water level can occur. When this happens, the passing flow is confined to the smaller cross-sectional area causing there to be an increase in velocity, resulting in even greater scour [12,19,20].

The Hydraulic Engineering Circular No. 18 (HEC-18) has identified that an ice formation or jam is a factor that affects the local scour depth around bridge piers, as a more severe scour condition can occur with a smaller flow rate (Q). HEC-18 further states that there are many examples of foundation scour from accelerated flow beneath an ice covering, but limited field measurements of scour induced by ice jams exist. When designing a bridge, HEC-18 requires that ice effects be considered when calculating the maximum scour depth, but the HEC-18 pier scour equation does not consider ice covers. Instead, HEC-18 suggests obtaining scour data from nearby bridges to estimate the expected scour [2]. This is not always an accurate approach. According to Wuebben, the resulting bathymetry after an ice covering has subsided may not represent the deepest scour, as a portion of the scour hole could have been refilled during the ice breakup period [16].

Various researchers have identified ice covers as a parameter that requires further research, e.g., Ettema et al. ranked ice covers at a medium-level priority in terms of bridge pier scour research needs [21]. Wu et al. performed a bridge pier scour study in the presence of a floating ice cover, which examined the effect of different pier diameters and water depths on scour depth and scour width [13]. The limitations of this study were that pressurized ice covers and different ice cover roughnesses were not considered. Ackermann et al. studied bridge pier scour with a floating ice cover possessing either a smooth or a rough surface [22]. Various velocities under both live-bed and clear-water conditions were tested. The limitations of this research were that pressurized conditions were not tested and only the maximum scour depth was recorded. Lastly, Hains and Zabilansky performed a thorough study of bridge pier scour under a floating and pressurized ice cover [23]. The limitations of this study were that live-bed conditions were reached in a number of clear-water tests making the final scour data for these points unusable, and Q was changed for each test rather than being kept constant.

It is understood that ice covers influence the hydrodynamics of the flow passing beneath, which can in turn increase bed erosion. However, the full extent of the impacts an ice cover has on bridge pier scour are not known. Therefore, the intent of this study is to expand the existing knowledgebase pertaining to bridge pier scour in the presence of an ice cover. In addition to maintaining a constant flow rate and remaining in the clear-water regime amongst all tests, the objectives are to:

1. Examine the differences in scour for floating versus fixed ice covers;
2. Investigate the effect of different levels of ice cover submergence (flow pressurization) on scour;
3. Evaluate the influence of both smooth and rough ice covers on scour.

2. Materials and Methods

2.1. Flume

The research presented in this study was completed in the University of Ottawa's Civil Engineering Hydraulics Laboratory in Ottawa, Canada. The flume utilized for the research measured 30 m in length, by 1.5 m in width, by 0.5 m in depth. A sand section, extending the width of the flume and measuring 3.16 m in length and 0.2 m in depth, was installed near the outlet. To contain the sand, concrete false floors, the same height as the sand, were installed upstream and downstream of the sand section. The downstream false floor measured 1.25 m in length and ended at the outlet. The upstream false floor measured 4.16 m in length and included an additional gradual gravel slope, 2.25 m in length, extending upstream from the leading face to smoothly transition the passing flow onto the false floor. An acrylic cylindrical bridge pier, 0.09 m in diameter (D), was positioned in the center of the sand section and bolted to a large steel plate situated beneath the sand. The width of the pier with respect to the width of the flume was checked to ensure flow blockage would not occur. The sediment used in the sand section was uniformly graded silica sand with a median grain size diameter (d_{50}) of 0.001 m. This sand size was selected based on a Shield's curve analysis. To maintain a constant bed roughness before and after the sand section, the same sand was adhered to the surface of both concrete false floors. Lastly, to reduce turbulence, two flow straighteners were installed near the inlet. An illustration of the flume can be seen in Valela et al. [24].

2.2. Ice Cover Construction

One of the objectives of this paper was to study scour under a fixed ice cover that induced pressurized flow conditions. To generate pressurized flow conditions, the ice cover was partially submerged. In order to ensure rigidity of the ice cover, it was constructed of dimensional lumber with a plywood bottom surface. The ice cover spanned the entire width of the flume and measured 7.5 m in length, therefore covering the sand section and a majority of the false floors (Figure 1). By making the ice cover this length, the ends were far enough away from the pier that the flow entering the front and exiting the rear of the ice cover did not influence the scour around the bridge pier. In addition, a fully developed flow profile was achieved prior to reaching the pier. To help guide the flow under the ice cover and avoid flow over top, two 30° sloped sections were attached to either end of the ice cover.

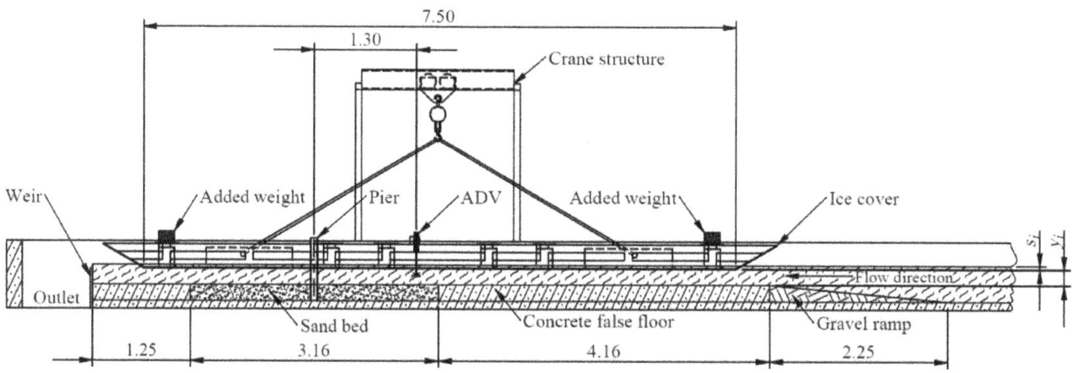

Figure 1. Cross-sectional drawing of the ice cover apparatus installed in the flume (dimensions in m). y_i is the flow depth under the ice and s_i is the ice cover submergence.

A hole, the diameter of the pier, was drilled through a bottom-flush viewing window in the ice cover to allow the pier to protrude through (Figure 2b). The hole was made so that the ice cover fit tight around the pier, limiting the flow passage in between. Due to the overall weight of the ice cover, it was constructed in three sections with a bolted connection between the sections. Once the ice cover was in the flume, it was bolted together and lowered into place for each test as one solid unit using a crane. Figure 2a displays the ice cover sections bolted together and resting on the side walls of the flume in preparation for a test.

The pressurized ice cover tests performed in this study were intended to replicate an ice cover that grows in the vertical direction, where the cross-sectional area under the ice decreases, while a constant Q is maintained. To recreate these conditions, the ice cover was submerged into the flow to different desired depths. Due to the large buoyancy forces that would otherwise be generated, water was allowed to enter the top side of the ice cover as added weight, as seen in Figure 2c. This was achieved using three fill holes at the front and three fill holes at the rear of the ice cover, as seen in Figure 2d. Once the water on top of the ice cover equilibrated with the water height in the flume, the fill holes were plugged for the duration of each test. Depending on s_i, additional weight was added in the form of concrete blocks to the top of the ice cover. An effort was made to reduce the transfer of water between the passing flow and the top side of the ice cover. This was achieved by using a rubber seal between the flume walls and the sides of the ice cover, as shown in Figure 2c. Furthermore, between each sheet of plywood contacting the passing flow, the edges were cut on a 45° degree angle and joined to eliminate the passage of water between the sheets.

Figure 2. Ice cover apparatus: (**a**) ice cover resting above the flume to allow for the flow conditions to be set, (**b**) pier protruding through the ice cover, (**c**) top view of the ice cover, and (**d**) fill holes used to allow the added water weight to enter above the ice cover.

2.3. Ice Cover Roughness

Another objective of this study was to compare scour under a smooth versus a rough ice cover. A smooth surface was created by first using plywood that possesses sanded faces. Then, any imperfections, joints, and screw heads were filled with a waterproof wood filler. Next, the entire surface was sanded to ensure a smooth finish, including the added wood filler. Lastly, a waterproofing product was applied to seal the surface using a paint brush. This resulted in a smooth and waterproof surface, as shown in Figure 3a, which closely resembles that of a thermally grown ice sheet.

Once all of the tests requiring a smooth ice cover were complete, the rough surface tests were performed. To achieve the desired roughness, 0.3 m by 0.3 m Polyvinyl Chloride (PVC) acoustic wall panels were used to cover the existing smooth bottom of the ice cover (Figure 3b), including the viewing window installed around the pier. The acoustic panels are waterproof, rigid, and possess the desired jagged shape which resembles the bottom of an ice jam. Each individual panel, shown in Figure 4a, has the same pattern which is symmetric about the diagonal axis. To ensure a more random configuration, the panels were staggered when installed. A topographic plot of an individual panel, Figure 4b, shows a maximum peak elevation of 0.025 m. The volumetric average elevation, for a given panel, was calculated to be located at 0.005 m from the base. When the ice cover was set at specific depths, the roughness's average elevation was used as the bottom of the ice cover.

Figure 3. Bottom surface of the ice cover: (**a**) smooth condition, and (**b**) rough condition.

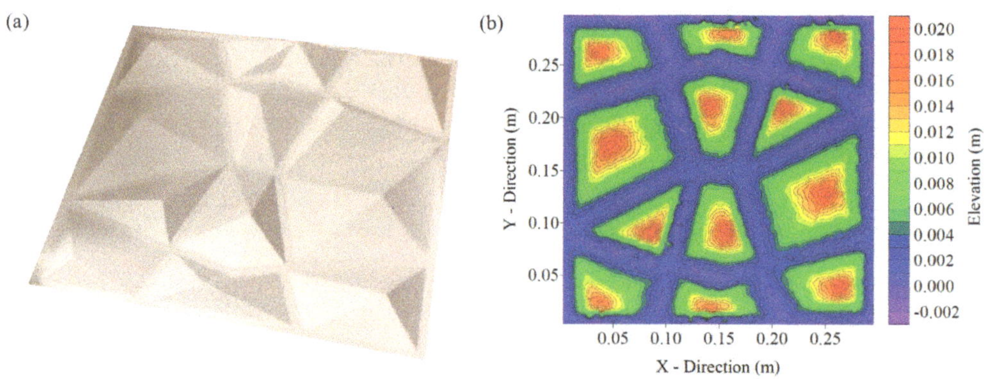

Figure 4. Rough ice cover: (**a**) individual panel, and (**b**) contour plot of the surface elevation of an individual panel.

To determine the roughness of the rough ice cover, a trial test was performed with the ice cover panels positioned flat on the flume floor facing upwards, without the presence of the ice cover. The flume was filled to a depth of 0.10 m, and a depth-averaged velocity (u_{avg}) of 0.37 m/s was maintained. A vertical u profile was collected over the ice cover panels where the flow was fully developed, and using the law of the wall, the ice cover roughness was found to be k_s = 0.013 m [25]. When using a scaling factor of 30, the maximum peak elevation and the roughness equate to 0.75 m and 0.39 m, respectively, which falls within the average range of an ice jam indicated by previous literature. Converting the k_s value to a Manning's coefficient (n) value, using the method presented by Li, yields an approximate average value of 0.0189 $s/m^{1/3}$ [15].

2.4. Instrumentation

The u measurements were collected using a Vectrino Acoustic Doppler Velocimeter (ADV). For each ice cover condition tested, a vertical u profile was collected 1.3 m upstream of the pier in the center of the flume. To collect the u measurements, a hole, the diameter of the ADV probe head, was cut in the ice cover so the probe head could protrude through. The ADV mounting setup can be seen in Figure 5. A side-looking, rather than a down-looking, probe head was utilized to ensure u measurements close to the ice cover were captured. Individual u time series were collected at 0.01 m intervals between the bed and the water surface or ice cover. Data were collected at each point for two minutes at a sampling frequency of 100 Hz. The measured data were filtered, despiked, and time-averaged [26].

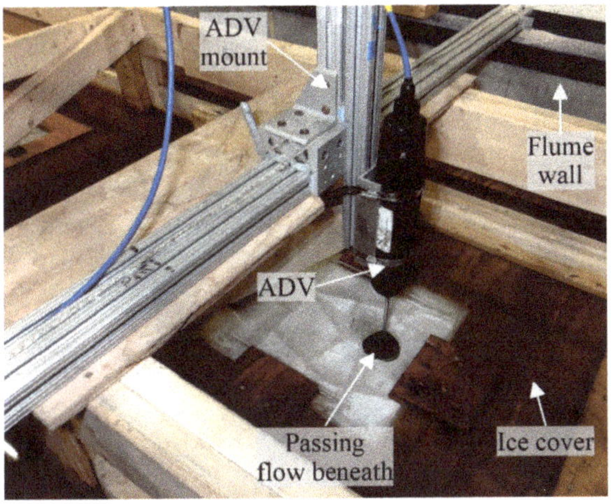

Figure 5. ADV setup with side-looking head protruding through the ice cover.

After each test was complete, the resulting bathymetry was recorded using a Leica ScanStation P50 Terrestrial Laser Scanner (TLS) (Figure 6). The accuracy of this device is +/− 0.0015 m in all three directions. To capture the sand bed with the TLS, the ice cover needed to be partly disassembled and raised after each test. To ensure accurate elevations, a very thin dusting of sodium bicarbonate was sprinkled upon the sand bed before scanning to increase opacity. Similarly, the Perspex pier was wrapped with fabric during scanning. To capture the entire bed, the TLS was setup at three locations around the pier, where a scan was performed from each location. Joining the scans together after was aided by four targets, which were placed in clear view of all three scanning locations. Once the scans were joined together, a detailed three-dimensional model of the entire scour region was created.

2.5. Protocol

A total of 13 tests were completed to examine the scour behaviour beneath an ice cover, which consisted of six tests with a smooth ice cover, six tests with a rough ice cover, and one test without any ice cover as a benchmark case. For both the smooth and rough ice covers, the same conditions were tested. One of the six tests was performed with the ice cover bottom touching the water surface, replicating the floating condition, as the flow was not pressurized below the ice cover. The remaining five tests were performed with the ice cover submerged to depths (s_i), which incremented by 0.015 m up to 0.075 m, achieving pressurized flow conditions under the ice cover. Setting the ice cover for each test was achieved by, first, lowering it into position with the crane. Then, depending on the required elevation, weight was added to the ice cover, in the form of water and concrete blocks, to reach the desired y_i. The ice cover remained attached to the crane throughout the duration of each test to maintain the desired elevation. The distance between each of the four ice cover corners and the bed was verified prior to commencing each test to ensure the ice cover was set at the required elevation and level in all directions. The experimental trials and the corresponding naming convention are presented in Table 1.

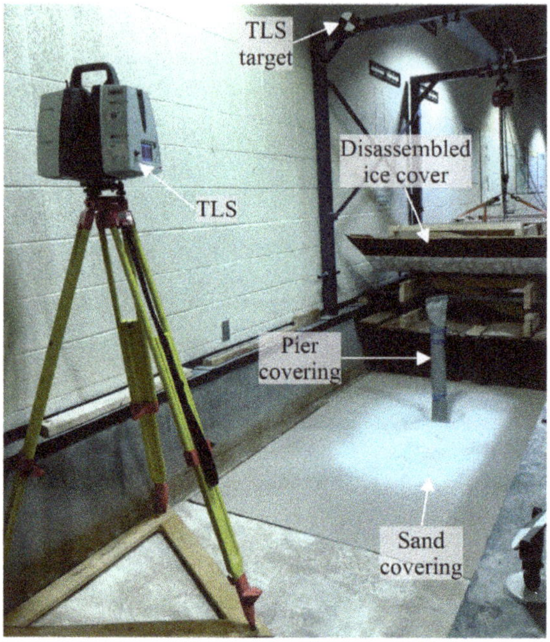

Figure 6. Post-test scanning procedure.

Table 1. Experimental trials.

Test	s_i (m)	y_i (m)	Ice Cover Roughness	Water Temperature (°C)	Reynolds Number	Stream Froude Number	Pier Froude Number
Open	-	-	-	20	41,237	0.163	0.055
SF	0.000 (Floating)	0.250	Smooth	20	91,187	0.104	0.052
S1	0.015	0.235	Smooth	20	90,329	0.112	0.056
S2	0.030	0.220	Smooth	20	99,949	0.135	0.077
S3	0.045	0.205	Smooth	20	99,960	0.148	0.088
S4	0.060	0.190	Smooth	20	100,299	0.164	0.101
S5	0.075	0.175	Smooth	20	107,527	0.196	0.134
RF	0.000 (Floating)	0.250	Rough	20	95,501	0.109	0.057
R1	0.015	0.235	Rough	20	97,901	0.121	0.066
R2	0.030	0.220	Rough	20	100,791	0.136	0.079
R3	0.045	0.205	Rough	20	102,657	0.152	0.092
R4	0.060	0.190	Rough	20	103,761	0.170	0.108
R5	0.075	0.175	Rough	20	106,497	0.195	0.132

This study is unique as each test was performed for the same duration (t) of 6 h and under the same Q (0.0825 m^3/s). Q was selected to be in the clear-water regime, for all ice cover conditions, determined from initial trial tests. To ensure the flow conditions were constant among all the tests, the flow was first set and verified without the ice cover. This was accomplished using a flow control valve on the pump and an outlet weir. Then, depending on the test that was being performed, the ice cover was either lowered into position at the desired elevation or left above the water surface for the duration of the test. By beginning each test in this manner, the same flow depth (y) of 0.25 m and u_{avg} of 0.22 m/s could be set for each test, whether or not the ice cover was being used. The time began once the ice cover was positioned at the desired elevation.

Prior to beginning each test, the ice cover was disassembled, and the bed was completely levelled. The ice cover was then reassembled and secured above the water surface. It was critical when beginning each test that the flume was slowly filled with water and that Q was gradually increased, as to avoid any initial scour.

3. Results

3.1. Velocity Data

During each of the experimental conditions tested, a vertical u profile was measured upstream of the pier. The u profiles beneath the smooth and rough ice covers are presented in Figure 7a,b, respectively. In addition, the u profile from the Open case (possessing no ice cover) is presented in both plots for comparison purposes.

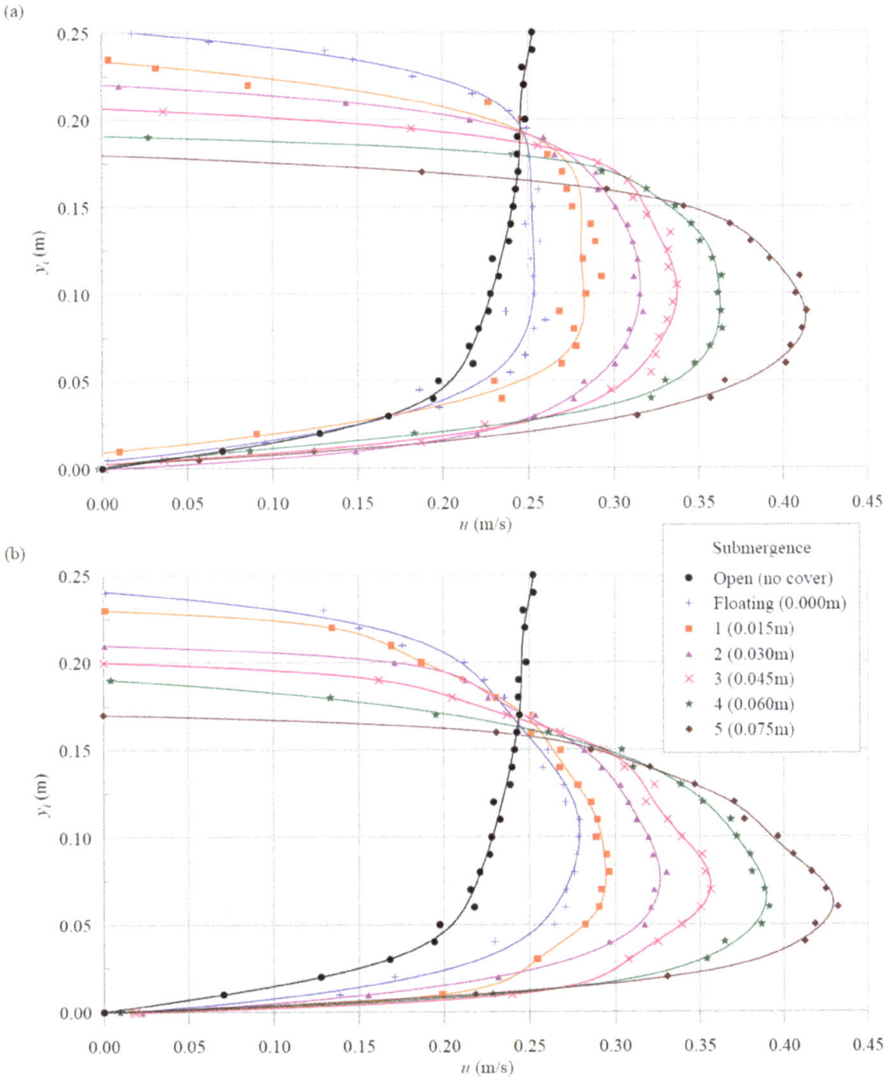

Figure 7. Velocity (u) profiles for Open, Floating, and submerged conditions: (**a**) smooth ice cover, and (**b**) rough ice cover.

The Open case u profile behaves as expected, with the maximum u occurring near the surface. The u gradient near the bed for the Open case is mild, which induces a low quantity of shear stress on the bed. Meanwhile, when any ice cover is present, the u profile changes to that resembling a pipe flow. The maximum u is reached towards mid-depth while the u at the top and bottom boundary approaches zero. As either ice cover becomes more submerged reducing the depth under the ice cover, the maximum u increases and the location of the maximum u shifts downwards to remain at approximately mid-depth. Furthermore, despite some anomalies, a general trend shows that the u gradient near the bed is lower for the Open condition than most of the ice covered conditions, especially the more submerged cases. This can result in the ice covered cases inducing greater shear stress on the bed.

The roughness of the ice cover is shown to have further influenced the shape of the u profile beneath. For the same s_i, the u profile under the rough ice cover possesses a greater maximum u and a sharper crest, when compared to the u profile under the smooth ice cover. In addition, the location of the maximum u under the rough ice cover occurs below mid-depth, closer to the bed, whereas the location of the maximum u under the smooth ice cover occurs at approximately mid-depth. This trend is consistent amongst all s_i levels. As a result, the u gradient near the bed is greater under the rough ice covers.

When comparing u profiles between the floating versus submerged (fixed) ice covers, little difference is present besides the influence of s_i. The shape of the u profiles for the floating cases match closely to those of s_i = 0.015 m. The main dissimilarity is the lower maximum u for the floating cases and that corresponds to the greater cross-sectional area.

3.2. Scour Data

After each test was performed and scans of the resulting bathymetry were collected, three-dimensional models of the bed were generated. Using these models, elevation plots were created to compare the differences in scour patterns between tests. Figure 8 displays the resulting scour for the Open case, where the flow travelled in the negative X direction. In the absence of an ice cover and under the given flow conditions, the quantity of scour around the pier is minimal. A small depression can be seen wrapping around the front and sides of the pier with a corresponding mound immediately behind the pier.

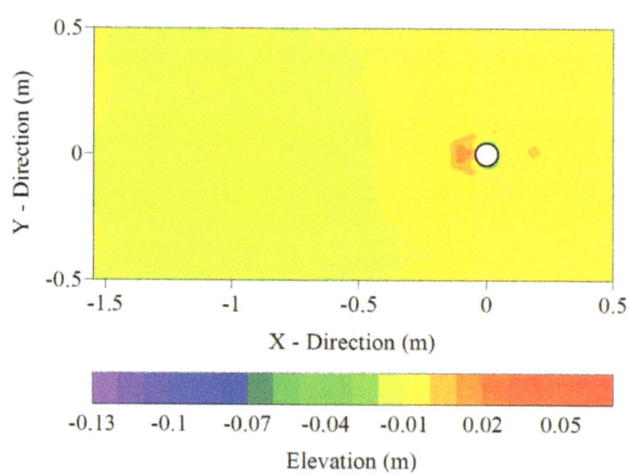

Figure 8. Scour contour plot for the Open trial.

Figure 9 displays the scour plots from each of the ice cover tests. The left and right column contain the smooth and rough ice cover results, respectively, while the rows contain increasing s_i levels. Regardless of the ice cover roughness or submergence, all tests possessing an ice cover yielded greater scour compared to the Open case. The floating cases (SF and RF) possess the least amount of scour, of the ice cover conditions tested, and the patterns look similar to that of the Open case. The difference is a magnified version of the Open case, where there is a deeper depression around the pier followed by a larger deposition pile behind the pier. As the ice cover becomes submerged, the quantity of scour and deposition increases steadily based on the level of s_i. The scour grows not only in depth but also proportionally in diameter. Once the deposition pile reaches approximately 0.05 m in height, two small scour holes begin to form immediately downstream of the deposition pile, as seen in test S2 and R2. These scour holes grow in size as the ice cover becomes further submerged and, as a result, create a second deposition pile further downstream.

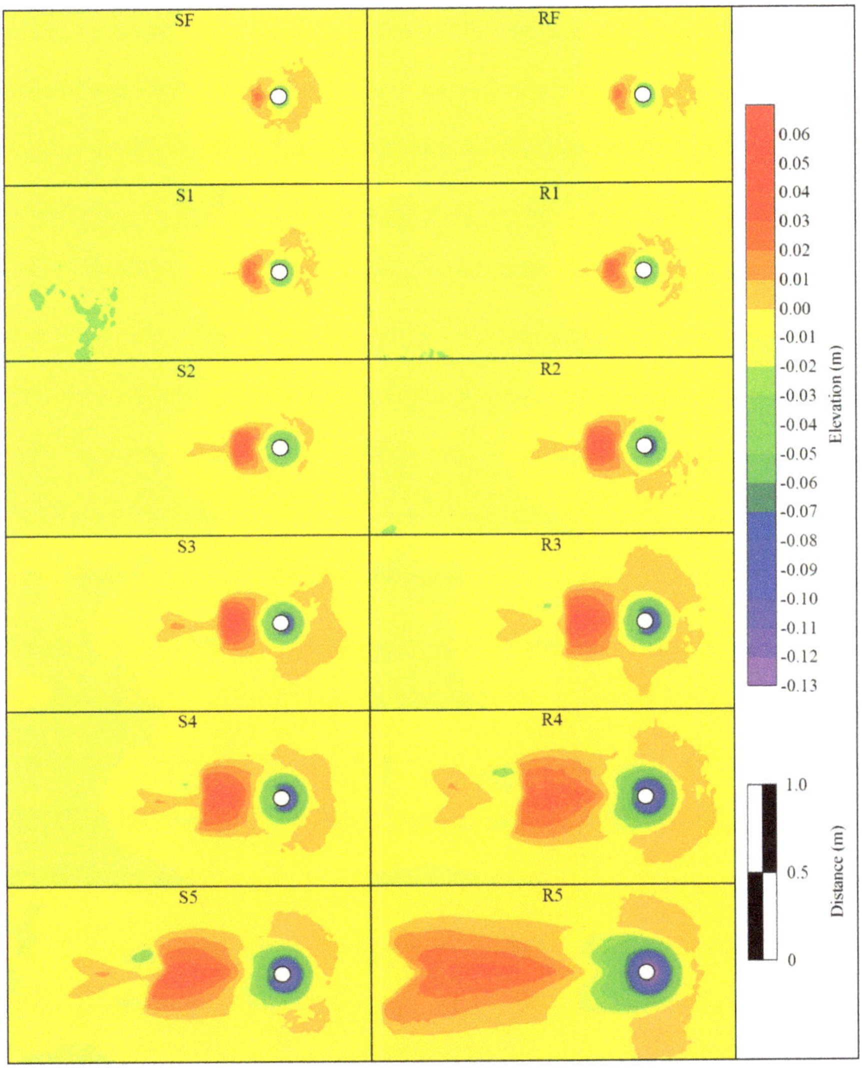

Figure 9. Scour contour plot comparison for the Floating and submerged trials.

The difference in scour between the smooth and rough ice covers is minimal for the floating and mildly submerged ice cover tests. As s_i increases, the ice cover roughness has a greater influence causing there to be a larger difference in scour between the smooth and rough ice covers. However, for all test conditions, the rough ice cover yielded more scour than the smooth ice cover. Test R5 contained the greatest scour with the upstream and side depression continuing around the rear of the pier and the deposition pile extending over double the length of that in test R4 and S5. Due to the deposition pile in test R5 nearing the end of the sand section, additional scour holes were not generated downstream of the deposition pile.

The maximum scour depth (d_s) for each test, as well as the percent increase in d_s when compared to the Open case, are presented in Table 2. All tests were within the clear-water regime, as live-bed conditions were not reached even under the most submerged ice cover. The Open case experienced a d_s of only 0.025 m, which increased drastically with the presence of any ice cover. The floating ice cover tests, SF and RF, yielded a 40% and 52% increase in d_s, respectively. However, when the ice covers became submerged for tests S2 and R2, the increase in d_s grew to 76% and 92%, respectively. The greatest d_s was achieved for test R5, reaching a depth of 0.128 m, implying a 412% increase over the Open case. The smooth and rough ice cover tests had a difference in d_s of 0.004 m between tests S1 and R1 and that grew to 0.019 m between tests S5 and R5.

Table 2. Scour results.

Test	d_s (m)	Increase in d_s (%)
Open	0.025	-
SF	0.035	40
S1	0.044	76
S2	0.062	148
S3	0.079	216
S4	0.090	260
S5	0.109	336
RF	0.038	52
R1	0.048	92
R2	0.074	196
R3	0.088	252
R4	0.107	328
R5	0.128	412

To better understand the difference between the effects of the smooth and rough ice cover on scour, the nondimensionalized ice cover scour depth (d_i) as a function of the nondimensionalized y_i, for both ice cover roughnesses, is plotted in Figure 10a. Note that d_i refers to the scour caused only by the ice cover, which is d_s minus the scour depth of the Open case (d_p). The purpose of plotting d_i, as opposed to d_s, was to highlight the additional scour induced by the presence of an ice cover.

Figure 10a confirms that the difference in scour between the smooth and rough ice covers is small for the floating condition but grows as the ice cover becomes more submerged. The equations of the smooth and rough fitted lines are presented in Equations (1) and (2), respectively:

$$\frac{d_i}{D} = -2.78\frac{y_i}{y} + 2.86 \quad (1)$$

$$\frac{d_i}{D} = -3.39\frac{y_i}{y} + 3.50 \quad (2)$$

where (y_i/y) represents the flow confinement (pressurization), such that $y_i \leq y$. Equations (1) and (2) were developed from y_i/y data ranging from 0.7 to 1.

The relationship between scour and flow confinement is linear for both smooth and rough ice covers. Due to a difference of 0.61 in the slopes, the data converge as y_i increases.

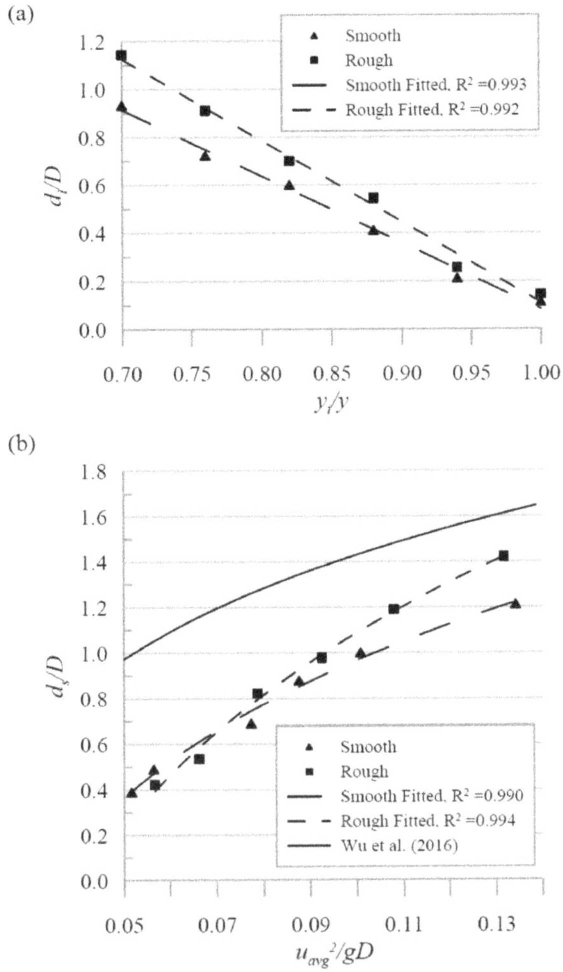

Figure 10. (**a**) Additional scour induced by the ice cover, and (**b**) the relationship between nondimensionalized maximum scour depth (d_s) and pier Froude number.

Another method of presenting the scour depth is as a function of the pier Froude number $\left(\frac{u_{avg}^2}{gD}\right)$, where g is the gravitational constant. According to Ettema et al., the pier Froude number can be seen as a way of conveying the similitude in the frequency and strength of vortices shed from the rear of a pier [27]. This parameter is often neglected when studying pier scour but has shown to be largely influential [13]. Figure 10b displays the nondimensionalized d_s as a function of pier Froude number. Similar to Figure 10a, this plot also indicates that the rate of change for the smooth and rough ice covers is not the same. For pier Froude numbers below approximately 0.07, the rough ice cover data points drop below that of the smooth ice cover. In other words, to achieve the same quantity of scour under both ice covers, greater pier Froude number values are required in the presence of a rough ice cover. However, above the pier Froude number of approximately 0.07, the rough ice cover yields greater scour for the same pier Froude number. The equations for the smooth and rough ice cover fitted lines, which are nonlinear, are shown in Equations (3) and (4), respectively:

$$\frac{d_s}{D} = 0.87 \ln\left(\frac{u_{avg}{}^2}{gD}\right) + 2.96 \tag{3}$$

$$\frac{d_s}{D} = 1.22 \ln\left(\frac{u_{avg}{}^2}{gD}\right) + 3.89 \tag{4}$$

The pier Froude number was altered through changing u_{avg} and leaving D constant, therefore implying that d_s is proportional to u_{avg}. Equations (3) and (4) are valid for $D > 0$ and $0 < u_{avg} < u_{cr}$, where u_{cr} is the critical velocity required for bed material motion.

4. Discussion

In this study, a number of conditions pertaining to the presence of surface ice around bridge piers, in regard to bridge pier scour, were analysed. One objective, which proved to be challenging to correctly replicate experimentally, was a floating ice cover. In the field, a floating ice cover can both freely adjust its elevation to accommodate the changing water levels, and it can also protrude into the passing flow to some degree, depending on the characteristics of the ice. Therefore, it is possible that a floating ice cover can generate accelerated flow beneath, which could result in similar scour behaviours to that of the mildly submerged fixed ice covers. To overcome this issue, a range of s_i levels were tested from just touching the surface to protruding into the flow 30% of y. Since the flow depth throughout each test remained constant and the ice cover was weighted to achieve the desired s_i, whether the ice cover was fixed in place or not would have no impact on the scour results. This implies that the submerged ice cover cases could represent a floating ice cover that protrudes into the flow.

Upon analysing the velocity and scour results with and without an ice cover, it is evident that the presence of any ice cover increases scour. Specifically, as an ice cover grows and becomes more submerged, the quantity of pier scour becomes greater. These scour results were confirmed by the general trend observed from the u data, which indicated an increase in the near bed u gradient as the ice covers became more submerged. Both a smooth and rough ice cover were tested, and this was to reach both ends of the possible roughness spectrum. It was discovered that the rough ice cover induced more scour than the smooth ice cover under all conditions. In addition, the quantity of scour also increased at a greater rate for the rough ice cover than the smooth ice cover, as y_i decreased. For these reasons, the rough ice cover should be used from a design perspective in order to achieve a conservative design. Since many pier scour equations do not take into consideration the effects of ice covers, the results presented in Figure 10a can be used in addition to such equations to help provide further insight.

Wu et al. presented equations that relate nondimensional scour to pier Froude number, under open-channel conditions, as shown in Equation (5), and smooth ice cover conditions, as shown in Equation (6) [13]:

$$\frac{d_s}{D} = 0.67 \ln\left(\frac{u_{avg}{}^2}{gD}\right) + 2.83 \tag{5}$$

$$\frac{d_s}{D} = 0.66 \ln\left(\frac{u_{avg}{}^2}{gD}\right) + 2.95 \tag{6}$$

Equation (6) from Wu et al. and Equation (3) presented in this paper both pertain to a smooth ice cover and possess a number of similarities (Figure 10b) [13]. Specifically, the constants are almost identical, and the coefficients are relatively close given the differences in the experimental conditions. This lends credence to both Equations (3) and (6), and indicates that the experimental results presented in this paper are reasonable. The experimental conditions of Wu et al. varied from this study in that y = 0.108 m, 0.150 m, 0.210 m, d_{50} = 0.00051 m, and u_{avg} = 0.24 m/s [13]. It can be noted that of the different experimental conditions, d_{50} and y are not considered in Equations (3) and (6), only u_{avg}.

When comparing the equations pertaining to a smooth ice cover (Equations (3) and (6)) to Equation (4) presented in this paper, which pertains to a rough ice cover, there are notable differences. The constant and the coefficient for Equation (4) are significantly greater than that of Equations (3) and (6). Wu et al. stated that the presence of a smooth ice cover, when compared to no ice cover, has only a limited influence on the pier Froude number [13]. Meanwhile, the results presented in this study (Equations (3) and (4)) show that the presence of a rough ice cover, in comparison to a smooth ice cover, has a substantial influence on the pier Froude number. Therefore, given the findings of this paper and that of Wu et al., the pier Froude number only becomes influenced with rough ice covers [13].

The work presented in this paper contains a number of limitations which warrants further research. First, bridge pier scour is a complex process which is influenced by numerous parameters, and due to the scope of the project, not all parameters could be examined to their full extent. Specifically, only one Q, y, D, and d_{50} were tested, and all the tests lied within the clear-water regime. The second limitation to this paper is that only one size of ice covering was used. The ice covering extended in the upstream and downstream direction a far distance from the pier, as to replicate an infinitely long ice cover, but perhaps shorter ice covers, such as ones localized around the pier, could have a different effect on the quantity of pier scour. The third limitation is that the ice cover itself was artificial as the smooth surface consisted of treated plywood and the rough surface consisted of PVC panels. While using artificial materials improves constructability, and the chosen materials were intended to mimic natural river ice, it is possible that natural river ice would induce different flow characteristics. The last limitation is that only one u profile was measured for each test condition and it was collected in the center of the flume upstream of the pier. Additional u profiles would be beneficial as they would create a more detailed flow field; however, this is difficult to achieve with an ADV in the presence of a solid ice cover.

A point worth noting is that when the ice cover was submerged, the flow depth beneath the ice cover consequently reduced. In the absence of an ice cover, the authors acknowledge that a change in flow depth could influence the scour depth. Therefore, the scour depth experienced under a submerged ice cover could be a combination of both the flow pressurization and the reduction in flow depth. Further research is required to distinguish the contribution of both factors.

5. Conclusions

Pier scour under increasing ice cover submergence was investigated. Velocity measurements collected under the ice cover for each experimental condition confirmed that the ice cover induced an upper boundary layer, resulting in maximum velocity near mid-depth. Increased ice cover submergence resulted in greater maximum velocity, and an associated increase in the near-bed velocity gradient. Furthermore, a rougher ice cover displaced the maximum velocity closer towards the bed, and thus the near-bed velocity gradient was even greater.

After each experiment, three-dimensional scans of the bed were performed. It was found that the presence of an ice cover, regardless of the submergence level or roughness, caused an increase in scour. The bathymetry measurements confirmed that the tests with an increased velocity gradient near the bed resulted in greater scour. Specifically, as the ice cover was submerged deeper into the flow, the quantity of scour grew, and for each level of submergence, the rough ice cover generated more scour than the smooth ice cover. The most scour occurred under the rough ice cover submerged to 30% of the flow depth (the largest submergence tested), as it resulted in a 412% increase in scour depth. Therefore, solid ice covers that span the water surface around bridge piers are an influential parameter in pier scour and additional research into the topic is a necessity. Importantly, when designing a bridge pier foundation to resist the additional erosive forces induced by an ice cover, focus should be put on rough ice covers that protrude into the passing flow, as such conditions yielded the most concerning scour depths.

Author Contributions: Conceptualization, C.V., D.A.B.S., I.N., C.D.R. and H.A.; methodology, C.V., D.A.B.S., I.N. and C.D.R.; software, C.V. and D.A.B.S.; validation, C.V. and D.A.B.S.; formal analysis, C.V., I.N. and C.D.R.; investigation, C.V. and D.A.B.S.; resources, I.N. and C.D.R.; data curation, C.V. and D.A.B.S.; writing—original draft preparation, C.V.; writing—review and editing, C.V., D.A.B.S., I.N., C.D.R. and H.A.; visualization, C.V.; supervision, I.N., C.D.R. and H.A.; project administration, C.V.; funding acquisition, I.N., C.D.R. and H.A. All authors have read and agreed to the published version of the manuscript. CRediT taxonomy for the term explanation.

Funding: This research was funded by the National Research Council (NRC) Canada and by the Natural Sciences and Engineering Research Council of Canada (NSERC).

Institutional Review Board Statement: Not applicable.

Informed Consent Statement: Not applicable.

Data Availability Statement: The data are contained within this article.

Conflicts of Interest: The authors declare no conflict of interest. The funders had no role in the design of the study; in the collection, analyses, or interpretation of data; in the writing of the manuscript, or in the decision to publish the results.

References

1. Chiew, Y.-M. Scour Protection at Bridge Piers. *J. Hydraul. Eng.* **1992**, *118*, 1260–1269. [CrossRef]
2. Arneson, L.A.; Zevenbergen, L.W.; Lagasse, P.F.; Clopper, P.E. *Evaluating Scour at Bridges*, 5th ed.; U.S. Department of Transportation: Washington, DC, USA, 2012.
3. Lagasse, P.F.; Clopper, P.E.; Zevenbergen, L.W.; Girard, L.G. *Countermeasures to Protect Bridge Piers from Scour*; Natl. Coop. Highw. Res. Progr. (NCHRP Rep. 593); Transportation Research Board: Washington, DC, USA, 2007.
4. Dargahi, B. Controlling Mechanism of Local Scouring. *J. Hydraul. Eng.* **1990**, *116*, 1197–1214. [CrossRef]
5. Ettema, R.; Constantinescu, G.; Melville, B.W. Flow-Field Complexity and Design Estimation of Pier-Scour Depth: Sixty Years Since Laursen and Toch. *J. Hydraul. Eng.* **2017**, *143*, 03117006. [CrossRef]
6. Raudkivi, A.J. Functional Trends of Scour at Bridge Piers. *J. Hydraul. Eng.* **1986**, *112*, 1–13. [CrossRef]
7. Stevens, M.A.; Gasser, M.M.; Saad, M.B.A.M. Wake Vortex Scour at Bridge Piers. *J. Hydraul. Eng.* **1991**, *117*, 891–904. [CrossRef]
8. Dey, S.; Raikar, R.V.; Roy, A. Scour at Submerged Cylindrical Obstacles under Steady Flow. *J. Hydraul. Eng.* **2008**, *134*, 105–109. [CrossRef]
9. Tafarojnoruz, A.; Lauria, A. Large Eddy Simulation of the Turbulent Flow Field around a Submerged Pile within a Scour Hole under Current Condition. *Coast. Eng. J.* **2020**, *62*, 489–503. [CrossRef]
10. Zare, S.G.A.; Moore, S.A.; Rennie, C.D.; Seidou, O.; Ahmari, H.; Malenchak, J. Estimation of Composite Hydraulic Resistance in Ice-Covered Alluvial Streams. *Water Resour. Res.* **2016**, 1306–1327. [CrossRef]
11. Beltaos, S. River Ice Jams: Theory, Case Studies, and Applications. *J. Hydraul. Eng.* **1983**, *109*, 1338–1359. [CrossRef]
12. Zabilansky, L.J.; Hains, D.B.; Remus, J.I. Increased Bed Erosion Due to Ice. In Proceedings of the 13th International Conference on Cold Regions Engineering, Orono, ME, USA, 23–26 July 2006; pp. 1–12. [CrossRef]
13. Wu, P.; Balachandar, R.; Sui, J. Local Scour around Bridge Piers under Ice-Covered Conditions. *J. Hydraul. Eng.* **2016**, *142*, 04015038. [CrossRef]
14. Beltaos, S. Hydraulic Roughness of Breakup Ice Jams. *J. Hydraul. Eng.* **2001**, *127*, 650–656. [CrossRef]
15. Li, S.S. Estimates of the Manning's Coefficient for Ice-Covered Rivers. *Proc. Inst. Civ. Eng. Water Manag.* **2012**, *165*, 495–505. [CrossRef]
16. Wuebben, J.L. A Preliminary Study of Scour under an Ice Jam. In Proceedings of the 5th Workshop on Hydraulics of River Ice/Ice Jams, Winnipeg, MB, Canada, 21–24 June 1988; pp. 177–192.
17. Zare, S.G.A.; Moore, S.A.; Rennie, C.D.; Seidou, O.; Ahmari, H.; Malenchak, J. Boundary Shear Stress in an Ice-Covered River during Breakup. *J. Hydraul. Eng.* **2016**, *142*, 04015065. [CrossRef]
18. Ambtman, K.E.D.; Hicks, F.E. Field Estimates of Discharge Associated with Ice Jam Formation and Release Events. *Can. Water Resour. J.* **2012**, *37*, 47–56. [CrossRef]
19. Zabilansky, L.J. Ice Cover Effects on Bed Scour: Case Studies. In Proceedings of the 11th International Conference on Cold Regions Engineering, Anchorage, AK, USA, 20–22 May 2002; pp. 795–803. [CrossRef]
20. Ambtman, K.E.D.; Hicks, F.E.; Steffler, P.M. Experimental Investigation of the Pressure Distribution beneath a Floating Ice Block. *J. Hydraul. Eng.* **2011**, *137*, 399–411. [CrossRef]

21. Ettema, R.; Constantinescu, G.; Melville, B.W. *Evaluation of Bridge Scour Research: Pier Scour Processes and Predictions*; Natl. Coop. Highw. Res. Progr. (NCHRP Proj. 24-27(01)); Transportation Research Board: Washington, DC, USA, 2011. [CrossRef]
22. Ackermann, N.L.; Shen, H.T.; Olsson, P. Local Scour around Circular Piers under Ice Covers. In *Ice in the Environment: Proceedings of the 16th IAHR International Symposium on Ice, Dunedin, New Zealand, 2–6 December 2002*; Department of Physics, University of Otago: Dunedin, New Zealand, 2002; pp. 149–155.
23. Hains, D.B.; Zabilansky, L.J. *Laboratory Test of Scour under Ice: Data and Preliminary Results*; ERDC/CRREL TR-04-9; US Army Corps Eng. Res. Dev. Center, Cold Reg. Res. Eng. Lab.: Hanover, NH, USA, 2004.
24. Valela, C.; Nistor, I.; Rennie, C.D.; Lara, J.L.; Maza, M. Hybrid Modelling for Design of a Novel Bridge Pier Collar for Reducing Scour. *J. Hydraul. Eng.* **2020**. [CrossRef]
25. Sirianni, D.A.B.; Valela, C.; Rennie, C.D.; Nistor, I.; Almansour, H. Effects of Developing Ice Covers on Local Bridge Pier Scour. *J. Hydraul. Res.* **2020**.
26. Jamieson, E.C.; Post, G.; Rennie, C.D. Spatial Variability of Three-Dimensional Reynolds Stresses in a Developing Channel Bend. *Earth Surf. Process. Landf.* **2010**, *35*, 1029–1043. [CrossRef]
27. Ettema, R.; Kirkil, G.; Muste, M. Similitude of Large-Scale Turbulence in Experiments on Local Scour at Cylinders. *J. Hydraul. Eng.* **2006**, *132*, 33–40. [CrossRef]

water

MDPI

Article

Local Scour around Tandem Double Piers under an Ice Cover

Liansheng Sang [1], Jun Wang [1,*], Tiejie Cheng [1], Zhixing Hou [1] and Jueyi Sui [2,*]

1 School of Civil and Hydraulic Engineering, Hefei University of Technology, Hefei 230009, China; 13665513421@163.com (L.S.); hfut_chengtj@126.com (T.C.); houhfut@163.com (Z.H.)
2 School of Engineering, University of Northern British Columbia, Prince George, BC V2N 4Z9, Canada
* Correspondence: junwanghfut@126.com (J.W.); jueyi.sui@unbc.ca (J.S.); Tel.: +86-137-0551-0008 (J.W.); +1-250-960-6399 (J.S.)

Abstract: Compared to the scour around a single pier, the local scour process around tandem double piers is much more complicated. Based on laboratory experiments in a flume, we conducted the scour process around tandem double piers under an ice-covered flow condition. The results showed that when the pier spacing ratio $L/D = 2$ (where L = the pier spacing distance, and D = the pier diameter), the rear pier (the downstream one) will intensify the horseshoe vortex process behind the front pier, and the scour depth around the front pier will increase by about 10%. As the pier spacing ratio L/D increases, the scour depth around the front pier will gradually decrease. When the pier spacing ratio $L/D = 5$, sediment scoured around the front pier begins to deposit between these two piers. To initiate a deposition dune between piers, the pier spacing distance under an ice-covered condition is about 20% more than that under an open flow condition. The results also showed that the existence of the rear pier will lead to an increase in the length of the scour hole but a decrease in the depth of the scour hole around the front pier. The local scour around the front pier interacts with the local scour of the rear pier. The maximum scour depth of the scour hole around the rear pier increases first, then decreases and increases again afterward. When the pier spacing ratio $L/D = 9$, the scour depth around the rear pier is the least. With an increase in the pier spacing ratio, the influence of the local scour around the front pier on the local scour around the rear pier gradually decreases. When the pier spacing ratio L/D is more than 17, the scour around the front pier has hardly any influence on that around the rear pier. The scour depth around the rear pier is about 90% of that around the front pier.

Keywords: ice cover; local scour depth; tandem double piers; pier spacing distance

Citation: Sang, L.; Wang, J.; Cheng, T.; Hou, Z.; Sui, J. Local Scour around Tandem Double Piers under an Ice Cover. *Water* **2022**, *14*, 1168. https://doi.org/10.3390/w14071168

Academic Editor: Maria Mimikou

Received: 27 February 2022
Accepted: 29 March 2022
Published: 6 April 2022

1. Introduction

The local scour around bridge piers/abutments has always been a concern of researchers and engineers. Accurately predicting the scour depth around bridge piers is important for the safety design of a bridge. To date, many studies regarding the local scour process around bridge piers under open flow conditions have been reported. Aksoy et al. [1] carried out experiments to study the scour around a circular pier. The mechanism of the local scour process has been studied. An empirical formula has been proposed to describe the relationship between the scour depth and flow intensity, water depth, and dimensionless time. Using both uniform and non-uniform sands, Chang et al. [2] conducted experiments to study the local scour around bridge piers, and proposed a model describing the local scour depth over a non-uniform sand bed under a steady flow condition. Kothyari et al. [3] carried out laboratory experiments to study local scour around circular piers under the conditions of both steady and unsteady flows. An empirical formula for calculating the time variation in the scour depth was proposed, and on this basis, the expression for the maximum scour depth was obtained by considering the influence of the inhomogeneity and stratification of sediment and the instability of water flow on the scour depth. Sheppard and Miller [4,5] conducted laboratory experiments to study the local scour around bridge piers under movable bed conditions, and proposed an empirical formula. Based on the experimental data, Pandey et al. [6] developed a new formula for predicting the maximum scour depth in front of piers. Compared to other formulas, the calculated results using their formula agree well with those of observed value.

Under an open flow condition, some studies on local scour around multiple piers or pile groups have been conducted. The results of an experimental study of local scour around pile groups conducted by Ataie-Ashitani and Beheshti [7] showed that the local scour depth around pile groups sometimes increases twice as much as that around a single pier. Using their experimental data and data of other researchers, the correction coefficient for predicting the maximum local scour depth around pile groups was proposed. Kim et al. [8] carried out numerical simulation of the local scour around both tandem double piers and side-by-side double piers, and pointed out that the maximum scour depth around tandem double piers increases with the pier spacing distance first, then decreases gradually with the pier spacing distance, and finally reaches a stable value. They also claimed that the maximum scour depth around side-by-side double piers decreases with the increase in the pier spacing distance. Wang et al. [9] conducted experiments to study the scour problem around tandem double piers by considering different pier spacing distances, flow velocities, and sediment particle sizes. They claimed that the scour depth around the front pier is nearly the same as that around a single pier. Moreover, depending on the flow velocity, the scoured channel bed around the rear pier can be divided into four different zones: a no scouring zone, a synchronous scouring zone, an excessive scouring zone, and a radial offset area. The relationship between the pier spacing distance and the deviation of the radial offset zone was obtained. Liu et al. [10] also studied the local scour around tandem double piers, and claimed that the presence of the transition zone is related to the sediment transportation process from the deposition dune downstream of the front pier to the scour hole around the rear pier. Formulas for determining the critical velocity and predicting the downstream scour depth have been derived. Khaple et al. [11] pointed out that when the pier spacing ratio L/D ranges from 4 to 8, the maximum scour depth around the rear pier of the tandem double piers decreases with the increase in the pier spacing ratio L/D. When $8 < L/D < 10$, the maximum scour depth around the rear pier increases with the pier spacing ratio, and the maximum scour depth around the rear pier is the smallest when $L/D = 8$. When $10 < L/D < 12$, the maximum scour depth around the rear pier approaches a constant which is about 0.80 to 0.87 times of the maximum scour depth around a single pier.

However, the presence of an ice cover in a river imposes a solid boundary to flow. The velocity profile under an ice-covered flow condition is totally different compared to that under an open channel flow. Under an ice-covered flow condition, the maximum velocity occurs between the channel bed and the bottom of the ice cover and is dependent on the relative roughness of these two boundaries [12,13]. Compared to the open flow condition, the scouring process around bridge piers under an ice cover is much more complicated. Bacula and Dargahi [14] carried out experiments to study the local scour around a bridge pier under an ice-covered flow condition. They claimed that the scour depth under an ice-covered flow condition is much greater than that under an open flow condition. Ackerman et al. [15] pointed out that the maximum scour depth under an ice-covered flow condition increases by about 25% to 35% compared to that under an open flow condition. Based on laboratory experiments, Hains [16] claimed that the presence of ice cover leads to about a 21% increase in the local scour depth around bridge piers compared to that under an open flow condition. Wu et al. [17] investigated the scour around a cylindrical bridge pier under ice-covered flow conditions. Empirical formulas have been proposed to calculate the depth and radius of scour holes around piers under both open flow and ice-covered flow conditions. Using three non-uniform natural sands, Wu et al. [18–21] studied the local scour process around both square and circular abutments under both smooth and rough ice-covered flow conditions. They reported that the maximum scour depth around abutments under an ice-covered flow condition depends mainly on the flow Froude number, $D50/H$, and H/D. An empirical formula was proposed to calculate the maximum scour depth around abutments. By introducing the densiometric Froude number, the influence of geometric shape on the maximum scour depth around abutment was analyzed. It was concluded that the influence of the shape coefficient on the maximum scour depth under ice-covered flow conditions is less than that under an open flow condition. The maximum scour depth increases with the ice cover roughness. Jafari and Sui [22] studied velocity field and turbulence structure around spur dikes with different angles of orientation under ice-covered flow conditions. They reported that the strongest velocity fluctuation occurs immediately above the scour hole surface and very close to the dike tip. With the increase in the dike angle toward upstream, the scour hole becomes larger. Wang et al. [23] compared the difference between the rate of scour depth under an ice-covered flow condition and that under an open flow condition. Their results show that the rate of the scour depth under an ice-covered flow condition is larger than that under an open flow condition. Namaee and Sui [24–28] investigated the local scour problem around side-by-side bridge piers under both open flow and ice-covered flow conditions. Their results showed that the maximum scour depth around piers increases with the ice cover roughness, densiometric Froude number, and pier size. The maximum scour depth decreases with the decrease in the pier size and the increase in pier spacing distance and the particle size of the armor layer in the scour hole. The results of the numerical model were also compared with those of laboratory experiments, and the feasibility of the sediment transport model was verified using the Meyer-Peter and Muller equation.

In summary, many research works have been carried out to study the local scour process around a single pier, multi-piers, and pile group under open flow conditions. To date, however, only a few studies have been conducted to investigate the local scour process around bridge piers or abutments under ice-covered flow conditions. Regarding the local scour process around multiple piers, to our knowledge, only five research works have been reported [24–28]. However, the setup of piers in their studies is a side-by-side installation. The scouring process around tandem double piers under ice-covered flow conditions have never been investigated. In the present study, laboratory experiments were carried out to study the impacts of flow velocities, water depth, and pier spacing distance on the mechanism of local scour around tandem double piers under an ice-covered flow condition.

2. Experiment Setup and Methodology

2.1. Experimental Setup

Laboratory experiments were conducted in the hydraulic laboratory of Hefei University of Technology. The flume used in this experimental study was 26.68 m long and 0.4 m wide. In total, 22 cross sections (CS) for measurements along the flume with an equal spacing distance of 1.2 m were set up. Water flowed out of the water holding tank located upstream of the flume. A triangle weir downstream of the water holding tank was used to determine the flow rate into the flume with the accuracy of 0.1%. A tailgate located at the end of the flume allowed the water to spill over into a downstream reservoir. The downstream reservoir was equipped with a sediment trap and was connected to the suction pipe of a centrifugal pump. The pumping system circulated the water between the downstream reservoir and the water holding tank located upstream of the flume. In the upstream tank, a multi-layer grid was installed to dissipate water energy and reduce water flow oscillation at the flume entrance. Along the left side flume wall, at each cross section, a set of pressure transducers was installed with the accuracy of 0.1 mm, and water levels along the flume were measured (see Figure 1). Between CS-2 and CS-22, a sand bed with an initial thickness of 10 cm was prepared. The median grain size (d50) of the sand bed material was 0.714 mm and the inhomogeneity coefficient η was 1.61. All experiment runs belonged to clear-water scouring. Styrofoam panels were used to model ice cover. The model pier was installed at CS-16. At CS-16, a transparent ruler was affixed to the flume side walls in order to record data such as the initial scouring depth around the piers. The experiment layout is shown in Figure 1.

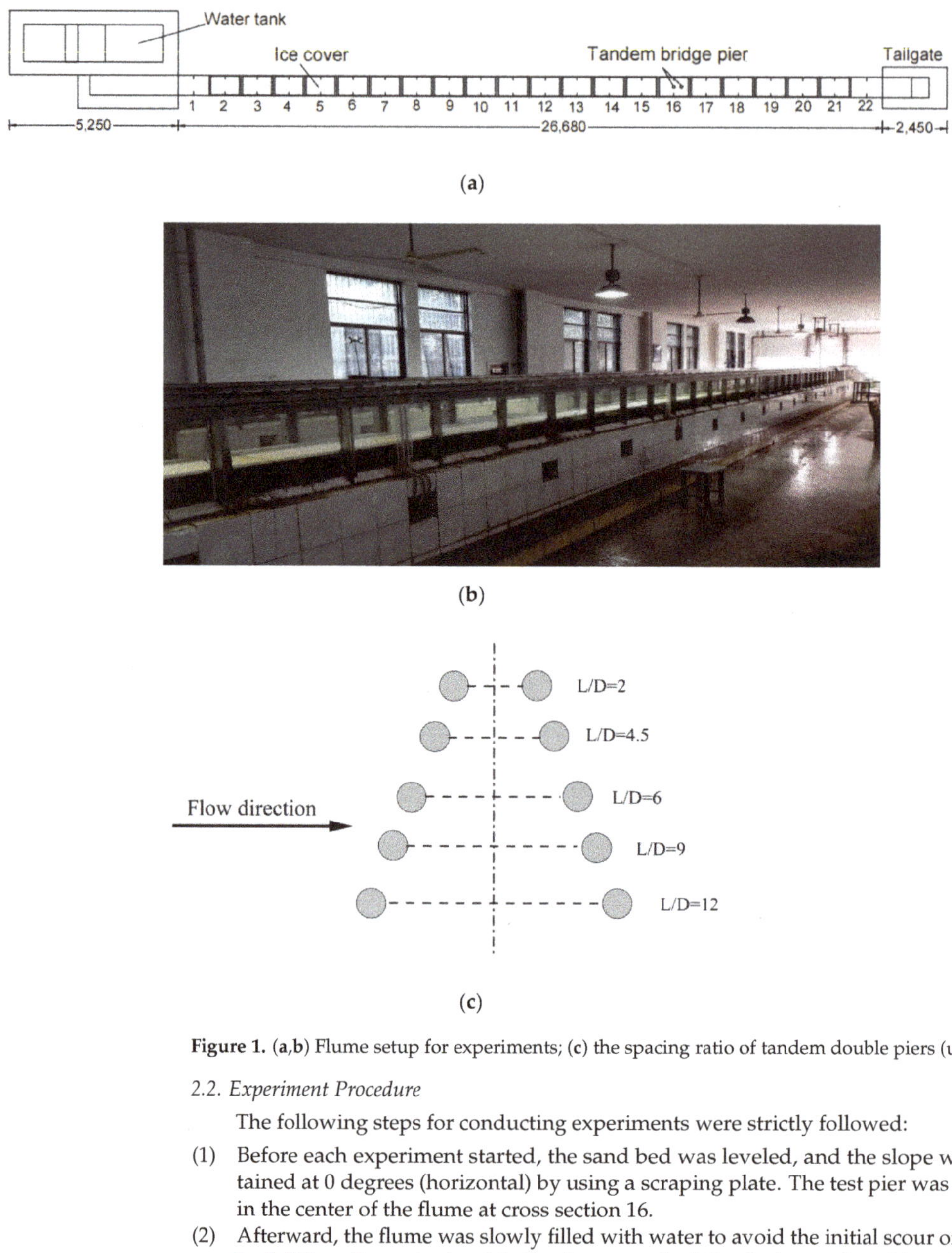

Figure 1. (**a**,**b**) Flume setup for experiments; (**c**) the spacing ratio of tandem double piers (unit: mm).

2.2. Experiment Procedure

The following steps for conducting experiments were strictly followed:

(1) Before each experiment started, the sand bed was leveled, and the slope was maintained at 0 degrees (horizontal) by using a scraping plate. The test pier was installed in the center of the flume at cross section 16.
(2) Afterward, the flume was slowly filled with water to avoid the initial scour of channel bed. When the water level in the flume reached the designated position, the water filling process was finished.
(3) Then, the model ice cover was carefully placed on the water surface. With this step, the preparation for this experiment was completed.
(4) To start each experiment run, a designated discharge through the triangle weir from the water holding tank was kept as a constant. The head (water surface) over the triangular weir was controlled by a point gauge with an accuracy of 0.1 mm. The water level in the flume was maintained at a designated level by adjusting the tailgate at the

downstream end of the flume. Under such a flow condition, this set of experiments was kept running.

(5) At CS-16 where the pier is located, both scour depth and water levels at each cross section were measured once every two minutes during the first half hour, once every five minutes during the 2nd half hour, once every 10 min during the 3rd half hour, and finally, once every 30 min until the end of each experiment run. Experiments showed that after about 6 h, no significant changes in the scouring process and the quasi-equilibrium depth of the scour hole were achieved. However, all experiment runs lasted 16 h to ensure that the local scour process around the tandem double piers reached an equilibrium state. Under such an equilibrium condition, the shape of the deposition dune does not change since the scour process at tandem double piers stops completely, although some sand particles reciprocate inside scour holes but cannot be washed out of scour holes. Then, the model ice cover was removed carefully. By reducing the incoming flow rate from the upstream water holding tank, and increasing the water level by raising the tailgate, the flow velocity in the flume was kept basically at a standstill. Thus, the shape of the scour hole did not change. Then, the bathymetry of the scour hole and deposition dune around the pier was measured by using a point gauge with the accuracy of 0.1 mm. In total, for each survey of the bathymetry of the scour hole and deposition dune, the measurements were conducted at 48 points, including inside the scour hole, along the middle ridge line of the mound, and the outer contour line of the mound. Some variables used in this study are defined and shown in Figure 2.

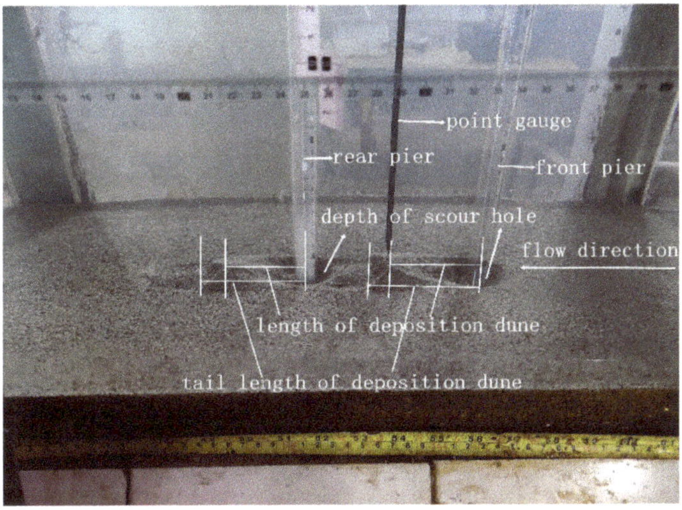

Figure 2. Measurement using a point gauge.

In total, 39 experiments were conducted under different flow conditions (flow depth, velocity) and pier spacing ratios, as shown in Table 1. The flow velocities in this experimental study range from 0.16 to 0.22 m/s.

Table 1. Experimental conditions and results.

Serial Number	V (m/s)	H (m)	D (m)	L (m)	Serial Number	V (m/s)	H (m)	D (m)	L (m)
A1	0.2	0.1	0.02	—	A21	0.2	0.2	0.02	0.09
A2	0.2	0.1	0.02	0.04	A22	0.2	0.2	0.02	0.10
A3	0.2	0.1	0.02	0.09	A23	0.2	0.2	0.02	0.12
A4	0.2	0.1	0.02	0.12	A24	0.2	0.2	0.02	0.17
A5	0.2	0.1	0.02	0.18	A25	0.2	0.2	0.02	0.18
A6	0.2	0.1	0.02	0.22	A26	0.2	0.2	0.02	0.19
A7	0.2	0.1	0.02	0.24	A27	0.2	0.2	0.02	0.24
A8	0.2	0.1	0.02	0.26	A28	0.2	0.2	0.02	0.32
A9	0.2	0.1	0.02	0.32	A29	0.2	0.2	0.02	0.34
A10	0.2	0.1	0.02	—	A30	0.2	0.25	0.02	—
A11	0.2	0.15	0.02	0.04	A31	0.2	025	0.02	0.04
A12	0.2	0.15	0.02	0.09	A32	0.2	0.25	0.02	0.09
A13	0.2	0.15	0.02	0.12	A33	0.2	0.25	0.02	0.12
A14	0.2	0.15	0.02	0.18	A34	0.2	0.25	0.02	0.15
A15	0.2	0.15	0.02	0.24	A35	0.2	0.25	0.02	0.18
A16	0.2	0.2	0.02	—	A36	0.2	0.25	0.02	0.24
A17	0.2	0.2	0.02	0.02	B1	0.16	0.2	0.02	0.18
A18	0.2	0.2	0.02	0.04	B2	0.18	0.2	0.02	0.18
A19	0.2	0.2	0.02	0.05	B3	0.22	0.2	0.02	0.18
A20	0.2	0.2	0.02	0.06	—	—	—	—	—

Note: The flow velocity (V) is the average approaching velocity and division of the discharge by the flow area. The value of H represents the approaching flow.

3. Results and Discussions

3.1. Analysis of Experiment Data

Ample data were collected for these 39 experiments. Table 2 summarizes some measurement results of these experiments. The variation in the depth of the scour hole with time around the front pier of the tandem double piers is compared to that around a single pier case. One can see from Figure 3 that under the same flow condition (approaching flow depth and velocity), the variation in the depth of the scour hole with time around the front pier of the tandem double piers is nearly the same as that around the single pier. When the spacing distance between piers is large enough, the existence of the rear pier hardly affects the local scour around the front pier, and the scour depth around the front pier of tandem double piers is nearly the same as that around a single pier.

As shown in Figure 4, when the pier spacing ratio L/D is less than 5, due to the horseshoe vortex between the front pier and the rear pier, the scoured particles around the front pier are delivered directly into the scour hole around the rear pier and are unable to deposit in the zone between these two piers. Therefore, the scour hole around the front pier is extended to the scour hole around the rear pier to form a large scour hole. With the increase in the pier spacing ratio $L/D > 5$, the effect of the "horseshoe" vortex becomes weak. Consequently, a deposition dune downstream of the front pier begins to develop (between the front pier and rear pier). With a further increase in the pier spacing ratio L/D, both the volume and height of the deposition dune gradually increase. Thus, the interaction between the deposition dune and the rear pier also gradually becomes strong. When the pier spacing ratio $L/D = 9$, the development of the deposition dune between two piers is complete, and the tail of the deposition dune (downstream of the front pier) has a ring shape surrounding the outer edge of the scour hole around the rear pier. When the pier spacing ratio $L/D > 9$, both the shape and size of the deposition dune between two piers hardly change. With a further increase in the pier spacing ratio, the interaction between the deposition dune downstream of the front pier and the local scour around the rear pier gradually becomes weak. When the pier spacing ratio L/D reaches 17, the deposition dune downstream of the front pier has hardly any impact on the local scour around the rear pier, and the local scour process around each of the tandem double piers can be treated as the local scour around a single pier.

Table 2. Experiment results.

Serial Number	Maximum Depth of Scour Hole		Maximum Length of Scour Hole		Maximum Height of Deposition Dune		Maximum Length of Deposition Dune	
	Front Pier	Rear Pier	Front Pier	Rear Pier	Front Pier	Rear Pier	Front Pier	Rear Pier
A1	2.63	—	10.4	—	1.94	—	14.6	—
A2	2.6	1.4	—	—	—	2.06	—	13
A3	2.61	1.6	—	—	—	2.15	—	13.5
A4	2.63	1.7	10	—	1.1	1.46	—	11
A5	2.58	1.54	11.7	—	2.02	1.53	16	12.2
A6	2.6	1.4	11.7	—	2	1.46	14.5	14
A7	2.6	1.5	11.5	—	1.63	1.37	18	14
A8	2.62	1.58	12.3	9.5	1.76	1.37	16.7	13
A9	2.6	2.25	10.4	8.7	1.9	1.74	15	12.7
A10	2.42	—	8.8	—	1.79	—	12.6	—
A11	2.5	1.27	—	—	—	1.73	—	14
A12	2.46	1.48	11	—	—	1.77	—	15
A13	2.4	1.66	10.8	7.1	0.93	1.53	—	16.3
A14	2.37	1.63	9.2	4.6	1.72	1.12	16.2	14.7
A15	2.41	1.7	10.2	8.2	1.71	1.1	14.8	11
A16	2.2	—	9.7	—	1.79	—	12	—
A17	2.43	1.1	—	—	—	1.77	—	11.4
A18	2.22	1.39	—	—	—	1.98	—	11.1
A19	2.22	1.36	—	—	—	2.21	—	10
A20	2.15	1.27	—	—	—	2.33	—	9
A21	2.2	1.52	—	—	—	2.11	—	9.3
A22	2.44	1.17	—	—	—	1.86	—	9.9
A23	2.35	1.26	9.3	6.1	1.32	1.45	8.7	9.7
A24	2.23	1.46	9.6	8.3	1.77	1.37	9.6	10.2
A25	2.2	1.28	9	6.3	1.62	1.04	10	9
A26	2.2	1.42	9.3	6.1	1.83	1.22	9.5	7.5
A27	2.2	1.91	8.9	7.3	1.7	1.39	13	10.3
A28	2.22	2	9.4	8.3	1.9	1.7	11.5	9.5
A29	2.38	0.81	9.1	8.1	1.96	1.78	11.3	9.6
A30	2	—	8.9	—	1.71	—	10.5	—
A31	2.09	0.93	—	—	—	1.75	—	6.7
A32	2.03	1.27	—	—	—	1.76	—	7.7
A33	1.95	1.4	9.4	—	1.38	1.33	8.2	9.1
A34	2	1.3	9.6	6.5	1.58	1.15	10.4	7.5
A35	2.02	1.45	9.3	7.3	1.67	1.16	10.4	7.5
A36	2.03	1.66	8.5	7.6	1.68	1.35	10.9	10.2
B1	0.99	0.87	6.2	6	0.78	0.63	8.1	7.6
B2	1.58	1.27	6.7	6.3	1.43	0.98	7.4	7.3
B3	2.83	1.75	12.3	7.9	1.35	1.65	13.5	18.3

Figure 5 presents the variation in tail length of a deposition dune (L_S) with time (T_i). One can see from Figure 5 that the tail length of a deposition dune around the front pier of tandem double piers is much greater than that for a single pier. This is because that the presence of the rear pier causes changes in the velocity field between the deposition dune downstream of the front pier and the scour hole around the rear pier, and thus, the effect of the "horseshoe" vortex causes more sediment erosion. The tail length of the deposition dune downstream of the rear pier decreases. This is because of the upward flow in front of the front pier and the deposition dune, which results in the decrease in shear stress near the channel bed around the rear pier. Thus, the local scour around the rear pier is weakened, and the tail length of the deposition dune downstream of the rear pier is reduced.

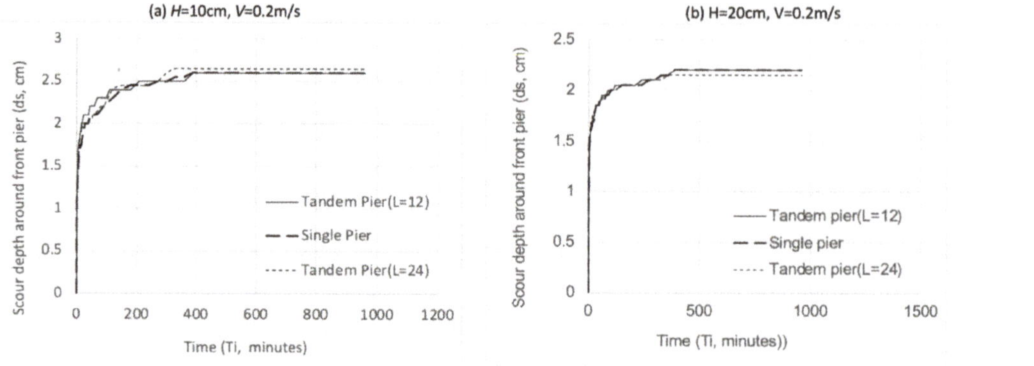

Figure 3. Variation in depth of scour hole with time around the front pier of the tandem double piers compared to that around a single pier condition. (**a**) H = 10 cm, V = 0.2 m/s; (**b**) H = 20 cm, V = 0.2 m/s).

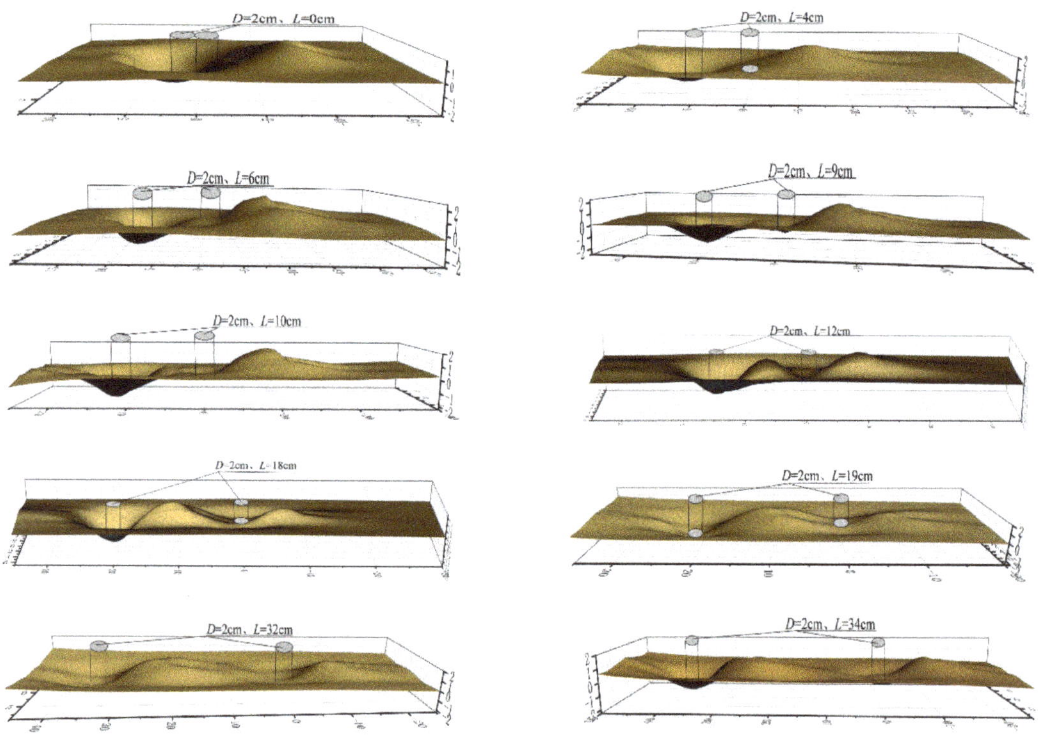

Figure 4. Three-dimensional local scour patterns around tandem double piers for different pier spacing ratio L/D (Note: H = 20 cm, D = 2 cm, and V = 0.2 m/s).

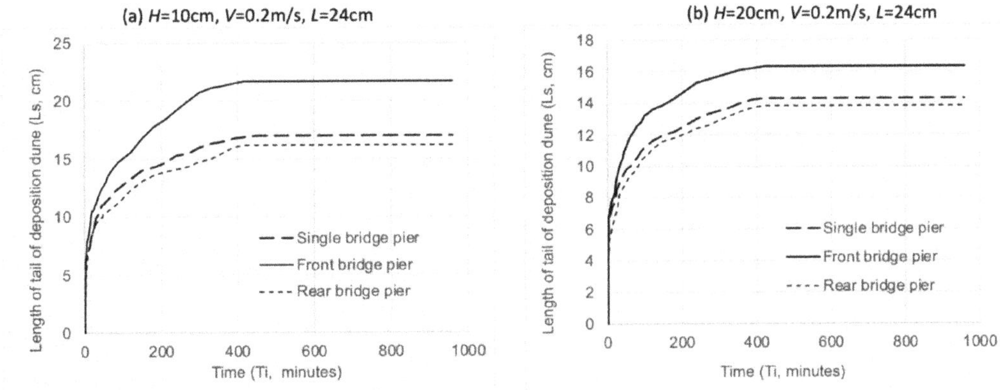

Figure 5. Variation in tail length of deposition dune (L_S) with time (T_i). (**a**) H = 10 cm, V = 0.2 m/s, L = 24 cm; (**b**) H = 20 cm, V = 0.2 m/s, L = 24 cm).

Both the height and length of deposition dunes downstream of tandem double piers have been compared to those of a single pier. As shown in Figure 6, compared to results for the single pier case, in the presence of tandem double piers, the height of the deposition dune downstream of the front pier is lower and its length is longer. This result is consistent with the conclusion from Figure 5. Obviously, the presence of the rear pier should be responsible for this change, since the presence of the rear pier changes the velocity field between the deposition dune in front of the rear pier and the scour hole around the rear pier. Therefore, the transport of sediment particles from the deposition dune became faster, which resulted in a lower but longer deposition dune downstream of the front pier. One can see from Figures 5 and 6 that both the height and length of the tail of the deposition dune downstream of the rear pier are smaller than those of the single pier case. This result reflects the impact of the deposition dune downstream of the front pier. When the spacing distance between piers is small, the deposition dune downstream of the front pier will be scoured and sediment particles will be delivered into the scour hole around the rear pier, which makes the scour process around the rear pier more complicated.

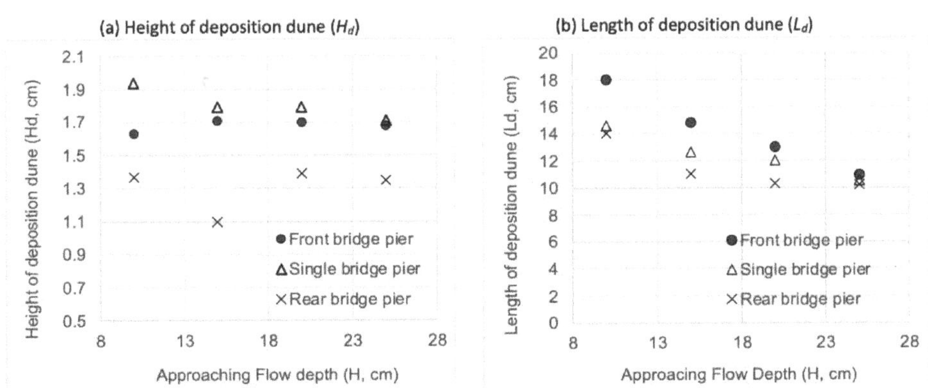

Figure 6. The height and length of deposition dunes downstream of tandem double piers compared to those of a single pier (V = 0.2 m/s, L = 24 cm).

One can see the variations in the length of the deposition dune (Figure 7a), and the variation in the length of the tail of the deposition dune downstream of the rear pier (Figure 7b). The results for the pier spacing distance of L = 12 cm are compared to those for the pier spacing distance of L = 24 cm. The results show that in the presence of tandem double piers, both the length of the deposition dune and the length of the tail of the deposition dune downstream of the rear pier are smaller than those for the single pier case. This is because the deposition dune downstream of the front pier lifts up the flowing current and thus weakens the local scour process around the rear pier. Due to the interaction between the rear pier and the deposition dune downstream of the front pier, the length of the deposition dune downstream of the rear pier for the case of the pier-spacing-distance of 12 cm is smaller than that for L = 24 cm.

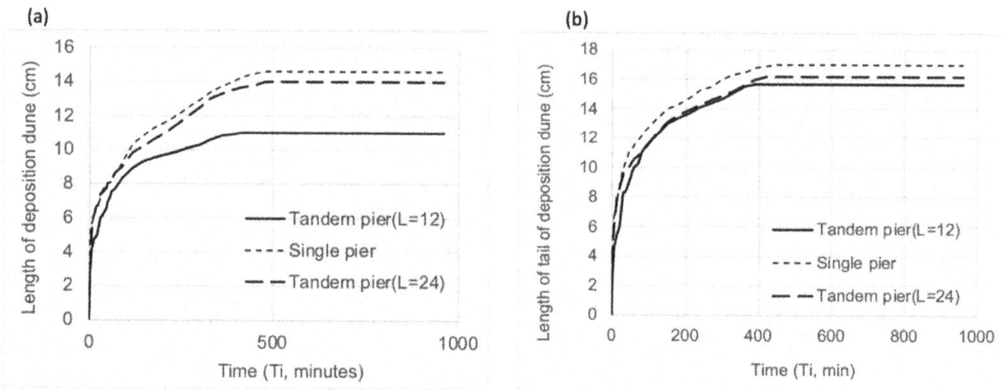

Figure 7. (**a**) Variations in the length of deposition dune downstream of rear pier; (**b**) the length of tail of deposition dune downstream of rear pier. (H = 10 cm, V = 0.2 m/s).

Figure 8 shows the relationship between scour depth and pier spacing ratio. The pier scour contour diagrams for different pier spacing distance are presented in Figure 9. It can be seen from Figure 8 that with the increase in the pier spacing distance (L/D), the dimensionless scour depth (ds/D) around the front pier does not change much and keeps as a constant depending on the flow condition (approaching flow velocity). When the pier spacing distance is L/D = 5, the deposition dune around the front pier begins to initiate between two piers. With the increase in the pier spacing distance, both the volume and height of the deposition dune between two piers gradually increase, and thus, the interaction between the deposition dune and rear piers also gradually increases. From Figures 8b and 9, one can see that when L/D = 9, the tail of the deposition dune downstream of the front pier forms a ring-shaped ridge surrounding the scour hole around the rear piers, and the depth of the scour hole around the rear pier reaches the minimum. At this time, downstream of the deposition dune around the front pier, there exists a ring-shaped deposition dune surrounding the outer edge of the scour hole around the rear pier, and the depth of the scour hole around the rear pier will reach the minimum. When L/D > 9, the shape of the deposition dune between two piers does not changes. With increasing pier spacing distance, the interaction between the deposition dune around the front pier and the local scour around the rear pier gradually weakens. When L/D = 17, there is hardly an interaction between the deposition dune around the front pier and the scour hole around the rear pier; the scouring process around the tandem double piers can be considered as a scour process of two individual single piers. However, due to the presence of an ice cover together with the sediment transport and local scour process, some energy is lost along the flow, including the flows through the scour hole around the front pier, so that the depth of the scour hole around the rear pier is about 90% of that around the front pier.

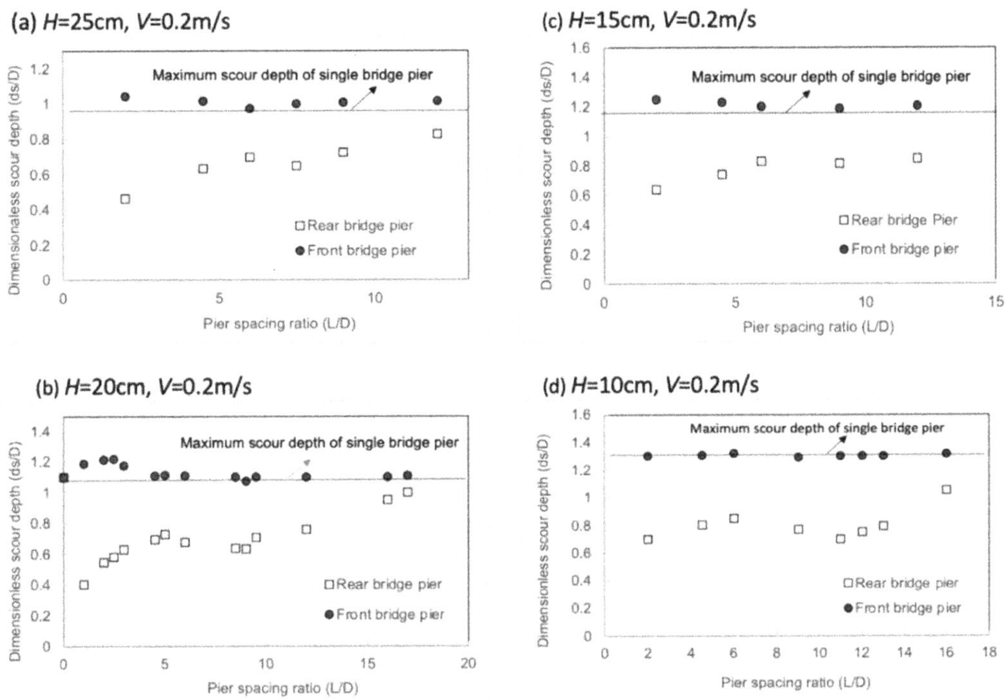

Figure 8. Relationship between scour depth and pier spacing ratio. (**a**) H = 25 cm, V = 0.2 m/s; (**b**) H = 20 cm, V = 0.2 m/s; (**c**) H = 15 cm, V = 0.2 m/s; (**d**) H = 10 cm, V = 0.2 m/s).

3.2. Empirical Expression for Calculating Scour Depth

The dimensionless number $(L_{sp}/D)*(D_{50}/D)^2*\eta$ is used to describe the influence of the pier spacing distance. The relationship between $(L_{sp}/D)*(D_{50}/D)^2*\eta$ and the flow Froude number (F_r) is presented in Figure 10. The data for open flow conditions were obtained from Liu (2018), who conducted experiments using tandem double piers. The results of the analysis show that under an ice-covered flow condition, the presence of the front pier causes a more significant effect with a larger influencing range compared to that under an open flow condition.

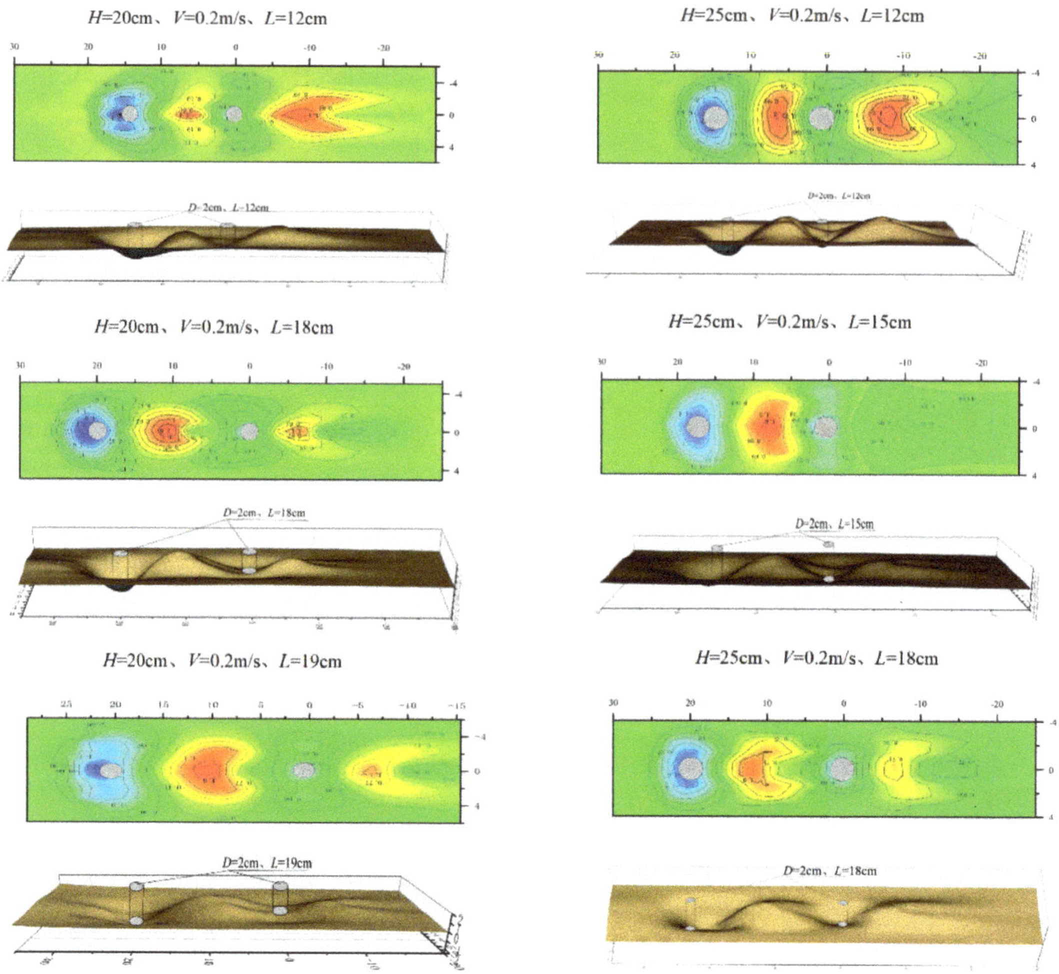

Figure 9. Pier scour contour diagrams for different pier spacing distances.

The depth of a scour hole around a pier is affected by hydraulic conditions, bed material, and the pier spacing distance, as shown in Figure 11. Considering a bridge pier in a river with steady and uniform flow, the following parameters may influence the depth of a scour hole (d_S) around the pier: (1) median grain size of the bed material (D_{50}); (2) flow Froude number (F_r); (3) pier spacing distance (L); (4) pier diameter (D). Thus, the relationship between dimensionless local scour depth under an ice-covered flow condition and factors affecting the depth of the scour hole around a pier can be expressed as follows:

$$\frac{d_s}{D} = kF_r{}^a \left(\frac{L}{D}\right)^b \left(\frac{D_{50}}{H}\right)^c \tag{1}$$

where k, a, b, and c are coefficients.

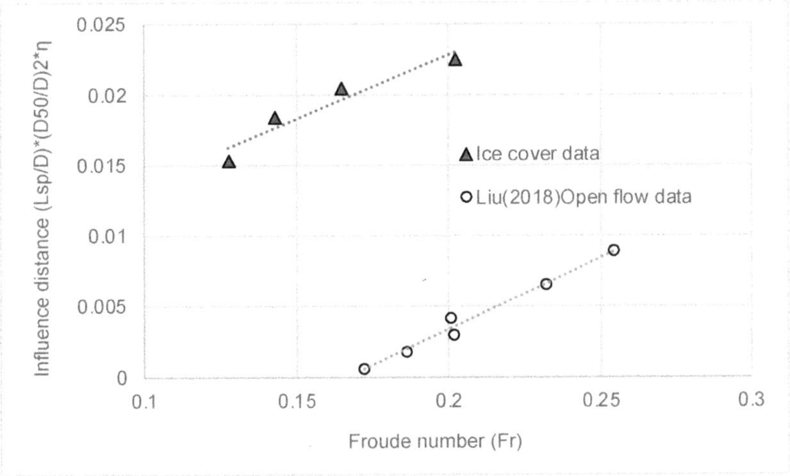

Figure 10. Relationship between the influence distance of the front pier on the rear pier and the flow Froude number.

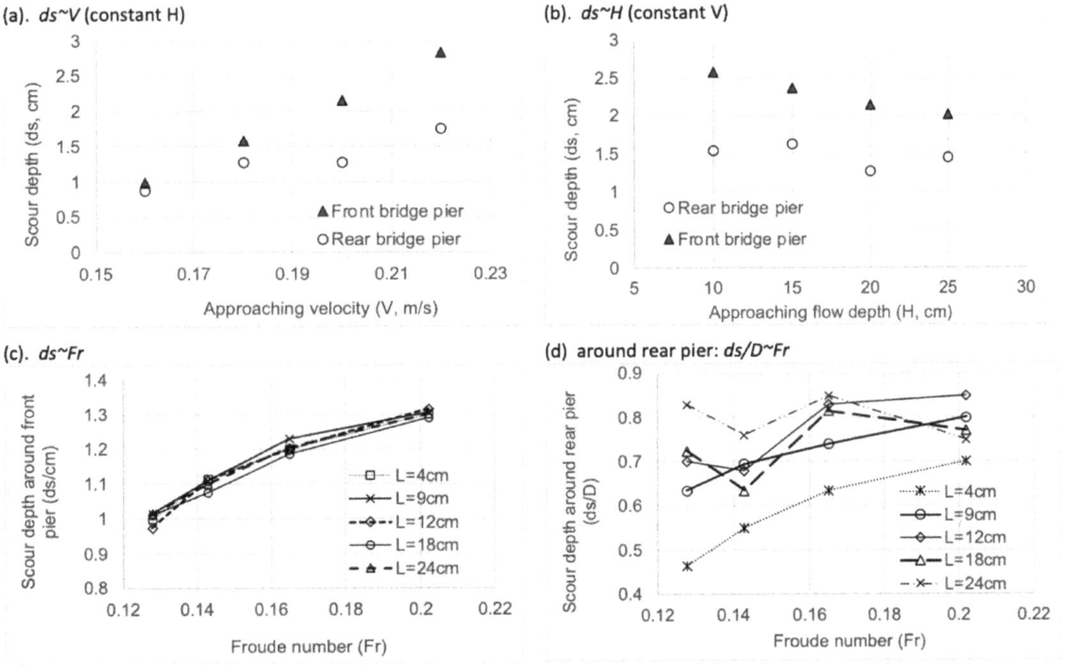

Figure 11. Relationship between scour depth (*ds*, or d_S/D) and hydraulic conditions (*V*, *H*, or *Fr*).

Based on data collected from laboratory experiments, the following relationships for determining the scour depth around both the front pier and the rear pier under an ice-covered flow condition are obtained, respectively:

Scour depth around the front pier:

$$\frac{d_s}{D} = F_r^{2.462}\left(\frac{L}{D}\right)^{-0.031}\left(\frac{D_{50}}{H}\right)^{-0.858} \quad (2)$$

Scour depth around the rear pier:

$$\frac{d_s}{D} = F_r^{1.462}\left(\frac{L}{D}\right)^{0.189}\left(\frac{D_{50}}{H}\right)^{-0.475} \quad (3)$$

One can see from Equations (2) and (3) that with the increase in the flow Froude number (F_r), the dimensionless depths (d_s/D) of the scour hole around both the front and rear piers increase. The scour depth (d_s/D) around the front pier increases with the decrease in the pier spacing ratio (L/D) between piers. However, the scour depth (d_s/D) around the rear pier increases with the increase in the pier spacing ratio (L/D) between piers. Figure 12 shows the observed scour depths around the front pier and rear pier compared to those calculated using Equations (2) and (3), respectively. One can see that the calculation results agree well with the observations.

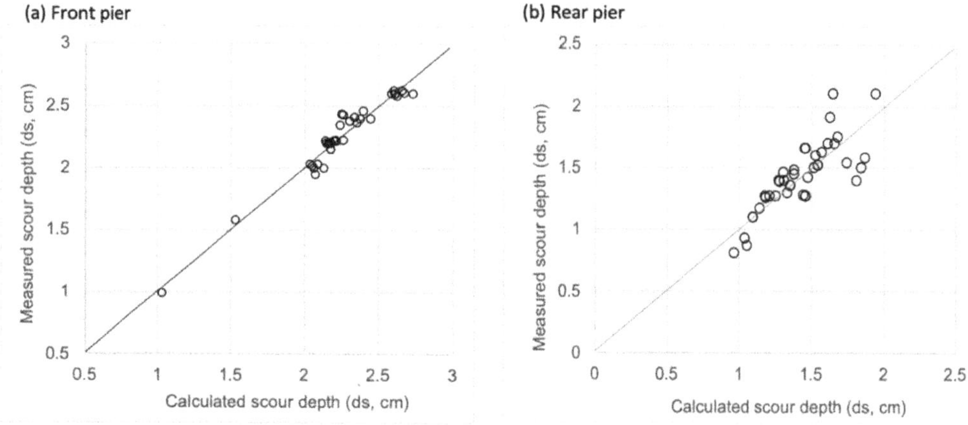

Figure 12. Comparison between calculated scour depths and those of observations. (**a**) Front pier; (**b**) Rear pier.

4. Conclusions

The present experimental study mainly focused on the influence of the pier spacing distance on the depth of scour holes and the height of deposition dunes downstream of piers in ice-covered flow conditions. The results of the present study have been compared to those of the local scour around a single pier under the same conditions. The following conclusions have been drawn from this experimental study:

1. Under an ice-covered flow condition, the patterns of local scour around the tandem double piers depends on the spacing distance between the piers. If the spacing distance between the piers is zero, the local scour process around the tandem double piers should be treated as a single pier scour. With the increase in the spacing distance between piers, the scour depth around the front pier under an ice-covered flow condition increases by about 10% compared to that under a single pier case. Moreover, with the increase in the pier spacing ratio, the scour depth around the front pier will gradually decrease. When the pier spacing ratio $L/D = 5$, sediment scoured at the front

pier begins to deposit between these two piers. To initiate a deposition dune between piers, the pier spacing distance under an ice-covered condition is about 20% more than that under an open flow condition. The existence of the rear pier will increase the length of the scour hole but reduce the depth of the scour hole around the front pier. The maximum scour depth around the rear pier increases first, then decreases and increases again afterward. When the pier spacing ratio $L/D = 9$, the local scour depth around the rear pier is the least. With the increase in the pier spacing ratio, the influence of the local scour around the front pier on the local scour around the rear pier gradually decreases. When the pier spacing ratio L/D is more than 17, the local scour around the front pier has hardly any influence on that around the rear pier. The scour depth around the rear pier is about 90% of that around the front pier.

2. Under the same experimental conditions, compared with the results of local scour around a single pier, when the pier spacing ratio $L/D > 4.5$, the height of the deposition dune downstream of the front pier is lower, but the length of the deposition dune downstream of the front pier is longer. Both the height and length of the deposition dune downstream of the rear pier are smaller than those of the single pier case.
3. Relationships for calculating the scour depth around both the front pier and rear pier under ice-covered flow conditions were obtained. With the increase in the flow Froude number (F_r), the depths (d_s/D) of the scour hole around both the front and rear piers increase. The scour depth (d_s/D) around the front pier increases with the decrease in the pier spacing ratio (L/D) between piers. However, the scour depth (d_s/D) around the rear pier increases with the increase in the pier spacing ratio (L/D) between piers. The calculation results using proposed equations agree well with those of observations.

Author Contributions: L.S.: Laboratory works, data curation, formal analysis, methodology, and writing—original draft preparation; J.W.: Conceptualization, laboratory supervision, methodology, funding acquisition, writing—original draft preparation; T.C.: Laboratory works, data curation, investigation; Z.H.: Laboratory works, data curation, investigation; J.S.: Conceptualization, formal analysis, methodology, writing—review and editing. All authors have read and agreed to the published version of the manuscript.

Funding: This research was supported by the National Natural Science Foundation of China (grant nos. 51879065). The authors are grateful for the financial support.

Institutional Review Board Statement: Not applicable.

Informed Consent Statement: Not applicable.

Data Availability Statement: The data are available in the case that it is required.

Conflicts of Interest: The authors declare no conflict of interest.

References

1. Aksoy, A.O.; Bombar, G.; Arkis, T.; Guney, M.S. Study of the time-dependent clear water scour around circular bridge piers. *J. Hydrol. Hydromech.* **2017**, *65*, 26–34. [CrossRef]
2. Chang, W.Y.; Lai, J.S.; Yen, C.L. Evolution of Scour Depth at Circular Bridge Piers. *J. Hydraul. Eng.* **2004**, *130*, 905–913. [CrossRef]
3. Kothyari, U.C.; Garde, R.C.J.; Raju, K.G.R. Temporal variation of scour around circular bridge piers. *J. Hydraul. Eng.* **1992**, *118*, 1091–1106. [CrossRef]
4. Sheppard, D.M.; Odeh, M.; Glasser, T. Large scale clear-water local pier scour experiments. *J. Hydraul. Eng.* **2004**, *130*, 957–963. [CrossRef]
5. Sheppard, D.M.; Miller, W. Live-bed local pier scour experiments. *J. Hydraul. Eng.* **2006**, *132*, 635–642. [CrossRef]
6. Pandey, M.; Sharma, P.K.; Ahmad, Z.; Singh, U.K. Experimental investigation of clear-water temporal scour variation around bridge pier in gravel. *Environ. Fluid Mech.* **2018**, *18*, 871–890. [CrossRef]
7. Ataie-Ashtiani, B.; Beheshti, A.A. Experimental investigation of clear-water local scour at pile groups. *J. Hydraul. Eng.* **2006**, *132*, 1100–1104. [CrossRef]
8. Kim, H.S.; Nabi, M.; Kimura, I.; Shimizu, Y. Numerical investigation of local scour at two adjacent cylinders. *Adv. Water Resour.* **2014**, *70*, 131–147. [CrossRef]

9. Wang, H.; Tang, H.W.; Liu, Q.S.; Wang, Y. Clear-water local scouring around three piers in a tandem arrangement. *J. Hydraul. Eng.* **2016**, *59*, 888–896. [CrossRef]
10. Liu, Q.S.; Tang, H.W.; Wang, H.; Xiao, J.F. Critical velocities for local scour around twin piers in tandem. *J. Hydrodyn.* **2018**, *30*, 1165–1173. [CrossRef]
11. Khaple, S.; Hanmaiahgari, P.R.; Gaudio, R.; Dey, S. Interference of an upstream pier on local scour at downstream piers. *Acta Geophys.* **2017**, *65*, 29–46. [CrossRef]
12. Sui, J.; Faruque, M.A.A.; Balachandar, R. Local scour caused by submerged square jets under ice cover. *ASCE J. Hydraul. Eng.* **2009**, *135*, 316–319. [CrossRef]
13. Sui, J.; Wang, J.; He, Y.; Krol, F. Velocity profiles and incipient motion of frazil particles under ice cover. *Int. J. Sediment Res.* **2010**, *25*, 39–51. [CrossRef]
14. Batuca, D.; Dargahi, B. Some experimental results on local scour around cylindrical piers for open and covered flow. In Proceedings of the 3rd International Symposium on River Sedimentation, Jackson, MS, USA, 31 March–4 April 1986.
15. Ackermann, N.L.; Shen, H.T.; Olsson, P. Local scour around circular piers under ice covers. Ice in the Environment. In Proceedings of the 16th IAHR International Symposium on Ice, Dunedin, New Zealand, 2–6 December 2002.
16. Hains, D.B. An Experimental Study of Ice Effects on Scour at Bridge Piers. Ph.D. Thesis, Lehigh University, Bethlehem, PA, USA, 2004.
17. Wu, P.; Balachandar, R.; Sui, J. Local scour around bridge piers under ice-covered conditions. *J. Hydraul. Eng.* **2016**, *2*, 115–121. [CrossRef]
18. Wu, P.; Hirshfield, F.; Sui, J. Further studies of incipient motion and shear stress on local scour around bridge abutment under ice cover. *Can. J. Civ. Eng.* **2014**, *41*, 892–899. [CrossRef]
19. Wu, P.; Hirshfield, F.; Sui, J.; Wang, J.; Chen, P.P. Impacts of ice cover on local scour around semi-circular bridge abutment. *J. Hydrodyn.* **2014**, *26*, 10–18. [CrossRef]
20. Wu, P.; Hirshfield, F.; Sui, J. Local scour around bridge abutments under ice covered condition-an experimental study. *Int. J. Sediment Res.* **2015**, *30*, 39–47. [CrossRef]
21. Wu, P.; Hirshfield, F.; Sui, J. Armour layer analysis of local scour around bridge abutments under ice cover. *River Res. Appl.* **2015**, *31*, 736–746. [CrossRef]
22. Jafari, R.; Sui, J. Velocity field and turbulence structure around spur dikes with different angles of orientation under ice covered flow conditions. *Water* **2021**, *13*, 1844. [CrossRef]
23. Wang, J.; Li, Z.Q.; Cheng, T.J.; Sui, J. Experimental study on local erosion of piers under ice cover with time. *J. Hydraul. Eng.* **2021**, *52*, 1174–1182.
24. Namaee, M.R.; Sui, J. Impact of armour layer on the depth of scour hole around side-by-side bridge piers under ice-covered flow condition. *J. Hydrol. Hydromech.* **2019**, *67*, 240–251. [CrossRef]
25. Namaee, M.R.; Sui, J. Effects of ice cover on the incipient motion of bed material and shear stress around side-by-side bridge piers. *Cold Reg. Sci. Technol.* **2019**, *165*, 102811. [CrossRef]
26. Namaee, M.R.; Sui, J. Local scour around two side-by-side cylindrical bridge piers under ice-covered condition. *Int. J. Sediment Res.* **2019**, *34*, 355–367. [CrossRef]
27. Namaee, M.R.; Sui, J. Velocity profiles and turbulence intensities around side-by-side bridge piers under ice-covered flow condition. *J. Hydrol. Hydromech.* **2020**, *68*, 70–82. [CrossRef]
28. Namaee, M.R.; Sui, J.; Wu, Y.; Linklater, N. Three-dimensional numerical simulation of local scour around circular side-by-side bridge piers with ice cover. *Can. J. Civ. Eng.* **2021**, *48*, 1335–1353. [CrossRef]

water

MDPI

Article

Velocity Field and Turbulence Structure around Spur Dikes with Different Angles of Orientation under Ice Covered Flow Conditions

Rahim Jafari and Jueyi Sui *

School of Engineering, University of Northern British Columbia, Prince George, BC V2N 4Z9, Canada; jafari@unbc.ca
* Correspondence: jueyi.sui@unbc.ca; Tel.: +1-250-960-6399

Abstract: Spur dikes are well-known structures that are widely used in rivers and coastal regions. Depending on their types, sizes, and orientation angles, spur dikes can substantially change flow characteristics. Results of previous studies indicate that the presence of an ice cover in rivers can cause complicated flow structures. The present experimental study investigates velocity fields and turbulence structures in the vicinity of spur dikes under ice cover with different roughness coefficients. The spur dikes were set up at the following three angles of orientation, 90°, 60°, and 45°. Our results show that the strongest velocity fluctuation occurs immediately above the scour hole surface and very close to the dike tip. The increase in the dike angle toward upstream, the velocity component values increase, leads to a larger scour hole. Results show that an increase in dike angle of each 10° (from 45° to 90°) increases the scour depth between 5% and 10%, depending on flow conditions. Furthermore, the increase in the cover roughness coefficient and the blockage ratio of a spur dike leads to a further increase in turbulence kinetic energy and 3D velocity components values. The findings of this study imply that the appearance of an ice cover can increase turbulence intensities up to nearly 30%.

Keywords: spur dike; ice cover; acoustic doppler velocimeter (ADV); instantaneous velocity component; turbulence intensity; local scour

Citation: Jafari, R.; Sui, J. Velocity Field and Turbulence Structure around Spur Dikes with Different Angles of Orientation under Ice Covered Flow Conditions. *Water* **2021**, *13*, 1844. https://doi.org/10.3390/w13131844

Academic Editors: Yong G. Lai and Achim A. Beylich

Received: 10 May 2021
Accepted: 27 June 2021
Published: 1 July 2021

Publisher's Note: MDPI stays neutral with regard to jurisdictional claims in published maps and institutional affiliations.

1. Introduction

Spur dikes are extended structures where one end is at the riverbank, and the other one is projected toward the river flow [1]. These constructions have been widely used for many purposes, such as river bank protection, flood control, improvement of a navigational course, control of scour process, landscape improvement, and ecosystem restoration [2]. Regardless of the different types of spur dikes, they redirect flow from the river bank and affect the flow regime, flow velocity, sediment transportation, and consequently scour process [3].

If a spur dike allows the flow to pass through it, that is called permeable. If the dikes block and then repel the river flow, it is called impermeable. Since the channel width is reduced by impermeable spur dikes, the energy gradient becomes steeper, and flow velocity increases. Consequently, the riverbed will be eroded. The scour process will be affected by the shape and size of the spur dike, features of bed material, and flow conditions [4]. Zhang et al. [5] investigated bed morphology, sediment distribution, and flow field in a channel with a series of impermeable spur dikes on both sides of the channel. They claimed that the flow field and local scour around the most upstream pairs of spur dikes are affected the most. Teraguchi et al. [6] studied flow field and scour patterns around spur dikes based on laboratory experiments and numerical simulations. They pointed out that the scour holes on the upstream side of the impermeable spur dikes are larger and deeper than those of permeable ones. Mizutani et al. [7] investigated the impact of the spur dike height and grain size of riverbed material on the morphology and topography around an impermeable spur dike. It was found that the maximum scour depth is reduced with the increase of the median particle size of bed materials. Furthermore, field observations regarding the bed deformation around stone-lined spur dikes in the Akashi river indicate that the maximum scoured depth around the permeable spur dikes is reduced to 40% comparing to that around the impermeable ones [8]. Clearly, the local scour depth around the impermeable spur dikes is much deeper than the permeable ones.

The spur dikes in channels can be classified as submerged and non-submerged spur dikes depending on the flow conditions and water depth. Normally, the impermeable spur dikes are designed as non-submerged spur dikes because the overflow on the top of a dike can form a vertical jet just behind the spur dike. This vertical jet moves toward the downstream and causes a significant erosion of the downstream bank along with the body of the spur dike itself [9]. Moreover, in terms of the orientation angle of a spur dike which is defined as the angle between the dike axis and the river flow direction, spur dikes are divided into three groups: Attracting, deflecting, and repelling dikes. The attracting spur dike which points downstream averts the river flow toward the middle of the channel and protects the region on its downstream side. Deflecting spur dike is a well-known type that changes the river flow direction by not repelling it; this type of spur dike is normally used to locally protect the banks against erosion. They create a turbulent flow, cause more sediment transportation in the middle of the rivers and make the main channel deeper [2]. Muto et al. [10] conducted two experimental studies about the effects of the opening ratio and water depth on the velocity distributions. Their first study was based on experiments in the Yodo river in Japan using large-scale particle image velocimetry. The second one was conducted in a laboratory flume with a downscaled spur dike. They pointed out that the flow was highly unsteady inside the scour hole based on the experiment in the field. However, laboratory results did not show the same characteristics. As the authors concluded that the difference in the result of the field experiment from that of the laboratory was caused by the complex bathymetry of the natural river and lower Reynolds numbers of flows in the laboratory.

Further research shows that models developed from a steady, uniform flow cannot be applied for nonuniform flows [11,12]. Duan et al. [13] construe that dikes increase shear stress and turbulence intensities at the riverbed by converging water flow which initiates the local scour process around the dikes. A two-dimensional (2D) experimental model has been used to evaluate the effect of turbulence intensities on the entrainment of bed material (coarse sand and gravel). It appears that the instantaneous streamwise velocity has a higher rate compared to the instantaneous vertical velocity [14]. An experimental study on a three-dimensional (3D) flow illuminates that the local scour around the dike causes more complexity of the flow field [15]. Dey and Barbhuiya [16] assessed the turbulent flow field around a vertical-wall abutment and reported that the near-bed Reynolds stress played a significant role in transporting sediment and scour process.

The turbulence structure in an open channel flow has been investigated by many researchers, since it plays an important role in the flow characteristics and the local scour process around the dike. However, nearly all reported research has been conducted under the open channel flow condition. The impact of an ice cover on the local scour process around the spur dikes has not been investigated. In cold regions, such as Canada, ice covers appear on rivers' surfaces and may last for about six months, such as the upper reach and middle reach of the Fraser River. During an ice-covered flow condition, the river hydrology will be dramatically changed comparing to that under an open flow condition. An ice cover adds an extra hydraulic boundary to the flow in a river, which leads to considerable changes in the velocity profile, flow rate, bed shear stress distribution, sediment transport, and consequently scour pattern [17].

Under an ice-covered flow condition, the location of the maximum flow velocity gets closer to the channel bed, which increases the bed shear stress [18–20]. In conclusion, the presence of ice on the water surface makes the flow condition much more complicated than an open flow condition. Namely, the existence of an ice cover leads to the increase in the turbulence kinetic energy at the riverbed. Thus, the Reynolds shear stress under an ice-covered flow condition is different from that under an open flow condition. This change affects the incipient motion of bed material, rate of sediment transportation, and sediment suspension loads which make evaluating (predicting) local scour more complex [18,21,22]. To our knowledge, there is no research regarding the effect of an ice cover on the local scour process around the spur dikes. Moreover, measurements of the turbulence flow fields inside scour holes around the spur dikes under an ice-covered flow condition have never been conducted.

Many factors affect the design of a spur dike, such as the river width and depth, flow velocity, channel sinuosity, grain size of bed material, sediment transport rate, bank cohesiveness, the length and shape of a spur dike, the orientation angle of the spur dike to the flow, and construction materials. Understanding the characteristics of the flow in the vicinity of a spur dike under an ice cover will help to develop formulas that can accurately estimate the scour depth and elaborate models for the design of spur dikes. Additionally, it can provide a better understanding of the turbulence structure around obstacles equivalent to spur dikes, such as boulders. The specific objectives of the present experimental study are summarized as follows:

- The impact of an ice cover, including ice cover roughness on the turbulence intensities, 3D flow fields, shear velocity, and Reynolds stress around the spur dikes.
- The dependence of the scour morphology and turbulence structure around spur dikes on the orientation angle of the spur dike, ice cover roughness, and hydraulic condition.

2. Materials and Methods

2.1. Site Description

The present experimental study was carried out in a large-scale outdoor flume located in the Quesnel River Research Center, BC, Canada. The flume was 38.2 m long, 2.0 m wide, and 1.3 m deep. The longitudinal slope of the flume bed was 0.2%. To have a constant discharge throughout each experimental run, a holding tank feeds water into the flume. The holding tank located upstream of the flume had a volume of 90 cubic meters, was 40 m long, 2 m wide, and 1.3 m deep. Three valves provided a wide range of flow rates for various experimental runs (Figure 1). By adjusting three valves, the following three different flow rates were generated in this experimental study, 0.055 m^3/s, 0.105 m^3/s, and 0.12 m^3/s.

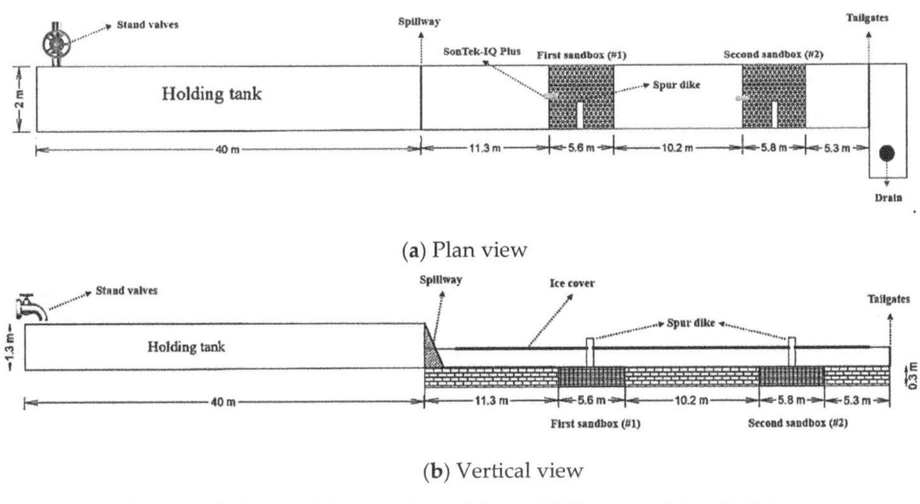

Figure 1. The layout of the experimental flume: (**a**) Plan view; (**b**) vertical view.

Due to the slope of the flume, water levels varied gradually from upstream to downstream. Therefore, the flow was classified as a nonuniform flow. Another important feature that influences the flow is the aspect ratio B/H (where *B* is the channel width and *H* is the flow depth), which is used to classify the channel (flume) either as a narrow (B/H < 5) or wide (B/H > 5) channel. A flow in a narrow channel is affected by the secondary currents from the banks (flume side walls) and causes the dip phenomena (the maximum velocity occurs below the flow surface). However, in the wide channel, the strength of the secondary currents reduces in the lateral direction of the flume [18]. In this experimental study, since the deepest flow depth was 35 cm and the flume width was 200 cm, all flows in this flume belong to wide channel flows (B/H > 5).

There were two sandboxes which are spaced 10.2 m from each other. These sandboxes were 2 m wide and 0.3 m deep. The upstream sandbox was 5.6 m long, and the downstream one was 5.8 m long (Figure 2). Three types of nonuniform sands with the median grain size (D_{50}) of 0.9 mm, 0.6 mm, and 0.48 mm were used in the sandboxes. One side of the flume wall along each sandbox was made of plexiglass to have a clear view of the scour process. A staff gauge was located in the middle of each sandbox to measure the water depth during experimental runs. The impermeable model spur dike, made of marine plywood (had a dimension of 80 cm height, 5 cm width, and 50 cm length), was installed in the middle of each sandbox. The model dike was placed at the bottom of each sandbox (namely, 30 cm buried in the sand) so that 50 cm of the model dike was exposed to the flow. Of note, the model spur dike was non-submerged for all experimental runs in this study.

Figure 2. The view of the original site (experimental flume).

The tailgate at the downstream end of the flume was used to control the water depth and flow velocity in the flume (Figure 2). Since the flume was very long with a longitudinal slope of 0.2%, water depths were different in these two sandboxes for each experimental run, which creates a wide range of water depths for this experimental study.

Since the ice cover roughness was one of the main factors affecting the local scour process, two types of model ice cover (smooth and rough) were used. Styrofoam panels were used to model the smooth cover, while the rough cover was made by attaching small 2.5 cm Styrofoam cubes to the bottom of the Styrofoam panels. Sixteen panels with dimensions of 1.99 × 2.4 m were used to cover almost the entire surface of the flow. For the experiments with ice cover, before each experimental run, the sixteen Styrofoam panels were put side by side on the flume bed. By opening the valves very slowly, water feeds inside the flume gradually. The Styrofoam started to float on the surface of the water. There were some ropes at the end of the flume to keep the Styrofoam panels floating side by side on the water surface through the experiment. Li [23] proposed the equation for calculating the roughness coefficient of an ice cover. He found that Manning's coefficient for ice-covered rivers averages from 0.013 to 0.04. By using Equation (2), the roughness coefficient of the rough ice cover was determined averagely as 0.03. The roughness coefficient for the smooth ice cover was considered 0.013, derived from Manning's value for the smooth concrete [24].

$$k_s = 30\ y_i \exp[-(1 - v_i/v_{max})] \quad (1)$$

where, y_i is the thickness of the ice-affected layer, v_i is the averaged velocity of the ice-affected layer, v_{max} is the maximum velocity of the velocity profile.

$$n_i = 0.039\ k_s^{1/6} \quad (2)$$

where, k_s is the average roughness height of the ice underside in meters.

2.2. Apparatuses for Measurements

There are many different apparatuses or devices for measuring flow velocities, such as Laser Doppler velocimeter (LDV), Particle Image velocimeter (PIV), Particle Tracking velocimeter (PTV), Acoustic Doppler velocimeter (ADV), Electro Magnetic velocimeter (EMV), Pitot Static Tube (PST), etc. [25–27]. LDV, ADV, EMV, and PST are all point instruments. Some of them, such as the LDV and PIV, have some limitations in their working range [28]. As pointed out by researchers in the literature, the ADV is accurate for measuring the turbulence properties of flows [29–31]. Thus, for 3D measurements of instantaneous velocity components, a 10-MHz SonTek ADV (SonTek-A Xylem Brand, San Diego, CA, USA) was used in this research. The ADV includes one probe (transmitter) and three receivers. It acquires the instantaneous velocity components by the sampling volume positioned at the intersection of the transmitted and received acoustic beams, which is located 10 cm beneath the probe head. The sample volume was a cylinder with a diameter of 0.61 cm and a height of 0.72 cm. The ADV's sampling volume is larger than those of the Laser-based velocimetry devices, such as the LDV and PIV [13]. The ADV cannot acquire the velocity very close to the flow surface, and the velocity profiles were not continuous to the top layer of the flow. However, this limitation does not affect the velocity analysis because the significant change of the velocity distribution occurs from the mid-water depth toward the flume bed and particularly inside of the scour hole.

The ADV measures the scattering particles' velocities in the flow. Therefore, the accuracy of the measurements depends on the quality and quantity of particles inside the sampling volume. To acquire robust data from ADV measurements, two auxiliary parameters provided in the ADV files, the signal-to-noise ratio (SNR) and the correlation (COR), should be evaluated [31]. The SNR represents the relative density of the particulate matter in the flow. The COR, which value varies from 0 to 100, indicates the relative consistency of the particle velocity scattering within the sampling period [29]. According to the user manual for ADV and previous studies, to achieve the most accurate data in this experimental study, these two parameters were set as follows: SNR > 15, and COR > 70% [30,31]. The Water Resources Research Laboratory of the US Bureau of Reclamation developed a software program known as Win ADV for filtering the ADV data files. In this study, Win ADV was used for data filterings, such as SNR and COR. Regarding previous research, to obtain the most accurate data at each measuring location, the data sampling rate of ADV was set with the highest frequency (25 Hz) with a duration of 120 to 150 s [11,30,31]. Following the proper setting of ADV and filtering data spikes, the velocity measurements were reliable within the range of 0.25 cm/s, with an error of ±1.5% of the measurement scale [20,32].

In the present study, a SonTek-IQ Plus (SonTek-A Xylem Brand, San Diego, CA, USA) was used to measure the approaching flow rate, average velocity, and water depth. There are two types of SonTek-IQ (standard and plus); the Plus edition contains advanced post-processing functions, providing deeper insight into approaching flow and volume data, making it very precise and robust [33]. The SonTek-IQ Plus contains six measuring beams (sensors); four of these are velocity beams, which monitor the flow velocity along both the longitudinal and latitudinal axis to secure the best possible coverage and most accurate depiction of the velocity field. The remaining two measuring beams, the pressure sensor, and vertical acoustic beams work together to measure the water level precisely [33]. According to the SonTek-IQ Series User's Manual, the data collected using the SonTek-IQ Plus is subjected to a 1% error in the range of measurement scale. The SonTek-IQ Plus was installed on the flume bed in front of each sandbox (the location of the SonTek-IQ is depicted in Figure 1). The flume shape, size, and exact location of the device were set in the SonTek-IQ software. Then based on these inputs and using the data collected by sensors, the software calculates the average flow velocity, water depth, and flow rate during each experimental run.

2.3. Data Collection

Before each experimental run, the surfaces of the sandboxes were leveled. Then the flume was filled with water slowly to avoid initial scouring around the spur dikes. After reaching the desired water levels, the valves opened completely, and the experimental run began. The initiation of the local scour process at the dike tip was observed from the very first minute of each run. A scour hole slowly develops and surrounds the upstream side of the dike to the flume wall. Then, as time goes on, the scour hole becomes deeper and wider, and extends to the downstream of the dike as well.

One of the important factors affecting the scour process is the time needed to reach the equilibrium condition. Equilibrium scour is defined as the condition when the dimensions of scour hole do not change with time. Zhang et al. [34] studied the local scour process around a spur dike and reported that 90% of the equilibrium depth happened about 2 h after the experiment started. Other researchers claimed that 80% of the maximum scour depth occurred at the first two hours of the experiment [35]. Namaee and Sui [20] conducted experimental research on local scour around side-by-side piers. They indicate that the equilibrium depth of a scour hole could be achieved within the first 6 h, although all their experimental runs lasted 24 h. In the present study, no change in scour depth was observed after about 12 h, and the scour hole reached its equilibrium condition. Therefore, each experimental run lasts for 24 h to make sure that the scour process was entirely completed. Additionally, some experiments were conducted for 48 and 72 h to investigate the time needed for achieving the equilibrium condition. Results confirm that after 24 h, there was no change in the scour hole around the spur dike comparing to those of experiments that lasted for 48 and 72 h.

After the scour process reached the equilibrium condition, velocity components in all three dimensions around the dike and inside scour hole were recorded by using a 10-MHZ ADV. Regardless of the dike layout, water surface condition, and flow depths, the velocity measurements were taken from the bottom of the scour hole to the water surface with the intervals of 1 cm for shallower flow depths and 2 cm for other flow depths (Figure 3).

Results showed that the highest turbulence intensity of the flow occurred in the center of the scour hole, where the maximum scour depth occurred. Therefore, the ADV locale for the velocity measurements of all experimental runs was at the deepest part of the scour hole, which was very close to the tip of the spur dike. For this experimental study, U_x (x-axis) is used to describe the streamwise velocity in the downstream direction, U_y (y-axis) is used to denote the lateral velocity in the transverse direction pointing to the left bank, and the vertical velocity is expressed as U_z in the z-axis (towards the water surface). Under an ice-covered flow condition, a small part of the Styrofoam panel around the spur dike was cut for placing the ADV probe in the flow for acquiring velocity data. After 24 h, all valves were closed completely, and the flume was drained gradually. Then the scour pattern was measured (Figure 4). Some of the experimental runs were repeated to validate the recorded data.

Figure 3. 3D Velocity data collection using the 10-MHz SonTek ADV.

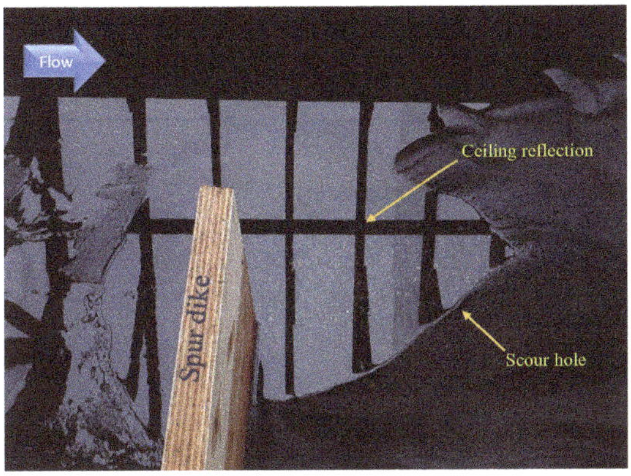

Figure 4. Plan view of the scour hole.

3. Results and Discussions

Throughout the scouring process, several horseshoe vortexes were detected inside of the scour hole. This vortex system has been observed in a clockwise direction. Some small bow waves on the surface of the upstream side of the dike were also observed (Figure 5). Since bow waves had opposite rotation directions to those of horseshoe vortexes, these two eddy systems were interfering with each other. Results indicate that these interferences become less noticeable with the increase in water depth. Moreover, by decreasing the dike orientation angle inclined to the downstream direction, the bow waves become smaller and less frequent. Moreover, for a dike with an orientation angle of 45° (toward downstream), these waves can hardly be observed, and their effect on the horseshoe vortex was negligible.

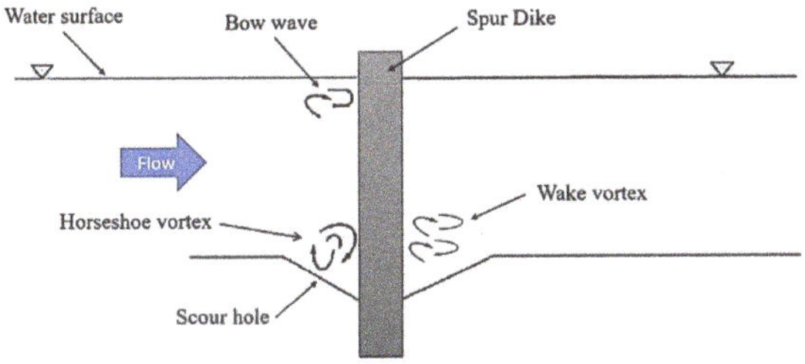

Figure 5. Development of vortex system around the spur dike (adapted from [36]).

Because of the flow separation at the dike tip and the formation of a powerful downflow at the upstream side of the dike, the horseshoe vortexes are created. As a consequence, a scour hole around the spur dike will be developed. The wake vortex system develops behind the spur dike, which results in the extension of the scour hole downstream of the dike. As pointed out by other researchers, wake vortexes are smaller and weaker than horseshoe vortexes; and they cannot carry sediment load that eroded from the scour hole [20,37]. This fact explains the development of the deposition ridge downstream of the dike (Figure 6).

Figure 6. Deposition ridge in the downstream side of the spur dike.

Inside the scour hole, horseshoe vortexes were created by the strong turbulence and a high level of instantaneous velocity fluctuation. Therefore, the deepest hole developed around the dike tip where the horseshoe vortex flow and downflow were stronger. By studying the 3D velocity inside a scour hole, the effects of different flow conditions and the dike setups on the local scour pattern should be assessed.

To better describe the laboratory findings which may be applicable in practical projects (rivers, spur dikes, bridge piers, etc.), the relationships between different parameters should be expressed using dimensionless variables. Furthermore, the normalization of variables (making them dimensionless) will enable us to compare results under different conditions of the experiment. Therefore, in all figures in the present study, both velocity components and water depth (h, the vertical distance at which 3D velocity data was collected) were normalized by average approaching flow velocity (U) and total water depth in the sandbox (H), respectively.

3.1. Streamwise Velocity Component (U_x)

Among all 3D velocity components, the streamwise velocity (U_x) plays a key role in developing the scour hole and the turbulence structure. Comparing to the other two velocity components (U_y, U_z), the streamwise velocity has the maximum value and highest fluctuation. Results show that regardless of the flow conditions and the orientation angle of the dike, velocity distributions inside the scour hole are less regular (or completely irregular). Moreover, the magnitudes of the streamwise velocity components inside the scour hole are smaller than those outside of the scour hole. However, in most cases, the highest level of fluctuation and the maximum velocity magnitude occur in the center of a scour hole, which is close to the dike tip.

In terms of the boundary conditions for the water surface, there were three different types, namely, open flow, smooth covered condition, and rough covered condition. As indicated in Figure 7, the streamwise velocity (U_x) inside the scour hole is minimum at the bottom of the scour hole and increases with the distance from the scour hole bottom.

The maximum streamwise velocity happens at the mid-water depth, and the streamwise velocity profile shows a convex shape. Of note, the ADV measuring volume is located 0.10 m from the probe head. Due to this limitation, the velocity profile cannot fully cover up to the water surface, as shown in Figure 7. Moreover, when a high level of turbulence existed at the measuring point using the ADV, the Doppler noise often appears [38]. These noises decrease with the data collection process and create data spikes. Moreover, sediment movement near the scour hole bed interferes with data collection during the ADV measurements [22]. By using the Win ADV software, these ambiguous data have been filtered. Therefore, sometimes velocity profiles cannot cover the bottom of the hole as well.

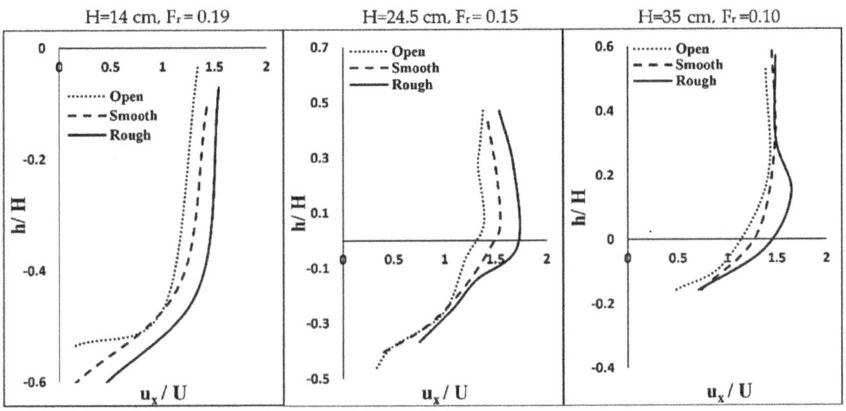

Figure 7. Streamwise velocity profile (u_x) under different surface cover conditions for the spur dike with an orientation angle of 90° for different water depths. ADV measurements subject to +/−0.25 cm/s error.

Regardless of the orientation angle of the dike and flow properties, under an ice-covered flow condition, the maximum velocity is located at the mid-depth of water. By increasing the roughness coefficient of the cover, the location of the maximum velocity is further shifted toward the channel bed. Additionally, cover conditions not only influence the location of U_{max}, but also affect the magnitude of the velocity as well (Figure 7). Results indicate that the rough ice cover can lead to an increase in the streamwise velocity values by nearly up to 25%. These findings are in good agreement with previous studies [17,18,20]. One can conclude that regardless of the shape and location of the barriers in an ice-covered river, the locale and magnitude of the maximum velocity depend on the features of an ice cover.

Under an ice-covered flow condition, the streamwise velocity (U_x) depends on the flow properties, such as water depth, cover roughness, and approaching velocity. In the present study, velocity profiles have been evaluated for all experiments. The flow Froud number (Fr) is one of the most important dimensionless parameters, and the effect of Fr on U_x was examined.

$$F_r = \frac{U}{\sqrt{gH}} \tag{3}$$

where, U is the average approaching velocity, g is the gravitational acceleration, and H is the water depth.

The result indicates that with the decrease in the flow Froud number, the streamwise velocity distributions under different surface conditions (open channel, smooth cover, and rough cover) become closer to each other, especially for velocity distributions under both open channel and smooth covered flow conditions. In Figures 7 and 8, the Froud numbers for water depths of 14 cm, 24.5 cm, and 35 cm are 0.19, 0.15, and 0.10, respectively. It appears that the effect of ice cover condition on the velocity distributions becomes intense with the increase in the flow Froud number. Moreover, as the dike orientation angle becomes smaller from 90° to 45°, this effect becomes more tangible (Figures 7 and 8).

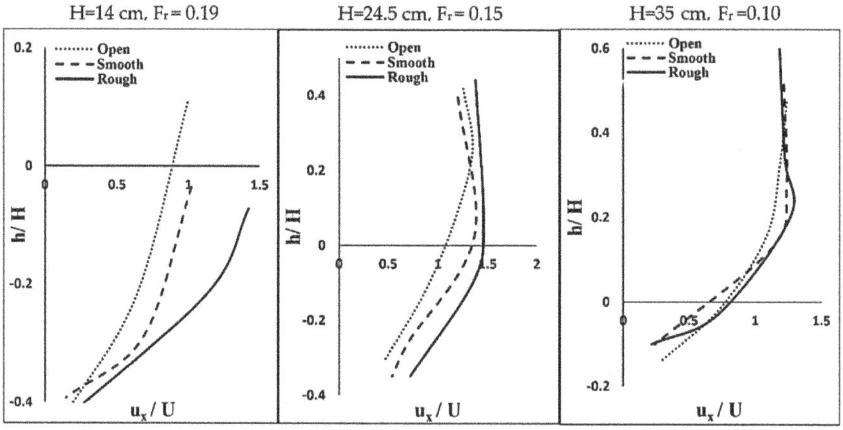

Figure 8. Streamwise velocity profile (u_x) inside and outside the scour holes under different surface cover conditions for the dike orientation angle of 45° for different water depths. ADV measurements subject to +/−0.25 cm/s error.

It is noted that as the dike orientation angle decreases, the magnitude of the streamwise velocity diminishes, and the scour hole becomes smaller. As indicated in Figures 9 and 10, regardless of the water depth and surface cover conditions, by decreasing the dike orientation angle from 90° to 45°, the velocity profiles are shifted upward, implying a decrease in the scour hole depth. These findings are consistent with the continuity theory (Equation (4)). Since the spur dike reduces the cross-section area of the flow, the velocity should increase. As the dike orientation angle decreases, the cross-section area increases, and velocity decreases.

$$(\rho AU)_{inlet} = (\rho AU)_{outlet} \tag{4}$$

where, ρ is the mass density of water, A is the cross-section area of the flow, and U is the average approaching velocity.

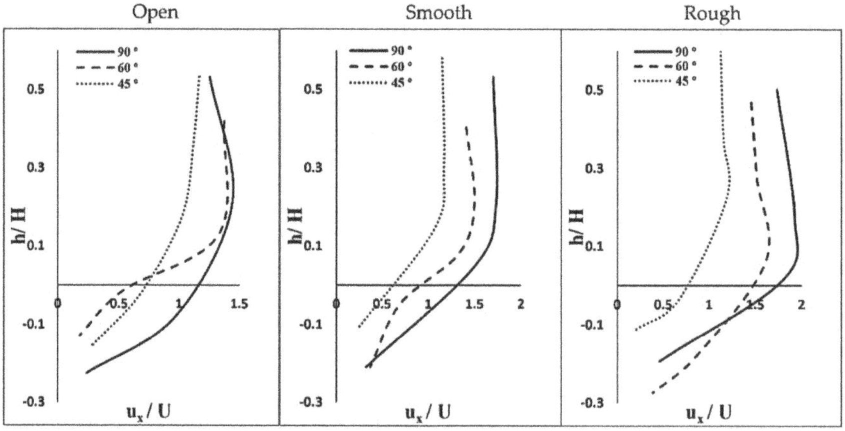

Figure 9. Streamwise velocity profile (u_x) inside and outside the scour holes for different dike orientation angles (water depth: 31 cm). ADV measurements subject to +/−0.25 cm/s error.

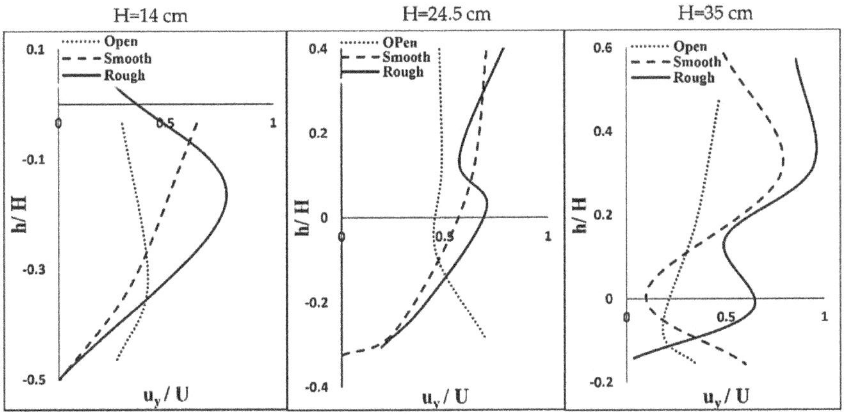

Figure 10. Lateral velocity profile (u_y) inside and outside the scour holes under different surface cover conditions for the dike orientation angle of 90° for different water depths. ADV measurements subject to +/−0.25 cm/s error.

With the decrease in the dike orientation angle (from 90° to 45°), the blockage ratio of the flow cross-section decreases. Therefore, both the downflow and horseshoe vortex become weaker. This leads to the formation of a smaller scour hole. Of note, these effects are independent of the surface cover condition, water depth, and flow rate (Figure 9).

3.2. Lateral Velocity Component (U_y)

Two significant characteristics of the lateral velocity component (U_y) are their irregularity and much smaller values compared to the streamwise velocity component (U_x). After scrutinizing U_y profiles (under different conditions of surface cover, flow rate, water depth, and dike orientation angle), results reveal that the vertical distribution of the lateral velocity component was nonmonotonic. Comparing to the streamwise velocity component (U_x), the lateral velocity component (U_y) is considerably smaller and mostly positive. Some non-negligible effects have been observed in the vertical distribution profiles of the lateral velocity component (U_y).

It is noted that the presence of an ice cover leads to an increase in the magnitude of the lateral velocity regardless of the flow rate, water depth, and dike orientation angle. With the increase in the cover roughness coefficient, the value of U_y increases (Figure 10). This result indicates the ice cover plays an important role in the value of U_y.

To determine the effects of the dike orientation angle on U_y, the vertical distribution profiles of the lateral velocity component (U_y) were evaluated. As indicated in Figure 11, with the decrease in the dike orientation angle, the magnitude of U_y decreases, similar to the effect of the dike orientation angle on the streamwise velocity. Furthermore, results indicate that only for the dike orientation angle of 90°, a semi-consistent pattern has been noticed for the vertical distribution profiles of the lateral velocity component (Figure 11). For the dike orientation angles of 60° and 45°, no clear or meaningful trend was detected (despite different flow rates, surface conditions, and bed materials). Nevertheless, in all cases, the vertical distribution profiles of the lateral velocity component (U_y) were positive inside and outside of the scour holes with an equivocal pattern. Thus, the locales of the maximum lateral velocity component in the vertical distribution profiles remain unknown (Figures 10 and 11).

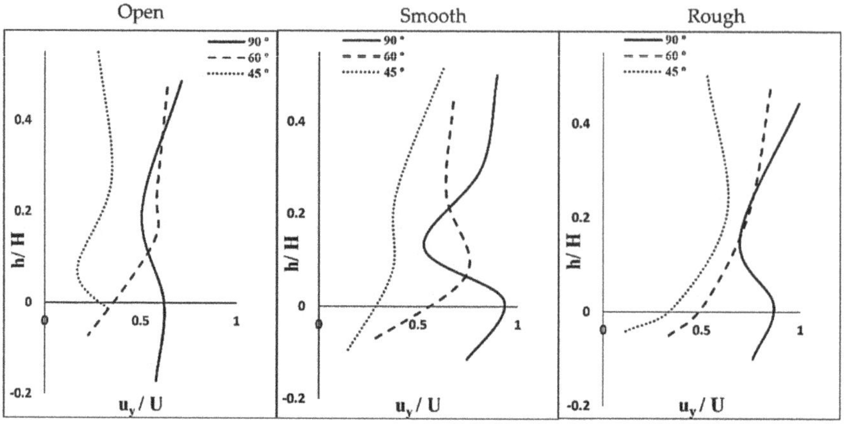

Figure 11. Lateral velocity profile (u_y) inside and outside the scour holes under different surface cover conditions for different dike orientations (water depth: 35 cm). ADV measurements subject to +/−0.25 cm/s error.

3.3. Vertical Velocity Component (U_z)

To evaluate the vertical velocity component (U_Z) under different experimental conditions, its vertical distribution in the center of the scour hole (the location of the maximum scour depth) has been examined. Results show that the values of the vertical velocity component are mostly negative inside and outside of the scour holes (Figure 12). These negative values signify the existence of the powerful downflow around the dike, which is created, due to the obstruction of the flow by the spur dike. This downflow plays an essential role in the local scour process because the downward velocity increases and strengthens the horseshoe vortex system, which leads to a deeper scour hole. This fact is consistent with the local scour morphology around the spur dike. As illustrated in Figure 13, for the dike, which is perpendicular to the flume wall (with the orientation angle of 90°), the scour hole is the deepest comparing to those of dikes with other orientation angles (60° and 45°). Thus, it can be concluded that by increasing the blockage ratio of the obstacle (such as piers and spur dikes), downward velocity will increase, and create a more powerful vertical velocity component inside the scour hole.

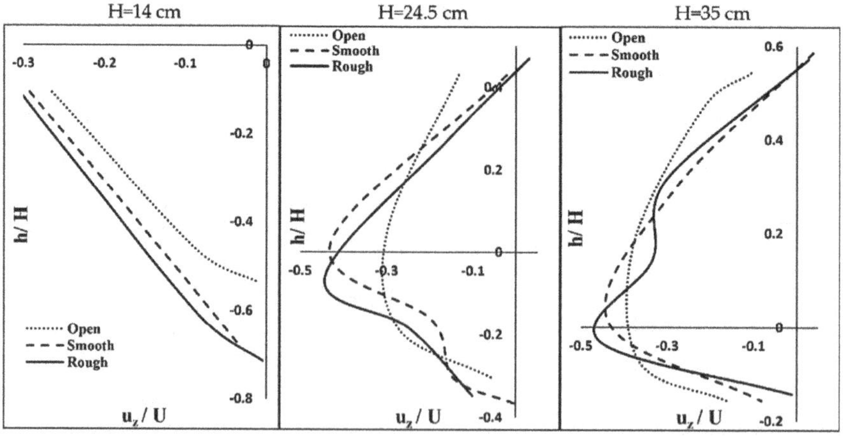

Figure 12. Vertical velocity profile (u_z) inside and outside the scour holes under the open channel, smooth covered, and rough covered flow conditions for the spur dike with an orientation angle of 90° for different water depths. ADV measurements subject to +/−0.25 cm/s error.

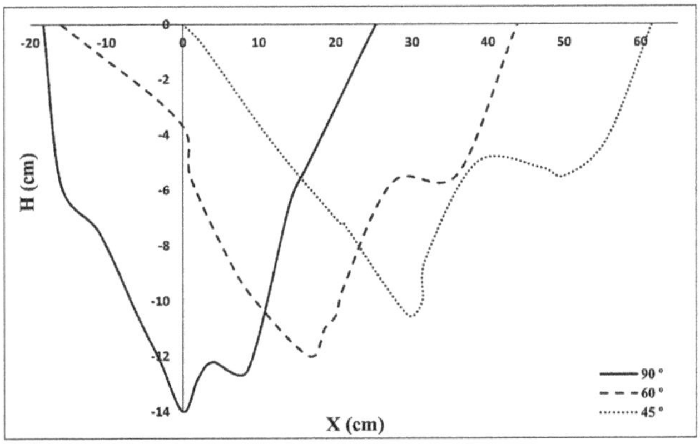

Figure 13. Scour hole profiles around the dike with different orientation angles (water depth: 11 cm, open channel flow). Scour hole measurements subject to +/−0.03 cm error.

Considering the absolute value of the vertical velocity, the U_Z values reach their minimum at the bottom of the scour hole, and increase with the distance from the bottom of scour holes. The maximum value of U_Z occurs around the initial level of the sand bed (before the scouring process started), then it reduces toward the flow surface. Close to the water surface, U_Z becomes very small (close to zero), and in some cases, it turns positive. The distribution of the vertical velocity shows a parabolic shape (Figure 12). Results show that vertical velocity vectors change their direction near the flow surface.

As shown in Figure 12, the ice cover has a significant impact on the value of U_Z. The presence of an ice cover on the surface of the flow results in obvious changes in the shape of the distribution of the vertical velocity. One can also notice, under the open flow conditions, the minimum values of U_Z are always negative. However, by adding an ice cover to the flow surface, the minimum values of U_Z are closer to zero and are positive in some cases. Furthermore, the absolute magnitude of the U_Z increases with the increase in the roughness coefficient of the cover. Therefore, it can be concluded that under an ice-covered flow condition, the vertical velocity component (U_Z) is the most important velocity component for developing scour holes comparing to other velocity components (U_X and U_Y). Moreover, results indicate that the presence of an ice cover influences the locale of the maximum vertical velocity. As the ice cover becomes rougher, the location of the maximum vertical velocity moved closer toward the channel bed. However, comparing to the streamwise velocity component (U_X), this effect on the location of the maximum vertical velocity component is less noticeable (Figures 7 and 12).

Results reveal that by increasing flow rate, the value of U_Z increases, and velocity profiles shift toward the water surface. For the water depths of 14 cm, 24.5 cm, and 35 cm, the average approaching velocity (U) is 21, 24, and 17 cm/s, respectively. One can see that the differences in the average approaching velocity are small. However, as shown in Figure 14, the difference in the vertical velocity distribution is remarkable. Therefore, one can say that the flow rate is one important factor responsible for the change in the vertical velocity component (U_Z) profile. Results also showed that under a rough covered flow condition, the maximum value of U_Z is a bit higher than those under the open flow and smooth covered flow conditions.

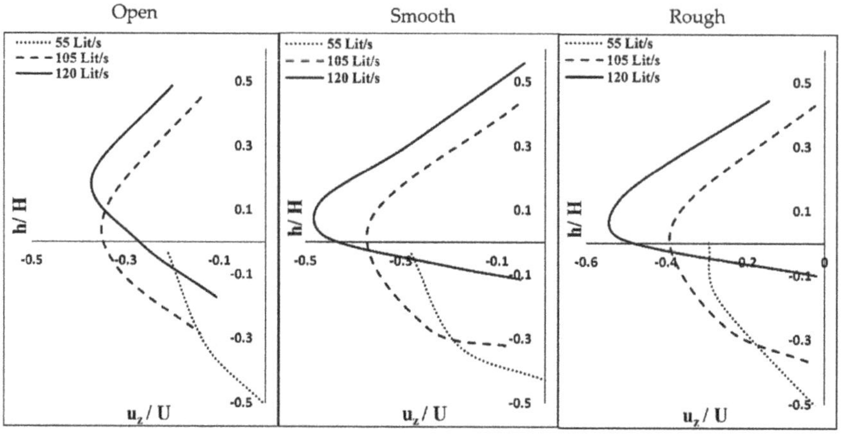

Figure 14. Vertical velocity profile (u_z) inside and outside the scour holes under conditions of different flow rates and surface cover (the dike orientation angle is 90°). ADV measurements subject to +/−0.25 cm/s error.

Further studies have been conducted to assess the impact of the dike orientation angle on the U_z distributions. Results reveal that by reducing the blockage ratio of the dike (namely, by decreasing the dike orientation angle), the effect of the ice cover on the vertical velocity profile became less noticeable. The presence of an ice cover on the flow surface leads to a considerable change in the U_z profile shape. However, by reducing the dike orientation angle from 90° to 45°, the vertical velocity profiles for a flow under an open channel become more similar to that under an ice-covered flow condition. Likewise, by reducing the dike orientation angle, the absolute value of U_z decreases (Figure 15). Results indicated that when the dike orientation angle is reduced, the vertical velocity component will be significantly affected in the following two ways, (1) the effect of an ice cover will be counteracted, (2) the downward velocity vectors will be weakened. Of note, this effect is more noticeable regarding scour hole depth comparing to the surface area of the scour holes.

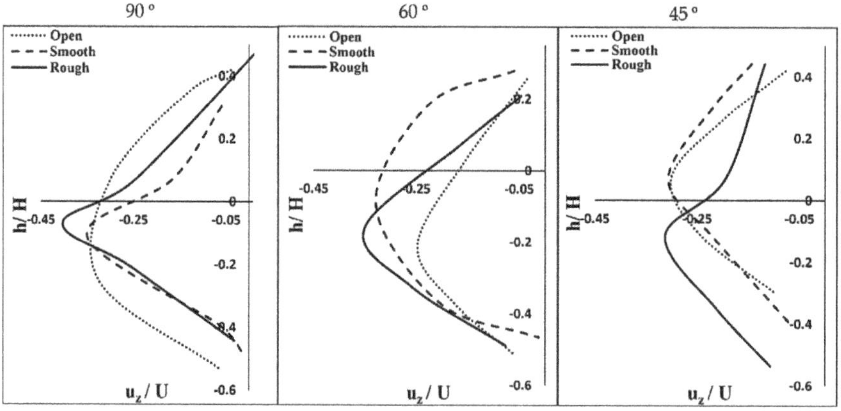

Figure 15. Vertical velocity profile (u_z) inside and outside the scour holes under conditions of different surface covers and different dike orientation angles (water depth: 21.5 cm). ADV measurements subject to +/−0.25 cm/s error.

3.4. Turbulence Intensities and Reynolds Shear Stress

It is noted that inside the scour holes, the flow is a combination of the downflow and horseshoe vortexes which cause a complex turbulence structure. Turbulent eddies generate velocity fluctuations which are referred to as turbulence strength (u_{rms}). u_{rms} was defined as the standard deviation (root mean square (RMS)) of the instantaneous velocity fluctuations. As described in Equation (5), u_{rms} was calculated based on 3D velocity components measured by ADV. A larger u_{rms} signifies a higher level of turbulence.

$$u_{rms} = \sqrt{\frac{1}{N}\sum_{i=1}^{N}(u_i')^2} \tag{5}$$

where, u_{rms} is the root-mean-square of the turbulent velocity fluctuations, u_i' is the instantaneous fluctuations of velocity components.

Because turbulent bursts are the main mechanism that entrains sediment and initiates the scouring process [11,18], the distribution patterns of the turbulence strength of 3D velocity components (u_x', u_y', u_z') have been examined. To accurately assess different instantaneous velocity fluctuations, the turbulence strength has been normalized by the average approach velocity (U). Results reveal that in most cases, the highest level of fluctuation occurs in the center of scour holes (very close to the dike tip) and the exit of the hole (flume bed), due to the exposure of a larger flow field. As indicated in Figure 16, generally, the streamwise turbulence intensity (u_x') is the largest, and the vertical turbulence intensity is the least, namely, $u_x' > u_y' > u_z'$. Moreover, the streamwise and vertical turbulence intensity generally followed a distinct pattern—they are small (close to zero) at the scour hole bottom and increase with the distance from the scour hole bottom. In most cases, u_x' and u_z' become maximum slightly above the scour hole (flume bed) and decrease toward the water surface, creating the reverse C shape profiles. Nonetheless, no consistent trend has been observed for the lateral turbulence intensity (Figure 16).

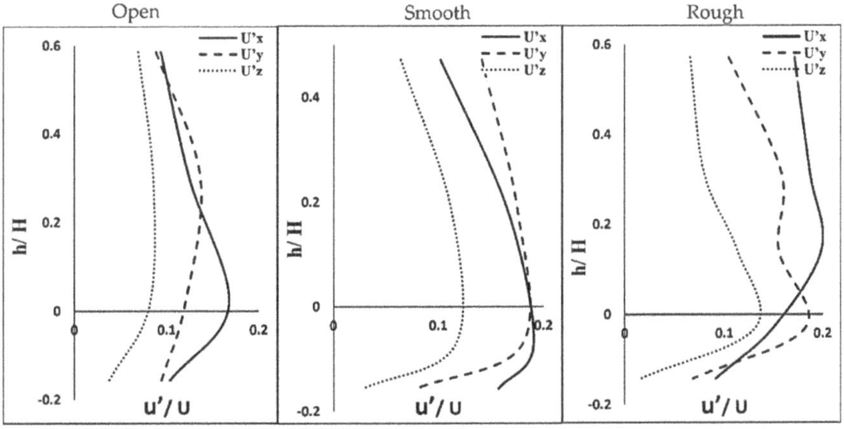

Figure 16. Turbulence intensity in three directions under different surface cover conditions with the dike orientation of 90° (water depth: 35 cm). ADV measurements subject to +/−0.25 cm/s error.

In terms of the impact of an ice cover on the turbulence strength, results indicate that the presence of an ice cover increases the turbulence intensities. With the increase in the roughness coefficient of an ice cover, the maximum fluctuation amount increases (Figure 16). Additionally, the approaching velocity plays an important role in the turbulence intensity. Regardless of the surface cover condition or the dike orientation angle, the maximum level of the instantaneous velocity fluctuations happened when the approaching velocity is the highest. Experiments clearly show that with a high approaching velocity, the maximum depth of the scour hole will be reached in a shorter amount of time. Moreover, the scour hole becomes larger with a higher approaching velocity.

The local scour process starts with the increase in the shear stress resulting from the accelerating flow around the spur dike. The shear stress refers to the Reynolds stress (τ), which can be illuminated as the transport of the streamwise momentum through a surface normal to the z-axis. Based on instantaneous velocity fluctuations collected using the ADV, the Reynolds shear stress can be calculated (Equation (6)) [34]. The Reynolds stress plays a key role in the entrainment and movement of sediment particles.

$$\tau = -\rho\left\langle u_x{}'u_z{}'\right\rangle \quad (6)$$

where, τ is the Reynolds stress, ρ is the mass density of water, $u_x{}'$ is the streamwise turbulence strength, and $u_z{}'$ is the vertical turbulence strength.

To better understand the Reynolds stress distribution inside scour holes, the Reynolds stress values were normalized by the shear velocity (friction velocity) u_*^2. The Shear velocity was calculated based on the boundary layer characteristic method (BLCM) [39], as illustrated in Equation (7).

$$\frac{\tau}{\tau_b} = \frac{-< u_x{}'u_z{}' >}{u_*^2} \quad (7)$$

where, τ is the Reynolds stress, $u_x{}'$ is the streamwise turbulence strength, $u_z{}'$ is the vertical turbulence strength, τ_b is the bed shear stress, and u_* is the shear velocity or friction velocity.

$$u_* = \frac{(\delta_* - \theta)\, u_{max}}{C\delta_*} \quad (8)$$

where, δ_* is the displacement thickness, θ is the momentum thickness which is defined by [40], C is a constant which was estimated as 4.4 for Canadian rivers [41].

As shown in Figure 17, the Reynolds stress is zero at the scour hole bottom, and gradually decreases to the least negative value, and it becomes zero again in the middle of scour holes; then it increases with the distance from the scour hole bottom, afterward it decreases toward the water surface. Inside the scour hole, the distribution of the Reynolds stress has a parabolic shape. Outside the scour hole, the value of the Reynolds stress becomes positive, and it reached its maximum slightly above the initial level of the sand bed (flume bed). Then it reduces again toward the flow surface and became negative close to the surface, creating a convex shape distribution (Figure 17).

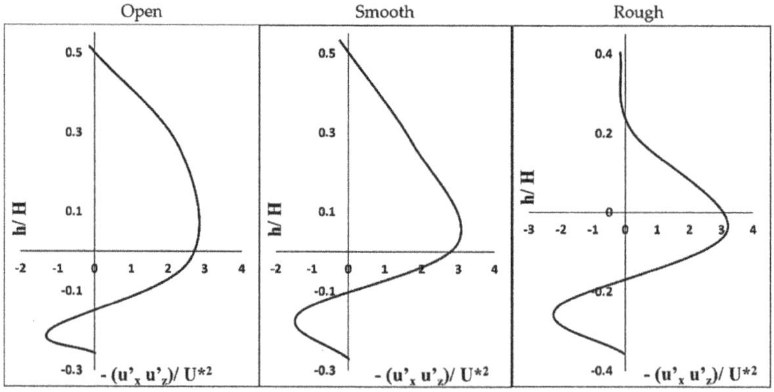

Figure 17. Reynolds stress, under different surface cover conditions for the dike with an orientation angle of 90° (water depth: 31 cm). ADV measurements subject to +/−0.25 cm/s error.

The negative Reynolds stress inside the scour hole demonstrates an upward vertical momentum transport caused by a negative velocity gradient ($\frac{du}{dz} < 0$). Similarly, the negative Reynolds stress close to the flow surface indicates the impact of the adverse pressure gradient at the upper portion of the flow, which has a negative velocity gradient as well.

Results show that the value of Reynolds stress under an ice-covered flow condition is greater than that under an open flow condition. With the increase in the roughness of an ice cover, the absolute value of the Reynolds stress rises (Figure 17). It can be concluded that under the ice-covered flow conditions, higher shear stress at the sand bed will be generated. This increase in the shear stress leads to more sediment movement around the dike and ultimately creates a larger scour hole.

4. Conclusions

The three-dimensional velocity components and turbulence structure inside and outside scour holes around the spur dikes in a large-scale flume have been investigated. The model spur dikes are impermeable and non-submerged, with orientation angles of 90°, 60°, and 45°, respectively. To have a comprehensive study on flow characteristics in the vicinity of a dike, experiments have been conducted under conditions of open channel, smooth covered, and rough covered flow condition. Based on data collected from laboratory experiments, the impacts of the spur dike on the 3D velocity distributions, Reynolds shear stress, and turbulence intensities have been investigated. Overall, the following results were obtained in this experimental study.

1. The presence of an ice cover on the water surface causes a considerable change in the bed shear stress and raises the turbulence intensities inside the scour holes which can have a significant effect on sediment transportation. To be more specific, the presence of an ice cover increases the maximum values of the 3D velocity components averagely by 10% to 25% for smooth and rough ice cover, respectively. The rough ice cover shifts the location of the maximum velocity further close to the sandbed, which leads to the increase in the Reynolds shear stress inside the scour hole and accordingly resulted in a deeper scour hole. These effects are independent of flow rates and dike orientations angle.
2. The dike with an orientation angle of 90° generates the strongest downflow around the dike comparing to those resulting from the orientations angle of 45° and 60°. Thus, the dike with an orientation angle of 90° creates high turbulence, and powerful horseshoe vortexes inside the scour holes, generate the deepest scour hole comparing to those around the dike with the smaller orientation angles. Results clearly show that by reducing the dike orientation angle from 90° to 45°, the velocity profiles are shifted upward, and scour hole depth decreases by 5–10% for each 10° angle decrease. These results imply that by changing the dike orientation angle, the blockage ratio of the cross-section area will reduce. Consequently, the vortex system in the vicinity of the spur dike will become weak. Thus, the turbulence kinetic energy inside the scour hole is reduced.
3. Generally, the 3D velocity distributions are less regular inside the scour holes. Besides, the velocity components inside the hole are smaller comparing to those outside of the scour holes. The highest level of the velocity fluctuation and turbulence intensity appears immediately above the scour holes. With the increase in the roughness coefficient of an ice cover, the maximum turbulence intensity increases. The smooth and rough ice cover raised turbulence intensity averagely by 15% and 30%, respectively. Moreover, with the increases in the approaching velocity, the instantaneous velocity fluctuation increases. The higher the flow velocity, the more powerful turbulence kinetic energy around the spur dike, and thus, the deeper the scour holes.
4. The streamwise velocity (U_x) is highest among all 3D velocity components, implying that U_x contributes more to the turbulence intensities, Reynolds shear stress, and consequently, the development of the scour holes. The lateral velocity component

(U_y) has the highest level of irregularities inside and outside the scour hole. Unlike the streamwise and vertical velocity (U_z) distributions, no meaningful trend has been observed for the lateral velocity component. Moreover, the presence of an ice cover on the water surface considerably affects the lateral velocity component. With the increase in the roughness coefficient of an ice cover, the lateral velocity increased.

5. Both the streamwise velocity component and lateral velocity component are almost always positive. However, the vertical velocity component is almost negative inside and outside the scour holes. The negative vertical velocity components indicate the powerful downflow and downward velocity in the vicinity of the dike. The absolute value of U_z increases proportionately with the approaching flow velocity. The higher the approaching velocity, the more the vertical velocity component absolute value. This effect has been intensified with the increase in the roughness coefficient of an ice cover. Under an ice-covered flow condition, the distribution pattern of U_z differs completely from that under an open flow condition. Moreover, the maximum vertical velocity has been observed when the dike has an orientation angle of 90°. With an increase in dike angle of each 10° (from 45° to 90°), the absolute value of vertical velocity relatively increases by up to almost 10%, implying that the dike orientation angle is one of the important controlling factors for U_z.
6. The Reynolds shear stress is negative inside the scour holes and becomes positive toward the flow surface. It reaches its maximum slightly above the scour holes. The negative values of the Reynolds stress are caused by the upward vertical momentum transport generated by a negative velocity gradient. Additionally, with the increase in the roughness coefficient of an ice cover, the absolute value of the Reynolds stress increases. It can be concluded that the presence of an ice cover creates more powerful shear stress at the sandbed, which causes a deeper scour hole.

Author Contributions: Conceptualization, R.J. and J.S.; methodology, R.J. and J.S.; software, R.J.; validation, R.J. and J.S.; formal analysis, R.J. and J.S.; investigation, R.J. and J.S.; resources, R.J. and J.S.; data curation, R.J.; writing—original draft preparation, R.J.; writing—review and editing, R.J. and J.S.; supervision, J.S.; project administration, R.J. and J.S.; funding acquisition, J.S. All authors have read and agreed to the published version of the manuscript.

Funding: The Natural Sciences and Engineering Research Council of Canada (NSERC) under the Discovery Grant Program RGPIN-2019-04278.

Institutional Review Board Statement: Not applicable.

Informed Consent Statement: Not applicable.

Data Availability Statement: The data used in this manuscript are available by writing to the first author.

Conflicts of Interest: The authors declare no conflict of interest.

References

1. Kuhnle, R.A.; Alonso, C.V.; Shields, F.D. Geometry of scour holes associated with 90 spur dikes. *J. Hydraul. Eng.* **1999**, *125*, 972–978. [CrossRef]
2. Sadat, S.H. Modification of Spur-Dike with Footing or Pile-Group to Stabilize River Morphology and Reduce Local Scour. Ph.D. Thesis, Nagoya Institute of Technology, Nagoya, Japan, 2015.
3. Giglou, A.N.; Mccorquodale, J.A.; Solari, L. Numerical study on the effect of the spur dikes on sedimentation pattern. *Ain Shams Eng. J.* **2018**, *9*, 2057–2066. [CrossRef]
4. Pandey, M.; Ahmad, Z.; Sharma, P. Scour around impermeable spur dikes: A review. *ISH J. Hydraul. Eng.* **2018**, *24*, 25–44. [CrossRef]
5. Zhang, H.; Nakagawa, H.; ISHIGAKI, T.; MUTO, Y. Prediction of 3D flow field and local scouring around spur dykes. *Proc. Hydraul. Eng.* **2005**, *49*, 1003–1008. [CrossRef]
6. Teraguchi, H.; Nakagawa, H.; MUTO, Y.; BABA, Y.; ZHANG, H. Flow and sediment transport around impermeable or permeable groins. *Proc. Hydraul. Eng.* **2008**, *52*, 175–180. [CrossRef]
7. Mizutani, H.; Nakagawa, H.; Kawaike, K.; Baba, Y.; Zhang, H. Study on local scour and variation of bed composition around non-submerged spur dyke. *J. Jpn. Soc. Civ. Eng. Ser. B1 Hydraul. Eng.* **2011**, *67*, I_829–I_834. [CrossRef]

8. Kanda, K.; Samoto, Y.; Li, Z. Study on morphodynamics around stone-lined spur dikes in the Akashi river. In *River, Coastal and Estuarine Morphodynamics: RCEM 2007, Two Volume Set*; CRC Press: Boca Raton, FL, USA, 2019; pp. 1176–1182.
9. Yossef, M.M.; Klaassen, G.J. Reproduction of groynes-induced river bed morphology using LES in a 2-D morphological model. *River Flow* **2002**, *2*, 1099–1108.
10. Muto, Y.; Baba, Y.; Aya, S. Velocity measurements in open channel flow with rectangular embayments formed by spur dikes. *Annu. Disaster Prev. Res. Inst. Kyoto Univ. B* **2002**, *45*, 2.
11. Jennifer, D.; Li, H.; Guangqian, W.; Xudong, F. Turbulent burst around experimental spur dike. *Int. J. Sediment Res.* **2011**, *26*, 471–523.
12. Nelson, J.M.; Shreve, R.L.; McLean, S.R.; Drake, T.G. Role of near-bed turbulence structure in bed load transport and bed form mechanics. *Water Resour. Res.* **1995**, *31*, 2071–2086. [CrossRef]
13. Duan, J.G.; He, L.; Fu, X.; Wang, Q. Mean flow and turbulence around experimental spur dike. *Adv. Water Resour.* **2009**, *32*, 1717–1725. [CrossRef]
14. Papanicolaou, A.; Diplas, P.; Dancey, C.; Balakrishnan, M. Surface roughness effects in near-bed turbulence: Implications to sediment entrainment. *J. Eng. Mech.* **2001**, *127*, 211–218. [CrossRef]
15. Huai, W.; Zeng, Y.; Xu, Z.; Yang, Z. Three-layer model for vertical velocity distribution in open channel flow with submerged rigid vegetation. *Adv. Water Resour.* **2009**, *32*, 487–492. [CrossRef]
16. Dey, S.; Barbhuiya, A.K. Velocity and turbulence in a scour hole at a vertical-wall abutment. *Flow Meas. Instrum.* **2006**, *17*, 13–21. [CrossRef]
17. Sui, J.; Wang, D.; Karney, B.W. Suspended sediment concentration and deformation of riverbed in a frazil jammed reach. *Can. J. Civ. Eng.* **2000**, *27*, 1120–1129. [CrossRef]
18. Sui, J.; Wang, J.; He, Y.; Krol, F. Velocity profiles and incipient motion of frazil particles under ice cover. *Int. J. Sediment Res.* **2010**, *25*, 39–51. [CrossRef]
19. Wu, P.; Hirshfield, F.; Sui, J.; Wang, J.; Chen, P.-P. Impacts of ice cover on local scour around semi-circular bridge abutment. *J. Hydrodyn.* **2014**, *26*, 10–18. [CrossRef]
20. Namaee, M.R.; Sui, J. Velocity profiles and turbulence intensities around side-by-side bridge piers under ice-covered flow condition. *J. Hydrol. Hydromech.* **2020**, *68*, 70–82. [CrossRef]
21. Ettema, R.; Kempema, E.W. River-Ice Effects on Gravel-Bed Channels. *Gravel-Bed Rivers Process. Tools Environ.* **2012**, 523–540. [CrossRef]
22. Muste, M.; Braileanu, F.; Ettema, R. Flow and sediment transport measurements in a simulated ice-covered channel. *Water Resour. Res.* **2000**, *36*, 2711–2720. [CrossRef]
23. Li, S.S. Estimates of the Manning's coefficient for ice-covered rivers. In Proceedings of the Institution of Civil Engineers-Water Management, Montreal, QC, Canada, October 2012; Thomas Telford Ltd.: London, UK, 2012; Volume 165, pp. 495–505. [CrossRef]
24. Mays, L.W. *Hydraulic Design Handbook*; McGraw-Hill Education: New York, NY, USA, 1999.
25. Weitbrecht, V.; Kühn, G.; Jirka, G. Large scale PIV-measurements at the surface of shallow water flows. *Flow Meas. Instrum.* **2002**, *13*, 237–245. [CrossRef]
26. Kuhnle, R.A.; Jia, Y.; Alonso, C.V. Measured and simulated flow near a submerged spur dike. *J. Hydraul. Eng.* **2008**, *134*, 916–924. [CrossRef]
27. Kadota, A.; Suzuki, K. Local scour and development of sand wave around T-type and L-type groynes. In *Scour and Erosion, Proceedings of the International Conference on Scour and Erosion (ICSE-5), San Francisco, CA, USA, 7–10 November 2010*; ASCE: Reston, VA, USA, 2010; pp. 707–714.
28. Telionis, D.; Yang, Y.; Rediniotis, O. Recent developments in multi-hole probe (mhp) technology. In Proceedings of the 20th International Congress of Mechanical Engineering, Gramado, Brazil, 15–20 November 2009.
29. Wahl, T.L. Analyzing ADV data using WinADV. In *Building Partnerships, Proceedings of the Building Partnerships, Minneapolis, MI, USA, 30 July 30–2 August*; ASCE: Reston, VA, USA, 2000; pp. 1–10.
30. Martin, V.; Fisher, T.; Millar, R.; Quick, M. ADV data analysis for turbulent flows: Low correlation problem. In Proceedings of the Hydraulic Measurements and Experimental Methods 2002, Estes Park, CO, USA, 28 July–1 August 2002; pp. 1–10.
31. Rehmel, M. Application of acoustic Doppler velocimeters for streamflow measurements. *J. Hydraul. Eng.* **2007**, *133*, 1433–1438. [CrossRef]
32. Sontek, A.D.V. *Operation Manual, Firmware Version 4.0*; Sontek: San Diego, CA, USA, 1997.
33. *SonTek-IQ Series Intelligent Flow Featuring Smart Puls ^{HD} User's Manual*; Xylem Inc: San Diego, CA, USA, 2017; p. 135.
34. Zhang, H.; Nakagawa, H.; Kawaike, K.; Yasuyuki, B. Experiment and simulation of turbulent flow in local scour around a spur dyke. *Int. J. Sediment Res.* **2009**, *24*, 33–45. [CrossRef]
35. Vijayasree, B.; Eldho, T.; Mazumder, B.; Ahmad, N. Influence of bridge pier shape on flow field and scour geometry. *Int. J. River Basin Manag.* **2019**, *17*, 109–129. [CrossRef]
36. Ghaderi, A.; Abbasi, S. CFD simulation of local scouring around airfoil-shaped bridge piers with and without collar. *Sādhanā* **2019**, *44*, 1–12. [CrossRef]
37. Melville, B.W.; Coleman, S.; Hoe, D. Countermeasures for Scour at Spill-Trough Bridge Abutments. In Proceedings of the First International Conference on Scour of Foundations, College Station, TX, USA, 17–20 November 2002; pp. 749–763.

38. McLelland, S.J.; Nicholas, A.P. A new method for evaluating errors in high-frequency ADV measurements. *Hydrol. Process.* **2000**, *14*, 351–366. [CrossRef]
39. Afzalimehr, H.; Maddahi, M.R.; Naziri, D.; Sui, J. Effects of non-submerged boulder on flow characteristics—A field investigation. *Int. J. Sediment Res.* **2019**, *34*, 136–143. [CrossRef]
40. Schlichting, H.; Gersten, K. *Boundary-Layer Theory*; Springer: Berlin, Germany, 2016.
41. Afzalimehr, H.; Rennie, C. Determination of bed shear stress using boundary layer parameters in a gravel-bed river. *Hydrol. Sci. J.* **2009**, *54*, 147–159. [CrossRef]

water

MDPI

Article

Analytical Models of Velocity, Reynolds Stress and Turbulence Intensity in Ice-Covered Channels

Jiao Zhang [1], Wen Wang [1,*], Zhanbin Li [1], Qian Li [2], Ya Zhong [3], Zhaohui Xia [4] and Hunan Qiu [1]

1 State Key Laboratory of Eco-Hydraulics in Northwest Arid Region of China, Xi'an University of Technology, Xi'an 710048, China; jiaozhang@whu.edu.cn (J.Z.); zbli@xaut.edu.cn (Z.L.); hunanqiu1999@126.com (H.Q.)

2 Yellow River Engineering Consulting Co., Ltd., Zhengzhou 450003, China; liqianyrec@163.com

3 Southwest Branch of China Construction Third Engineering Bureau Group Co., Ltd., Chengdu 610041, China; hljoanna@whu.edu.cn

4 PowerChina Northwest Engineering Corporation Limited, Xi'an 710065, China; xzh911911@163.com

* Correspondence: wangwen1986@xaut.edu.cn

Abstract: Ice cover in an open channel can influence the flow structure, such as the flow velocity, Reynolds stress and turbulence intensity. This study analyzes the vertical distributions of velocity, Reynolds stress and turbulence intensity in fully and partially ice-covered channels by theoretical methods and laboratory experiments. According to the experimental data, the vertical profile of longitudinal velocities follows an approximately symmetry form. Different from the open channel flow, the maximum value of longitudinal velocity occurs near the middle of the water depth, which is close to the channel bed with a smoother boundary roughness compared to the ice cover. The measured Reynolds stress has a linear distribution along the vertical axis, and the vertical distribution of measured turbulence intensity follows an exponential law. Theoretically, a two-power-law function is presented to obtain the analytical formula of the longitudinal velocity. In addition, the vertical profile of Reynolds stress is obtained by the simplified momentum equation and the vertical profile of turbulence intensity is investigated by an improved exponential model. The predicted data from the analytical models agree well with the experimental ones, thereby confirming that the analytical models are feasible to predict the vertical distribution of velocity, Reynolds stress and turbulence intensity in ice-covered channels. The proposed models can offer an important theoretical reference for future study about the sediment transport and contaminant dispersion in ice-covered channels.

Keywords: ice-covered channel; longitudinal velocity; Reynolds stress; turbulence intensity

Citation: Zhang, J.; Wang, W.; Li, Z.; Li, Q.; Zhong, Y.; Xia, Z.; Qiu, H. Analytical Models of Velocity, Reynolds Stress and Turbulence Intensity in Ice-Covered Channels. *Water* **2021**, *13*, 1107. https://doi.org/10.3390/w13081107

Academic Editors: Paweł M. Rowiński and Jueyi Sui

Received: 28 February 2021
Accepted: 14 April 2021
Published: 17 April 2021

1. Introduction

Most rivers at high altitude in cold northern regions always freeze in the winter and form ice sheets. The ice sheet in some Canadian rivers is at least 0.6 m thick and lasts for at least 4 months [1]. The wetted perimeter of the cross section and flow resistance in ice-covered flows increases with the presence of the ice sheet, which significantly affects the hydraulic characteristics of river and topographical features and greatly change the flow velocity distribution, flow transport capacity and sediment transport rate [2–11]. Therefore, it is necessary to study the ice-covered flows.

Unlike the open channel flow, flows in the ice-covered channel have asymmetric forms. The presence of the ice sheet makes the maximum streamwise velocity appear at the inner center of the flow, and the location of the maximum velocity is generally considered the division point in the asymmetrically distributed flow [12,13]. The asymmetric distribution mainly depends on the roughness of the ice sheet and the riverbed. The division point of the velocity tends to be away from the rougher surface [14]. Previous studies have shown that the main effects of the ice sheets on alluvial channels can be summarized as: they increase the water level (compared to open channels at the same flow rate), reduce the average flow velocity, increase the channel drag force and reduce the bed sediment transport rate [15].

Previous investigations on the flow characteristics of ice-covered flows are mainly obtained through experiments [16–18]. Many laboratory experimental data and field observations show that the vertical distribution of streamwise velocity forms a double-layer, which is characterized by the plane of maximum velocity. Shen and Harden (1978) [19] and Lau and Krishnappan (1981) [20] applied this double-layer theory in the ice-covered flows. They divided the ice-covered flows into two separate layers: upper ice layer and lower channel bed. Parthasarathy and Muste (1994) [13] found that the zero shear stress plane in the ice-covered flows was inconsistent with the maximum velocity plane. Chen et al. (2015) [21] proposed that the horizontal plane of zero shear stress should be determined as the dividing plane of sublayers and modified the double-layer assumption.

Some researchers have studied the vertical profiles of the longitudinal velocity by numerical simulation methods and proposed two-dimensional and three-dimensional models [22,23], which may require sensitive hydraulic parameters. However, these hydraulic parameters do not have explicit expressions to be calculated, and they have uncertainty that cannot be ignored. Except the numerical methods, previous researchers also attempted to obtain the vertical distribution of streamwise velocity by theoretical methods [24,25]. Uzuner (1975) [26] separated the ice-covered flow into two layers based on the position of the maximum velocity and assumed that the velocity distributions in the upper ice layer and lower bed layer were consistent with the logarithmic distribution law, and Manning formula could be independently applied to each layer. In addition, a two-power-law function was adopted to calculate the vertical distribution of the longitudinal velocity. Teal et al. (1994) [27] reported a reasonable fit of streamwise velocity data to the two-power-law function. Compared to the two-power-law function, the logarithmic law appears to overestimate velocities near the location of maximum velocity. The two-power-law function is a reasonable extension of the power-law expression that is used to describe velocity profiles of open channel flow. The advantage for the two-power-law function is that it describes the entire flow with a single continuous curve. Hence, the two-power-law function deserves further study.

The objective of this study is to obtain the vertical profiles of longitudinal velocity, Reynolds stress and turbulence intensity in ice-covered channels. Hence, this study focuses on (1) adopting a two-power-law function to calculate the vertical profile of the longitudinal velocity, (2) simplifying the time-averaged momentum equation to analyze the vertical distributions of Reynolds stress, (3) improving the exponential model to calculate the turbulence intensity and (4) employing experimental data to validate the theoretical models and giving detail discussion on the coefficients in the theoretical models.

2. Material and Methods

Since the free water surface is covered with ice, the open channel flow changes to a closed conduit flow but retains the flow characteristics of the open channel flow. The vertical distribution of longitudinal velocity significantly changes in the open channel flow with ice-cover, which reflects that the plane of maximum longitudinal velocity shifts from the free water surface to the inner water (see Figure 1). Here, we used a Cartesian coordinate system with the x axis in the main flow direction and z axis in the water depth direction. Corresponding to the x and z axes, u and w are the longitudinal and vertical velocities, respectively.

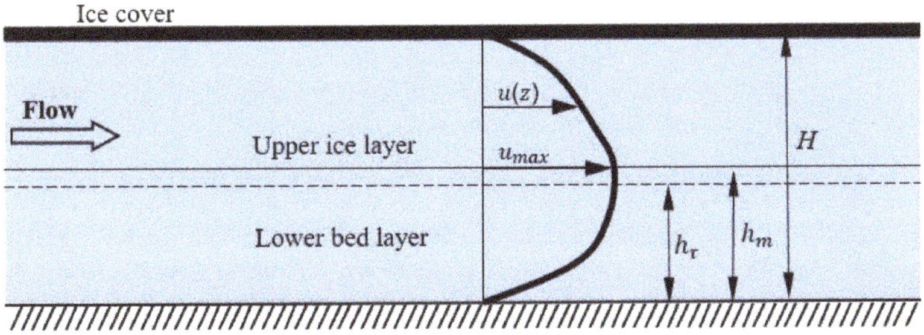

Figure 1. Schematic profile of the longitudinal velocity in an iced-covered open channel.

2.1. Vertical Distribution of Longitudinal Velocity

The presence of the ice cover causes the increase of the wetted perimeter, which increases the composite flow resistance. Owing to the different roughness of ice cover and channel bed, the velocity profile is vertically asymmetric (see Figure 1). In Figure 1, the vertical flow structure can be divided into two independent layers at the plane of maximum velocity, i.e., the upper ice layer and lower bed layer. The flow in the upper ice layer is mainly affected by the ice cover, and the lower flow is primarily affected by the channel bed. The vertical location of the maximum velocity is determined by the roughness of the ice cover and channel bed, i.e., the maximum velocity will not occur at the free water surface but near the middle of the water depth.

Tsai and Ettema (1994) [28] adopted a power-law function to predict the velocity profile in the open channel flow. For the ice-covered flow, a two-power-law function was used to obtain the vertical profile of the longitudinal velocity [27]. The first advantage of the two-power-law function is that it describes the flow just using a single continuous curve. The second is that the analytical solution of velocity is adjustable in accordance with the roughness change of channel bed and ice cover. The two-power-law is shown as

$$u = K_0\left(\frac{z}{H}\right)^{1/m_b}\left(1-\frac{z}{H}\right)^{1/m_i} \tag{1}$$

where z is the vertical axis with $z = 0$ at the channel bed; H is the water depth; K_0 is the flow parameter based on a given flow discharge per unit flow width and m_b and m_i are the parameters corresponding to the boundary roughness of the channel bed and ice cover. When parameter m_i approaches infinity, the velocity profile becomes equivalent to a single power law expression, which indicates that the ice cover disappears.

The velocity gradient obtained from Equation (1) is

$$\frac{du}{dz} = \frac{K_0}{H}\left[\frac{1}{m_b}\left(\frac{z}{H}\right)^{\frac{1}{m_b}-1}\left(1-\frac{z}{H}\right)^{1/m_i} - \frac{1}{m_i}\left(\frac{z}{H}\right)^{\frac{1}{m_b}}\left(1-\frac{z}{H}\right)^{\frac{1}{m_i}-1}\right] \tag{2}$$

By setting the velocity gradient $\frac{\partial u}{\partial z}$ to zero, we deduced the position of the maximum velocity as

$$\left(\frac{z}{H}\right)_{u_{max}} = \frac{h_m}{H} = \frac{m_i}{m_i + m_b} \tag{3}$$

where u_{max} is the maximum longitudinal velocity and h_m is the height of the maximum velocity from the channel bed.

2.2. Vertical Distribution of Reynolds Stress

To predict the vertical distribution of Reynolds stress in ice-covered channels for a steady uniform flow, the time-averaged momentum equation in the longitudinal direction can be simplified to

$$-\overline{u'w'}+\nu\frac{du}{dz}+\frac{u_{*i}^2 z}{H}=u_{*b}^2\left(1-\frac{z}{H}\right) \tag{4}$$

where u' and w' are turbulent fluctuations of the longitudinal and vertical velocities, $-\overline{u'w'}$ is Reynolds stress; ν is the flow kinematic viscosity; u_{*i} is the shear velocity at the ice cover and u_{*b} is the shear velocity at the channel bed.

Since the viscosity shear stress in the flow is much smaller than the Reynolds stress, the viscosity shear stress can be neglected in Equation (4) [29,30]. Then, there is a linear relationship between the shear stress (τ) and the vertical axis (z). Equation (4) has another form and is shown as

$$\tau(z)=\tau_b-(\tau_b-\tau_i)\frac{z}{H} \tag{5}$$

where $\tau(z)$ is the shear stress at distance z; τ_b and τ_i are the Reynolds stresses at the channel bed and ice cover; z is the vertical distance from the channel bed and H is the water depth.

Following the method of Rowinski and Kubrak (2002) [31], who combine the eddy viscosity and flow conditions by a mixing length concept, the shear stress can be given as

$$\tau(z)=\rho l^2\frac{\partial u}{\partial z}\left|\frac{\partial u}{\partial z}\right| \tag{6}$$

where ρ is the flow density and l is the proposed mixing length. Chen et al. (2015) [21] demonstrate that the mixing length theory can be applied at a certain distance from the fixed boundaries. Two mixing lengths, l_b and l_i, are proposed corresponding to the two fixed boundaries. The first one is the mixing length of channel bed and the second one is the mixing length of the ice cover. Hence, the linear relationships of two mixing lengths are approximately written as

$$\begin{cases} l_b=\kappa z,\ 0\le z\le\delta_b \\ l_i=\kappa(H-z),\ H-\delta_i\le z\le H \end{cases} \tag{7}$$

where δ_b and δ_i are the distances from the channel bed and ice cover, where the mixing length theory is valid. κ = 0.41 is the von Karman constant.

By substituting Equation (7) into Equation (6), we updated the expression of shear stress as

$$\begin{cases} \tau(z)=\rho\kappa^2 z^2\frac{\partial u}{\partial z}\left|\frac{\partial u}{\partial z}\right|,\ 0\le z\le\delta_b \\ \tau(z)=\rho\kappa^2(H-z)^2\frac{\partial u}{\partial z}\left|\frac{\partial u}{\partial z}\right|,\ H-\delta_i\le z\le H \end{cases} \tag{8}$$

The gradient of shear stress from Equation (8) is

$$\begin{cases} \frac{\partial\tau}{\partial z}=2\rho\kappa^2 z^2\frac{\partial u}{\partial z}\left(\frac{1}{z}\frac{\partial u}{\partial z}+\frac{\partial^2 u}{\partial z^2}\right),\ 0\le z\le\delta_b \\ \frac{\partial\tau}{\partial z}=2\rho\kappa^2(H-z)^2\frac{\partial u}{\partial z}\left(\frac{1}{(H-z)}\frac{\partial u}{\partial z}-\frac{\partial^2 u}{\partial z^2}\right),\ H-\delta_i\le z\le H \end{cases} \tag{9}$$

By substituting the velocity gradient (Equation (2)) into the Reynolds stress gradient (Equation (9)) and letting $\frac{\partial\tau}{\partial z}=0$, we obtained the two maximum Reynolds stresses, τ_{bmax} and τ_{imax}, near the channel bed and ice cover, respectively, which are presented as

$$\begin{cases} \tau_{bmax}=\rho\kappa^2 {z_{bmax}}^2\frac{\partial u}{\partial z_{bmax}}\left|\frac{\partial u}{\partial z_{bmax}}\right|,\ 0\le z\le\delta_b \\ \tau_{imax}=\rho\kappa^2(H-z_{imax})^2\frac{\partial u}{\partial z_{imax}}\left|\frac{\partial u}{\partial z_{imax}}\right|,\ H-\delta_i\le z\le H \end{cases} \tag{10}$$

where z_{bmax} and z_{imax} denote the positions where the corresponding maximum shear stresses τ_{bmax} and τ_{imax} occurred, and they can be calculated as

$$\begin{cases} z_{bmax} = \frac{-A_b - \sqrt{A_b^2 - 4B_bC_b}}{2B_b}H \\ z_{imax} = \frac{-A_i - \sqrt{A_i^2 - 4B_iC_i}}{2B_i}H \end{cases} \tag{11}$$

where A_b, B_b, C_b, A_i, B_i, and C_i are the constants relevant to the roughness of the channel bed and ice cover. Their expressions are

$$\begin{cases} \begin{cases} A_b = -m_b - 2\frac{m_i}{m_b} - 2 \\ B_b = (m_b + m_i)\left(\frac{1}{m_b} + \frac{1}{m_i}\right) \\ C_b = \frac{m_i}{m_b} \end{cases} \\ \begin{cases} A_i = -m_i + 2\frac{m_i}{m_b} + 2 \\ B_i = -(m_b + m_i)\left(\frac{1}{m_b} + \frac{1}{m_i}\right) \\ C_i = m_i - \frac{m_i}{m_b} \end{cases} \end{cases} \tag{12}$$

The shear stresses at the fixed boundaries, i.e., the channel bed and ice cover, can be considered the maximum shear stresses in the lower bed layer and upper ice layer. Hence, the linear relationship (Equation (5)) between shear stress and vertical position can be rewritten as

$$\tau(z) = \tau_{bmax} - (\tau_{bmax} - \tau_{imax})\frac{z}{H} \tag{13}$$

2.3. Vertical Distribution of Turbulence Intensity

The flow turbulence characteristics can be presented by the turbulence intensity, which can be calculated by the root mean square of the fluctuating longitudinal velocity, i.e., $u_{rms} = \sqrt{(u')^2}$. In the open channel flow, the term u_{rms} reaches its maximum value near the channel bed and it has a linear relationship with the vertical distance. Nezu and Rodi (1986) [32] established an exponential model to describe the vertical distribution of the turbulence intensity. However, in the ice-covered channel flow, it first decreases from the channel bed and reaches its minimum value near the middle flow depth, after which it gradually increases towards to the ice cover. For the profile of the turbulence intensity in the ice-covered channel, we adopted the division scheme to investigate the profile of the turbulence intensity. Based on the location of zero shear stress, its profile is separated into two regions. Then, the exponential model can be applied in each region as follows

$$\begin{cases} \frac{u_{rms}}{u_{*b}} = D_b e^{-E_b z/h_\tau}, 0 < z < h_\tau \\ \frac{u_{rms}}{u_{*i}} = D_i e^{-E_i z/h_\tau}, h_\tau < z < H \end{cases} \tag{14}$$

where D_b, E_b, D_i and E_i are constant parameters.

3. Experimental Verification

Experiments were performed in a rectangular glass flume with a total length of 20 m, a width of 1 m and a depth of 0.5 m in the State Key Laboratory of Water Resources and Hydropower Engineering Science in Wuhan University to investigate the longitudinal velocity profile in the ice-covered flow (Figure 2). The flow is circulated by a pump system and adjusted to be steady by a tailgate at the end of the flume. A plastic foam board was used to simulate the ice with the length of 15 m and a width of 1 m [33].

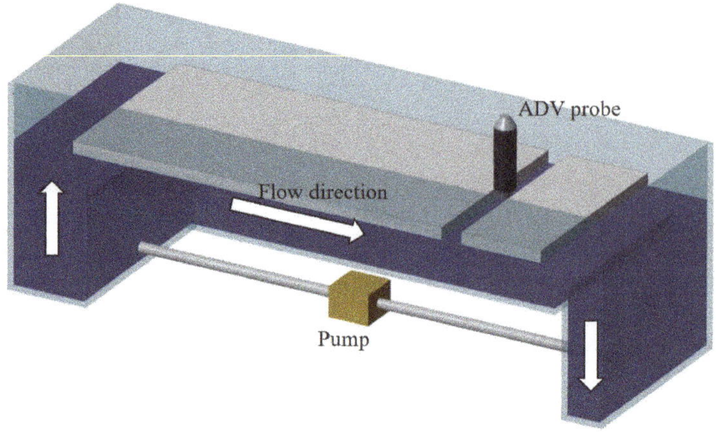

(**a**)

(**b**)

Figure 2. (**a**) Schematic of the experimental flume, the dark blue denoting the flow in the flume and (**b**) images of the experimental site.

The rectangular coordinates are as follows: x is the main flow direction with x = 0 at the beginning of the ice cover; y is the lateral direction with y = 0 at the sidewall and z is the vertical direction counting from the channel bed. The velocity was measured at the cross section with x = 9 m, which is sufficiently far from the ice cover entry to form a fully developed flow. In the measured cross section, 14 measured lines are arranged, where the measured points are also arranged with a uniform vertical distance 1 cm. The detail layout is shown in Figure 3.

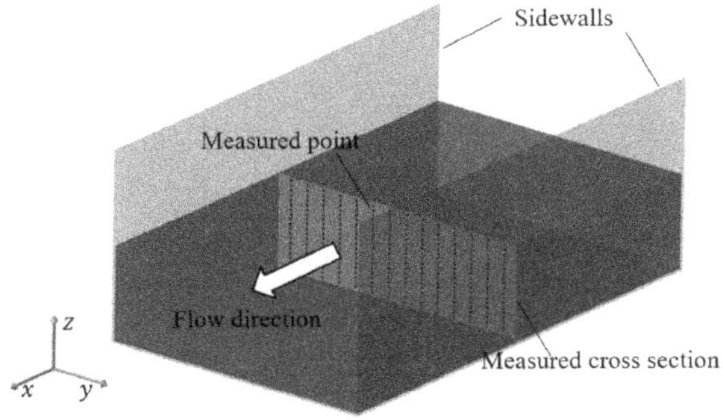

Figure 3. Layout of the measured cross section, lines and points.

A 3D acoustic Doppler velocimeter (ADV) with a precision of ±0.25 cm/s was adopted to measure the instantaneous velocity. The maximum sampling frequency was 50 Hz, and the sampling time for every measured point was 120 s, which resulted in 6000 velocity data for each point. Five experimental conditions were considered here, and the variables among these five cases were the cover conditions and water depth H, which changes from 15 to 20 cm. The details of the experimental parameters are listed in Table 1.

Table 1. Basic details of the characteristic parameters. B is the width of the flume, H is the water depth, S is the bed slope and Re is the Reynolds number.

Cases	Cover Condition	B (m)	H (m)	S	Re
1	Full ice cover	1	0.15	0.001	25590
2	Full ice cover	1	0.185	0.001	33725
3	Symmetrical shore cover	1	0.16	0.001	32480
4	Symmetrical shore cover	1	0.20	0.001	49160
5	Asymmetrical shore cover	1	0.16	0.001	31888

4. Model Parameters

To predict the profiles of velocity, Reynolds stress and turbulence intensity in the ice-covered flow, the model parameters, i.e., m_b and m_i, n_b and n_i, K_0, etc., should be given first.

4.1. m_b and m_i

In the single-power law for the free surface flow, exponent m is affected by Darcy–Weisbach resistance coefficient f, i.e., $m = \kappa\left(\frac{8}{f}\right)^{0.5}$ [30]. For the two-power expression, exponents m_b and m_i can also be related to Darcy–Weisbach resistance coefficients of channel bed f_b and ice cover f_i. Specifically, the value of one exponent m can be related to the resistance coefficient of one fixed boundary. Hence, exponents m_b and m_i can be expressed as

$$\begin{cases} m_b = \kappa\left(\frac{8}{f_b}\right)^{0.5} \\ m_i = \kappa\left(\frac{8}{f_i}\right)^{0.5} \end{cases} \tag{15}$$

The resistance coefficients can be calculated as [34]

$$\begin{cases} f_b = \frac{8n_b^2 g}{R_b^{1/3}} \\ f_i = \frac{8n_i^2 g}{R_i^{1/3}} \end{cases} \tag{16}$$

where n denotes Manning's roughness coefficients; R indicates the hydraulic radiuses and the subscript b and i in them represent the channel bed and ice cover. g is the gravity acceleration. Here, the ratio of channel width to water depth was approximately greater than 5, which indicates that the flow could be considered a shallow flow. Hence, the hydraulic radius of each sublayer could be simplified as the corresponding sublayer depth, i.e., $R_b = h_b$ and $R_i = h_i$. The flow depths of the lower bed layer and upper ice layer have been defined from the location of the maximum velocity and are represented by h_m and $H - h_m$. Hence, we easily obtained

$$\begin{cases} h_m = h_b \\ H - h_m = h_i \end{cases} \tag{17}$$

Substituting Equation (16) into Equation (15), we obtained

$$\begin{cases} m_b = \kappa\left(\frac{R_b^{1/3}}{n_b^2 g}\right)^{0.5} \\ m_i = \kappa\left(\frac{R_i^{1/3}}{n_i^2 g}\right)^{0.5} \end{cases} \tag{18}$$

Then, we substituted Equations (17) and (18) into Equation (3) ($\frac{h_m}{H} = \frac{m_i}{m_i+m_b}$) and obtained

$$h_b = \frac{Hn_b(H-h_b)^{1/6}}{n_b(H-h_b)^{1/6} + n_i h_b{}^{1/6}} \tag{19}$$

Once Manning's coefficients n_b and n_i are known, the depth of the lower bed layer (h_b) is iteratively solved. Then, exponents m_b and m_i were calculated by Equation (18).

4.2. n_b and n_i

The basic calculation model of Darcy–Weisbach friction coefficient f is

$$f = \frac{8u_*^2}{U^2} \tag{20}$$

where $U(= Q/BH)$ is the averaged velocity in a cross section. Substituting Equation (16) into Equation (20), we obtained

$$n = \frac{u_*}{U\sqrt{g}} R^{1/6} \tag{21}$$

For the ice-covered flow, the flow is divided as two layers, and two depth-averaged velocities, U_b and U_i, are averaged from the channel bed and ice cover to the position of the maximum velocity. Hence, Manning's coefficient in each sublayer can be expressed as

$$\begin{cases} n_b = \frac{u_{*b}}{U_b\sqrt{g}} h_b^{1/6} \\ n_i = \frac{u_{*i}}{U_i\sqrt{g}} h_i^{1/6} \end{cases} \tag{22}$$

These two Manning's parameters can be determined by the measured velocity profile.

For the fully developed asymmetric flow, the method of logarithmic law can be adopted to predict the vertical distribution of the longitudinal velocity. However, because the velocity gradient of the maximum velocity is discontinuous, the method of logarithmic law may not be completely feasible for the entire water depth. For the ice-covered flow, the method of logarithmic law can be applied in each sublayer, i.e., the lower bed layer and upper ice layer. Bonakdari et al. (2008) [35] further demonstrate that the turbulent boundary layer consists of the inner region near the sidewall and outer region far from the sidewall and propose that the flow velocity inside the inner region can better satisfy the distribution form of the logarithmic law. In this study, we define that the inner region inside the lower bed layer starts from the channel bed (z = 0) to $0.2H$, and the inner region inside the upper ice layer is from $0.8H$ to the ice cover (H). Therefore, we used the following logarithmic law to describe the velocity profile in the two inner regions [36]

$$\begin{cases} \frac{U_b}{u_{*b}} = \frac{1}{\kappa} ln\left(30\frac{z}{k_{sb}}\right) \\ \frac{U_i}{u_{*i}} = \frac{1}{\kappa} ln\left(30\frac{z}{k_{si}}\right) \end{cases} \tag{23}$$

where k_{sb} and k_{si} are the roughness heights of the turbulence boundary in the lower bed layer and upper ice layer. The logarithmic law can be simplified to

$$\begin{cases} U_b = a_b ln(z) + F_b \\ U_i = a_i ln(z) + F_i \end{cases} \tag{24}$$

where $a_b = u_{*b}/\kappa$, $a_i = u_{*i}/\kappa$, $F_b = a_b\left(ln\frac{30}{k_{sb}}\right)$ and $F_i = a_i\left(ln\frac{30}{k_{si}}\right)$. These parameters can be obtained by regression analysis based on the measured longitudinal velocity data; then, shear velocities u_{*b} and u_{*i} and roughness heights k_{sb} and k_{si} can be obtained.

4.3. K_0

According to the previous work [28,37], K_0 is the flow model parameter for a known flow rate per unit flow width and is

$$K_0 = U/K_1 \tag{25}$$

where K_1 is the normalized depth-averaged velocities for total flow and is calculated by

$$K_1 = \int_0^1 \left(\frac{z}{H}\right)^{1/m_b} \left(1 - \frac{z}{H}\right)^{1/m_i} d(z/H) \tag{26}$$

5. Results and Discussion

5.1. Model Verification

With the above model parameters, the theoretical model can be applied to the experimental data. Figure 4 compares the predicted velocity from the model with the measured ones. Both profiles followed the assumed one, where the maximum velocity occurs near the middle of the water depth, and the minimum velocity appears near the fixed boundaries. The analytical results were consistent with the experimental velocities, so the proposed model and corresponding parameters are reasonable and reliable to be used to predict the velocity profile in the straight open channel flow covered by the ice cover.

Figure 5 presents the profiles of measured Reynolds stress and analytical Reynolds stress for all experimental cases. The Reynolds stress in the longitudinal direction demonstrates a linear distribution along the water depth. Zero shear stress occurred at the location of h_T near the middle of the water depth, after which the absolute value of Reynolds stress gradually increased and reached each peak near the fixed boundaries. The measured and predicted data in Figure 5 were basically consistent and had identical trends, which indicates that our proposed model for Reynolds stresses was feasible.

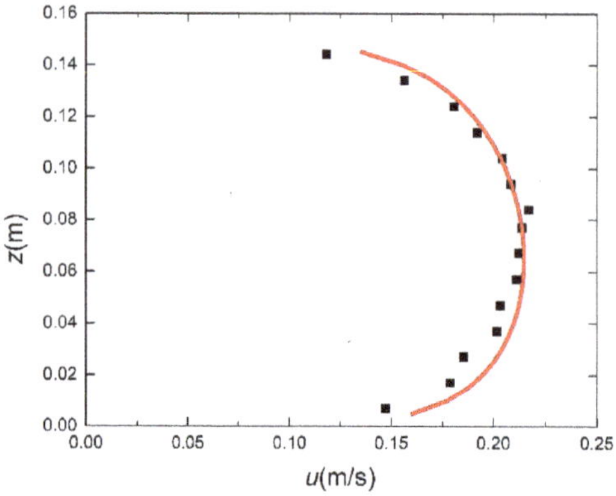

(**a**) Case 1

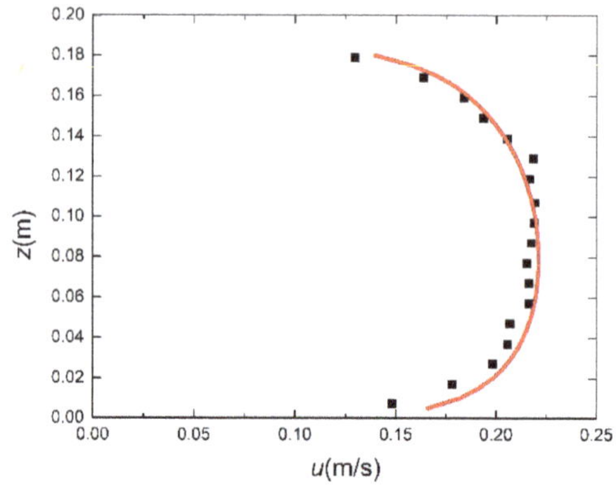

(**b**) Case 2

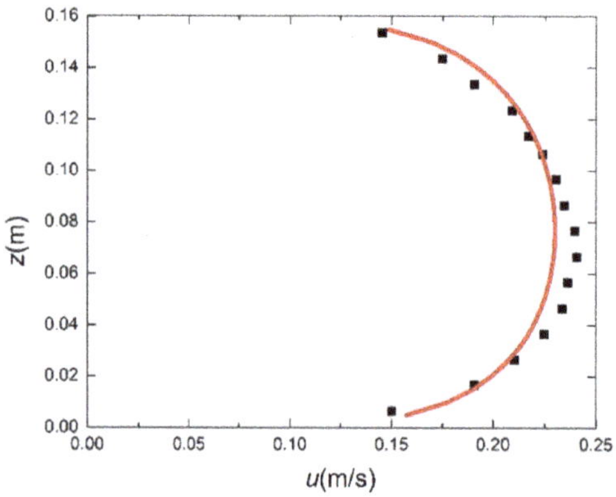

(**c**) Case 3

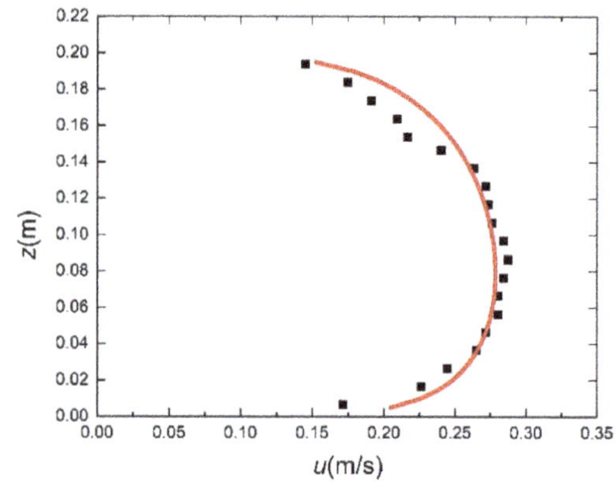

(**d**) Case 4

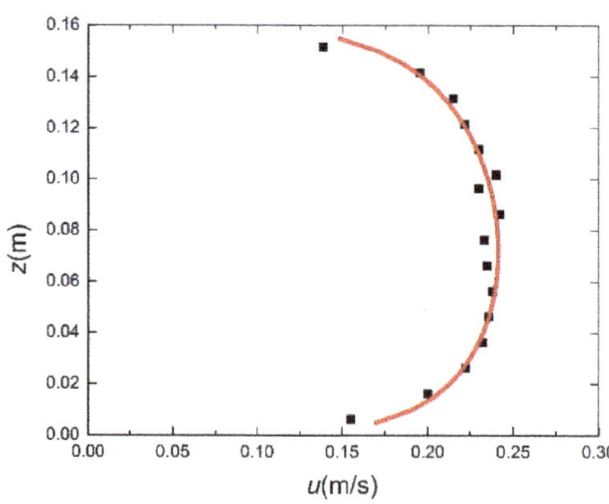

(**e**) Case 5.

Figure 4. Comparison of the measured and analytical velocities. Black squares denote the measured data and red lines denote the analytical ones.

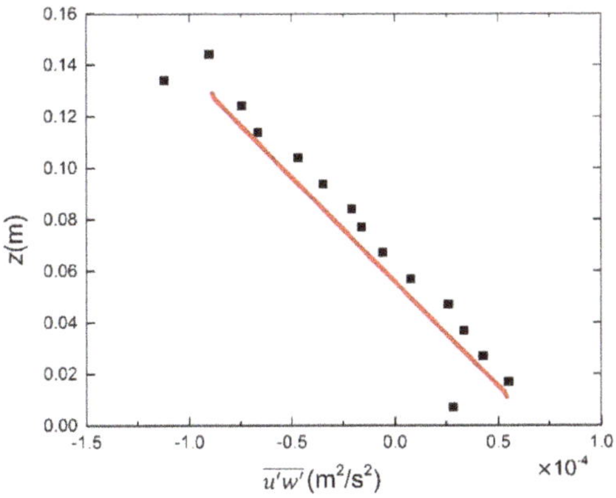

(**a**) Case 1

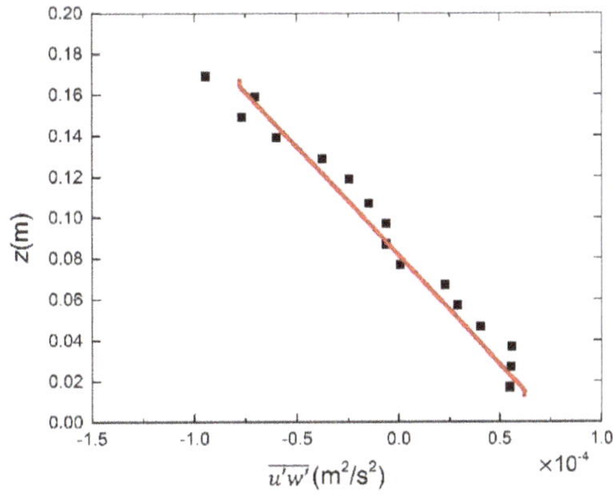

(**b**) Case 2

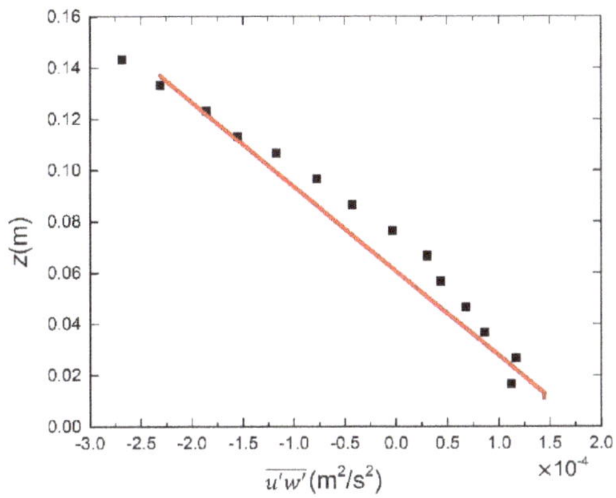

(**c**) Case 3

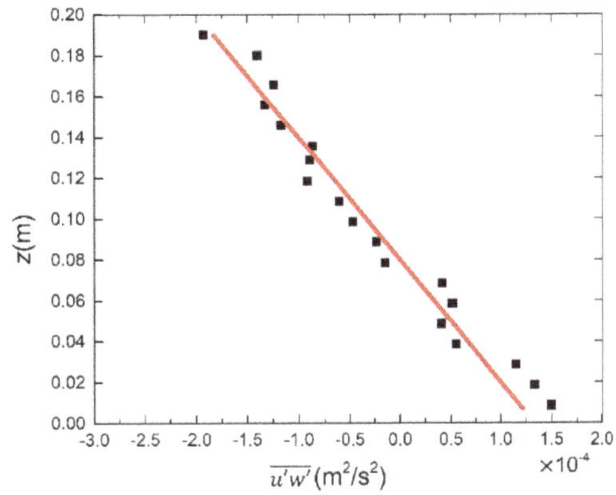

(**d**) Case 4

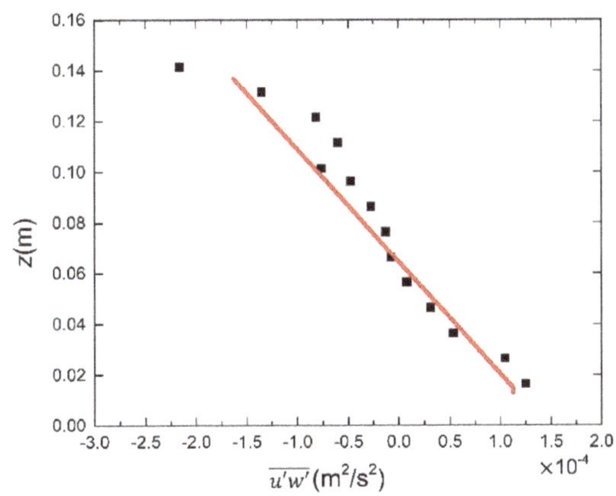

(**e**) Case 5

Figure 5. Comparison of the measured and analytical Reynolds stresses. Black squares denote the measured data and red lines denote the analytical ones.

Figure 6 shows the results of the turbulence intensity from the experiments and analytical model. The turbulence intensity reached its minimum near the middle water depth, from which it had an increasing trend towards the channel bed and ice cover. The profiles of turbulence intensity are consistent with the results from Papanicolaou et al. (2007) [38], who demonstrate that the turbulence production near the central region is small, and its diffusion effect is significant, but the turbulence production near the fixed boundaries reaches its maximum value. Overall, the analytical model can catch the general trend of the measured turbulence intensity, although the measured one has some fluctuations, which confirms that the model can be applied to calculate the turbulence intensity.

To find the difference between analytical and measured data, the error analysis was conducted. An absolute error is defined as the difference between analytical and measured time-averaged velocities. Hence, the average absolute error is calculated as

$$\overline{\Delta_a} = \frac{1}{N}\sum_{1}^{N}\left|(I)_{analytical} - (I)_{measured}\right| \tag{27}$$

where N is the number of measured points in each measured line for each case, $(I)_{analytical}$ and $(I)_{measured}$ are the analytical and measured values and I represents the variables, i.e., longitudinal velocity, Reynolds stress and turbulence intensity.

The average relative error is defined as

$$\overline{\Delta_r} = \frac{1}{N}\sum_{1}^{N}\left|\frac{(I)_{analytical} - (I)_{measured}}{(I)_{measured}}\right| \times 100\% \tag{28}$$

As shown in Table 2, the time-averaged longitudinal velocities obtained from the proposed model were reliable within an accuracy of 0.019 m/s in terms of the average absolute error $\overline{\Delta_a}$. The average relative error $\overline{\Delta_r}$ was 2.86–10.97%. The average absolute error and average relative error of the Reynolds stress were within 0.093 m/s and 13.63%, respectively. The average absolute error and relative error of the turbulence intensity were within 0.0019 m/s and 12.54%, respectively. All values in Table 2 were below 20%, which further confirmed that the proposed analytical model was reliable and feasible to predict the velocity and turbulence structure in the ice-covered flow.

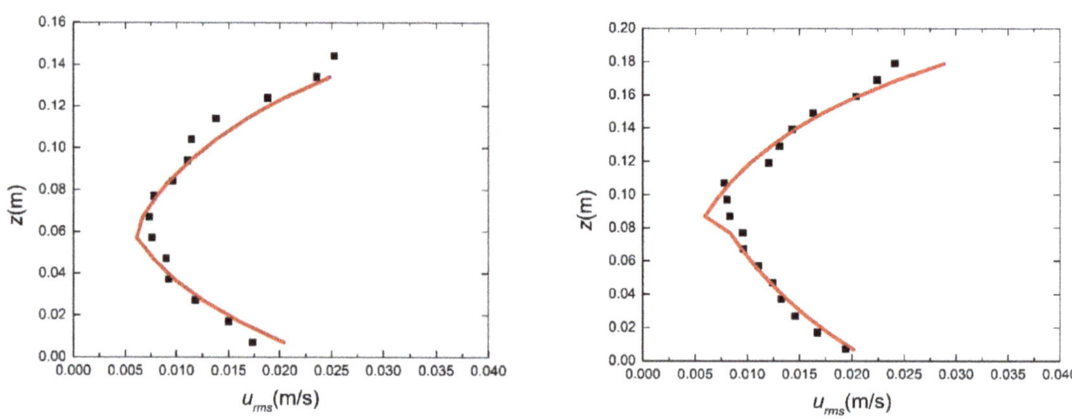

(**a**) Case 1 (**b**) Case 2

Figure 6. *Cont.*

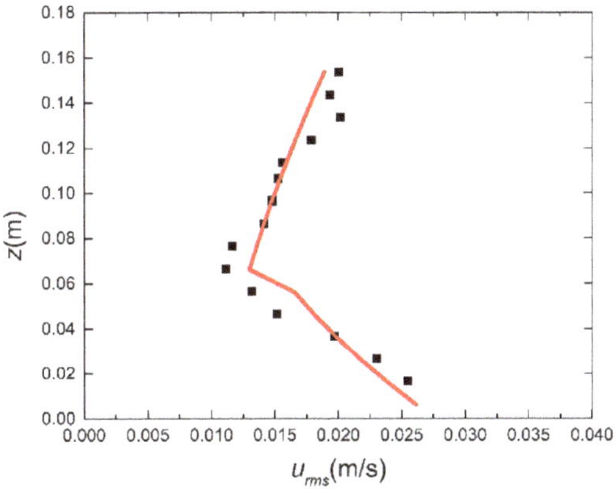

(**c**) Case 3

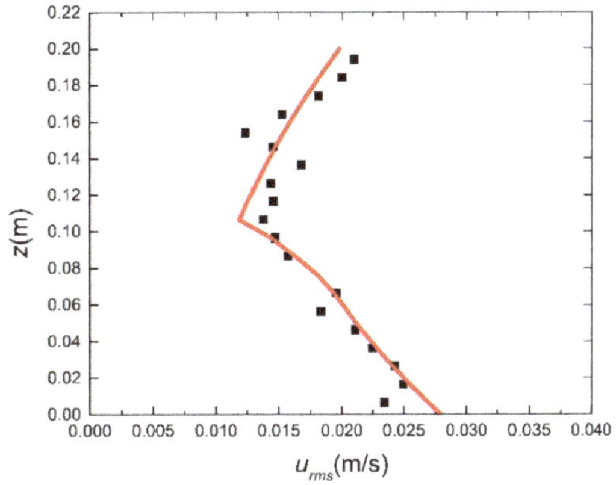

(**d**) Case 4

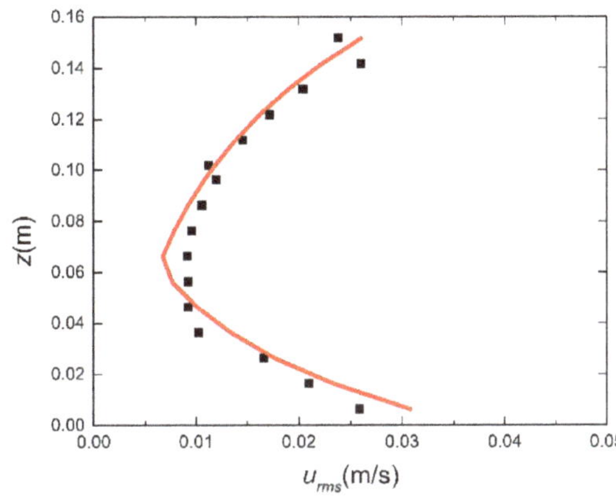

(**e**) Case 5

Figure 6. Comparison of the measured and analytical turbulence intensity. Black squares denote the measured data and red lines denote the analytical ones.

Table 2. Error statistics for the longitudinal velocity, Reynolds stress and turbulence intensity. $\overline{\Delta_a}$ is the average absolute error and $\overline{\Delta_r}$ is the average relative error.

Cases	Velocity		Reynolds Stress		Turbulence Intensity	
	$\overline{\Delta_a}$(m/s)	$\overline{\Delta_r}$(%)	$\overline{\Delta_a}$(m/s)	$\overline{\Delta_r}$(%)	$\overline{\Delta_a}$(m/s)	$\overline{\Delta_r}$(%)
1	0.0065	3.99	0.036	7.01	0.0013	10.52
2	0.0057	3.18	0.038	10.00	0.0011	8.09
3	0.0059	2.86	0.050	9.11	0.0012	8.05
4	0.011	5.25	0.082	8.42	0.0012	6.99
5	0.019	10.97	0.093	13.63	0.0019	12.54

5.2. Discussion

5.2.1. Manning's Coefficients n_b and n_i

Table 3 lists the calculated Manning's roughness coefficients of the channel bed and ice cover for each case, i.e., n_b and n_i. Manning's roughness coefficient of the channel bed was 0.012–0.015, and its mean value was 0.0138 with the standard deviation of 0.0012, which verified that n_b hardly changed among all studied cases. Manning's roughness coefficients of the ice cover in all cases were 0.017–0.02, which had small fluctuations and were lightly larger than those of the channel bed. The mean value of n_i was 0.0182 with a standard deviation of 0.0012. For the experimental channel, Manning's roughness coefficients of the channel bed and ice cover should be considered two specific constants. We took the mean Manning's roughness coefficient in each layer as the final Manning's roughness coefficient. Hence, n_b was set to be 0.0138, and n_i was equal to 0.0182.

Table 3. Values of the model parameters to predict the velocity and Reynolds stress.

Cases	n_b	n_i	m_b	m_i	h_m/H		h_τ/H	
					Calculated	Measured	Calculated	Measured
1	0.013	0.018	6.35	4.84	0.43	0.56	0.37	0.35
2	0.012	0.017	7.13	5.31	0.43	0.58	0.43	0.44
3	0.015	0.017	5.41	4.96	0.48	0.42	0.43	0.43
4	0.014	0.02	6.57	4.30	0.40	0.44	0.39	0.44
5	0.015	0.019	5.63	4.59	0.45	0.54	0.39	0.39

5.2.2. Flow Parameters m_b and m_i

The values of exponents m_b and m_i in this study are shown in Table 3. All values of m_b were $5.4 \leq m_b \leq 7.1$, and all m_i were $4.3 \leq m_i \leq 5.3$. Teal et al. (1994) [27] estimated m_b and m_i by nonlinear regression for more than measured 2300 vertical velocity profiles, and they found that these two parameters were 1.5–8.5, which includes the theoretical range (see Table 3). For an open channel flow, only exponent m_b is considered and is approximately 6–7, which is unlike the values used here. Hence, both m_b and m_i in the covered flow are affected by the roughness characteristics of both the channel bed and ice cover.

The shape of the vertical profile of the longitudinal velocity is determined by exponents m_b and m_i. The ratio of them can be given from Equation (17) as

$$\frac{m_i}{m_b} = \left(\frac{h_i}{h_b}\right)^{1/6} \frac{n_b}{n_v} \tag{29}$$

According to Equation (29), the ratio of m_b and m_i is mainly influenced by the roughness coefficients of the channel bed and ice cover. This result is also confirmed by the measured data of Li et al. (2020) [29]. They demonstrate that the vertical distribution of the velocity remains constant with changing water depth and flow rate under the same m_b and m_i or the same ratio of m_b and m_i. In the asymmetric flow, the maximum velocity tended to be closer to the smooth boundary with smaller roughness coefficient. In Table 3, the channel bed had a smaller roughness coefficient than the ice cover, which corresponded to the close maximum velocity to the channel bed, which was verified by $h_m < H - h_m$. In general, both open channel flow and symmetry flow can be considered as the special cases of asymmetric flow. For the open channel flow, the exponent m_i tends to infinity. For the symmetry flow, $m_b = m_i$.

5.2.3. Comparison of h_m and h_τ

Table 3 presents the locations (h_m and h_τ) where the maximum velocity and zero shear stress occur for all cases. These two locations were not consistent and had distinct difference, i.e., calculated $h_m/H > h_\tau/H$. The location of zero shear stress is closer to the channel bed than that of maximum velocity [13,29,38]. By contrast, the locations of the maximum velocity and zero shear stress for the symmetry and open channel flows were the same. Specifically, the location of maximum velocity and zero shear stress for the symmetry flows was at the middle water depth, and the location for the open channel flows was at the free water surface.

Considering the locations of the maximum velocity and zero shear stress for cases 1, 3 and 5, when the ratio of n_i to n_b increased, h_m/H decreased, so the location of the maximum velocity approached the channel bed, and the vertical inhomogeneity of the velocity profile was strengthened. Meanwhile, τ_i/τ_b increased with the increase in n_i/n_b; then, h_τ/H decreased, which indicates the location of the zero shear stress gets closer to the channel bed.

5.2.4. Empirical Constants D_b, E_b, D_i and E_i

The empirical constants of turbulence intensity for all cases are listed in Table 4. No remarkable changes of D_b and D_i were observed in any case. The mean D_b was 2.22 with a standard deviation of 0.05, and the mean D_i was 2.16 with a standard deviation of 0.06. Hence, it is reasonable to consider that $D_b = D_i$. The difference between E_b and E_i was not negligible, similar to the results of Li et al. (2020) [29].

Table 4. Parameters to predict the turbulence intensity in each case.

Cases	D_b	E_b	D_i	E_i
1	2.22	1.44	2.06	1.82
2	2.21	1.46	2.13	1.87
3	2.22	1.73	2.24	1.50
4	2.15	1.00	2.21	1.83
5	2.31	1.81	2.16	1.51

6. Conclusions

The existence of ice cover dramatically changed the flow velocity and turbulence structure. We here proposed theoretical models to describe the vertical distribution of longitudinal velocity, shear stress and turbulence intensity. By dividing the ice-covered flow into an ice-affected layer and a channel bed-affected layer, a two-power-law function was adopted to predict the vertical profile of velocity. The calculated velocity distribution presents that the maximum velocity occurred near the middle of the water depth close to the channel bed with smooth boundary. Theoretical analysis shows that the shear stress had a linear distribution form in the vertical direction, with the positive values in the lower bed layer and negative values in the upper ice layer. Moreover, the Manning's roughness coefficient of the ice cover was larger than that of the channel bed. The two exponents m_b and m_i were influenced by the roughness coefficients of the channel bed and ice cover. The location of zero shear stress was not the same as that of maximum velocity and was closer to the smooth fixed boundary than the plane of maximum velocity, namely, $h_m > h_\tau$. A comparison of the analytical and experimental velocities, the Reynolds stress and turbulence intensity displays that the theoretical models can provide satisfied predictions of the vertical distribution of these flow characteristics. This study expands our understanding of the effects of ice cover on the hydraulic characteristics in the open channels. However, we still need to do more research to explore the application of the proposed models in other conditions, like compound channels or confluence channels and we will involve comprehensive experiments to reveal detailed flow characteristics, such as vortex structure.

Author Contributions: Conceptualization, J.Z. and W.W.; methodology, J.Z.; software, Z.X., Y.Z.; resources, Q.L., Y.Z.; data curation, H.Q.; writing—original draft preparation, J.Z.; writing—review and editing, J.Z.; visualization, H.Q.; supervision, W.W.; project administration, Z.L.; funding acquisition, Z.L. All authors have read and agreed to the published version of the manuscript.

Funding: This research was funded by the National Key R&D Program of China (2017YFC0504704), the National Natural Science Foundation of China (Grant No. 51609198), the Science and Technology Project Funded by Shaanxi Provincial Department of Water Resources (2020slkj-10) and the Technology Project Funded by Clean Energy and Ecological Water Conservancy Engineering Research Center (QNZX-2019-03).

Conflicts of Interest: The authors declare no conflict of interest.

References

1. Davar, K.S.; Elhadi, N.A. Management of ice-covered rivers: Problems and perspectives. *J. Hydrol.* **1981**, *51*, 245–253. [CrossRef]
2. Chen, Y.; Wang, Z.; Zhu, D.; Liu, Z. Longitudinal dispersion coefficient in ice-covered rivers. *J. Hydraul. Res.* **2016**, *54*, 558–566. [CrossRef]
3. Chen, G.; Gu, S.; Li, B.; Zhou, M.; Huai, W. Physically based coefficient for streamflow estimation in ice-covered channels. *J. Hydrol.* **2018**, *563*, 470–479. [CrossRef]
4. Knack, I.; Shen, H.-T. Sediment transport in ice-covered channels. *Int. J. Sediment Res.* **2015**, *30*, 63–67. [CrossRef]
5. Lee, M.; Moser, R.D. Direct numerical simulation of turbulent channel flow up to Reτ ≈ 5200. *J. Fluid Mech.* **2015**, *774*, 395–415. [CrossRef]
6. Lotsari, E.; Tarsa, T.; Mri, M.K.; Alho, P.; Kasvi, E. Spatial variation of flow characteristics in a subarctic meandering river in ice-covered and open-channel conditions: A 2d hydrodynamic modelling approach. *Earth Surf. Process. Landf.* **2019**, *44*, 1509–1529. [CrossRef]
7. Turcotte, B.; Morse, B.; Bergeron, N.E.; Roy, A.G. Sediment transport in ice-affected rivers. *J. Hydrol.* **2011**, *409*, 561–577. [CrossRef]
8. Wang, F.; Huai, W.; Liu, M.; Fu, X. Modeling depth-averaged streamwise velocity in straight trapezoidal compound channels with ice cover. *J. Hydrol.* **2020**, *585*, 124336. [CrossRef]
9. Smith, B.T.; Ettema, R. Flow Resistance in Ice-Covered Alluvial Channels. *J. Hydraul. Eng.* **1997**, *123*, 592–599. [CrossRef]
10. Wang, J.; Wu, Y.; Sui, J.; Karney, B. Formation and movement of ice accumulation waves under ice cover—An experimental study. *J. Hydrol. Hydromech.* **2019**, *67*, 171–178. [CrossRef]
11. Namaee, M.R.; Sui, J. Velocity profiles and turbulence intensities around side-by-side bridge piers under ice-covered flow condition. *J. Hydrol. Hydromech.* **2020**, *68*, 70–82. [CrossRef]
12. Hanjalić, K.; Launder, B.E. Fully developed asymmetric flow in a plane channel. *J. Fluid Mech.* **1972**, *51*, 301–335. [CrossRef]
13. Parthasarathy, R.N.; Muste, M. Velocity Measurements in Asymmetric Turbulent Channel Flows. *J. Hydraul. Eng.* **1994**, *120*, 1000–1020. [CrossRef]
14. Tatinclaux, J.; Gogus, M. Asymmetric Plane Flow with Application to Ice Jams. *J. Hydraul. Eng.* **1983**, *109*, 1540–1554. [CrossRef]
15. Lau, Y.L.; Krishnappan, B.G. Sediment Transport Under Ice Cover. *J. Hydraul. Eng.* **1985**, *111*, 934–950. [CrossRef]
16. Muste, M.; Braileanu, F.; Ettema, R. Flow and sediment transport measurements in a simulated ice-covered channel. *Water Resour. Res.* **2000**, *36*, 2711–2720. [CrossRef]
17. Robert, A.; Tran, T. Mean and turbulent flow fields in a simulated ice-covered channel with a gravel bed: Some laboratory observations. *Earth Surf. Process. Landforms* **2012**, *37*, 951–956. [CrossRef]
18. Tao, L. Experimental Study on vertical velocity distribution of water flow under ice sheet. *Eng. Constr.* **2015**, *29*, 370–371.
19. Shen, H.T.; Harden, T.O. The effect of ice cover on vertical transfer in stream channels. *J. Am. Water Resour. Assoc.* **1978**, *14*, 1429–1439. [CrossRef]
20. Lau, Y.L.; Krishnappan, B.G. Ice Cover Effects on Stream Flows and Mixing. *J. Hydraul. Div.* **1981**, *107*, 1225–1242. [CrossRef]
21. Chen, G.; Gu, S.; Huai, W.; Zhang, Y. Boundary Shear Stress in Rectangular Ice-Covered Channels. *J. Hydraul. Eng.* **2015**, *141*, 06015005. [CrossRef]
22. Attar, S.; Li, S. Data-fitted velocity profiles for ice-covered rivers. *Can. J. Civ. Eng.* **2012**, *39*, 334–338. [CrossRef]
23. Sui, J.; Wang, J.; Yun, H.E.; Krol, F. Velocity profiles and incipient motion of frazil particles under ice cover. *Int. J. Sediment Res.* **2010**, *25*, 39–51. [CrossRef]
24. Larsen, P.A. Head losses caused by an ice cover on open channels. *J. Boston Soc. Civ. Eng.* **1969**, *56*, 45–67.
25. Sayre, W.W.; Song, G.B. Effects of Ice Covers on Alluvial Channel Flow and Sediment Transport Processes. IIHR Report No. 218, University of Lowa. Available online: https://apps.dtic.mil/dtic/tr/fulltext/u2/a066991.pdf (accessed on 16 April 2021).
26. Uzuner, M.S. The composite roughness of ice covered streams. *J. Hydraul. Res.* **1975**, *13*, 79–102. [CrossRef]
27. Teal, M.J.; Ettema, R.; Walker, J.F. Estimation of Mean Flow Velocity in Ice-Covered Channels. *J. Hydraul. Eng.* **1994**, *120*, 1385–1400. [CrossRef]
28. Tsai, W.; Ettema, R. Modified Eddy Viscosity Model in Fully Developed Asymmetric Channel Flows. *J. Eng. Mech.* **1994**, *120*, 720–732. [CrossRef]
29. Li, Q.; Zeng, Y.-H.; Bai, Y. Mean flow and turbulence structure of open channel flow with suspended vegetation. *J. Hydrodyn.* **2020**, *32*, 314–325. [CrossRef]
30. Zhang, J.; Lei, J.; Huai, W.; Nepf, H. Turbulence and Particle Deposition Under Steady Flow Along a Submerged Seagrass Meadow. *J. Geophys. Res. Oceans* **2020**, *125*, 2019–015985. [CrossRef]
31. Rowiński, P.M.; Kubrak, J. A mixing-length model for predicting vertical velocity distribution in flows through emergent vegetation. *Hydrol. Sci. J.* **2002**, *47*, 893–904. [CrossRef]
32. Nezu, I.; Rodi, W. Open-channel Flow Measurements with a Laser Doppler Anemometer. *J. Hydraul. Eng.* **1986**, *112*, 335–355. [CrossRef]
33. Zhong, Y.; Huai, W.; Chen, G. Analytical Model for Lateral Depth-Averaged Velocity Distributions in Rectangular Ice-Covered Channels. *J. Hydraul. Eng.* **2019**, *145*, 04018080. [CrossRef]
34. Huai, W.-X.; Xu, Z.-G.; Yang, Z.-H.; Zeng, Y.-H. Two dimensional analytical solution for a partially vegetated compound channel flow. *Appl. Math. Mech.* **2008**, *29*, 1077–1084. [CrossRef]

35. Bonakdari, H.; Larrarte, F.; Lassabatere, L.; Joannis, C. Turbulent velocity profile in fully-developed open channel flows. *Environ. Fluid Mech.* **2008**, *8*, 1–17. [CrossRef]
36. Zare, S.G.A.; Moore, S.A.; Rennie, C.D.; Seidou, O.; Ahmari, H.; Malenchak, J. Estimation of composite hydraulic resistance in ice-covered alluvial streams. *Water Resour. Res.* **2016**, *52*, 1306–1327. [CrossRef]
37. Han, L.; Zeng, Y.; Chen, L.; Li, M. Modeling streamwise velocity and boundary shear stress of vegetation-covered flow. *Ecol. Indic.* **2018**, *92*, 379–387. [CrossRef]
38. Papanicolaou, A.N.; Elhakeem, M.; Hilldale, R. Secondary current effects on cohesive river bank erosion. *Water Resour. Res.* **2007**, *43*, 497–507. [CrossRef]

water

MDPI

Article

Waved-Shape Accumulation of Ice Jam—Analysis and Experimental Study

Pangpang Chen [1], Jueyi Sui [2,*], Guangxue Cao [1,*] and Tiejie Cheng [1]

1 School of Civil and Hydraulic Engineering, Hefei University of Technology, Hefei 230009, China
2 School of Engineering, University of Northern British Columbia, Prince George, BC V2N 4Z9, Canada
* Correspondence: jueyi.sui@unbc.ca (J.S.); cao-gx@hfut.edu.cn (G.C.); Tel.: +1-250-9506399 (J.S.)

Abstract: Ice jam is a unique hydrological phenomenon in rivers in cold regions. The appearance of an ice jam in a river results in an increase in the wetted perimeter of the flow cross-section, and thus an increase in flow resistance as well as water level. It may cause ice flooding sometimes. Similar to the "sand wave" phenomenon in riverbed, it has been observed in laboratory experiments that the waved-shape accumulation of ice particles (termed as "ice wave") under an ice jam occurred. In this study, an Equation for describing the relationship between the approaching flow Froude number (Fr) and the ratio of ice jam thickness to flow depth (t/H) has been proposed. Taking the inflection point value of the equation under different flow depths, a characteristic curve has been developed to judge whether ice waves under an ice jam occurs. When the flow Froude number in front of an ice jam is below the value at the inflection point of the curve, the ice jam can maintain a mechanical stability within the ice jam thickness in a range from the lower limiting value to the upper limiting value, which were close to the ice wave trough thickness and the ice wave crest thickness, respectively. An Equation for calculating the ice wavelength has been derived and verified by using results of laboratory experiments. The relationship between the migration speed of ice wave and the ratio of ice discharge to water flow rate (Qi/Q) has been also analyzed. At last, case studies have been conducted with respect to ice accumulation in the St. Lawrence River, the Beauharnois Canal and the La Grande River. Results of case studies show that the shoving and ice dam have been dominated by mechanical factors, which would be accompanied by the ice wave phenomenon during the ice jam accumulation process. Results of case studies about ice accumulation in natural rivers also show that the relative thickness of an ice jam (t/H) of 0.4 is the criterion for assessing whether an ice jam in a river belongs to an ice dam.

Keywords: ice jam; ice wave; experiment; wave crest; wave trough; wavelength; migration speed

Citation: Chen, P.; Sui, J.; Cao, G.; Cheng, T. Waved-Shape Accumulation of Ice Jam—Analysis and Experimental Study. *Water* **2022**, *14*, 3945. https://doi.org/10.3390/w14233945

Academic Editor: Giuseppe Oliveto

Received: 15 October 2022
Accepted: 30 November 2022
Published: 4 December 2022

1. Introduction

River ice is a hydrological phenomenon in cold regions. River ice has a great impact on the global hydrological system, especially in the northern hemisphere [1]. Ice jam and ice dam can be developed when ice floes transported by flowing water is arrested by obstacles such as stationary ice cover or congested [2]. Due to a large aggregate thickness and a high hydraulic resistance comparing to those of a sheet ice cover, ice jams tend to disturb the riverbed, and cause high water level. River ice has many repercussions for operation and design including determining the overturning moment on structures, controlling the severity of spring flooding and assessing the associated water level—frequency relationships, or of predicting riverbed scour due to surges associated with suddenly rebreak-up of ice jams [3–5]. The Yellow River in China is one of the rivers with serious ice flooding disaster [6,7]. During winter periods, flooding resulted from ice jams happens very often in some river reaches including the Inner Mongolia Reach, Hequ Reach, and the Lower reach of the Yellow River [2,4,7–9].

A variety of research works regarding river ice hydraulics have been carried out based on prototype observations [2,4,7–9], laboratory experiments [10,11], theoretical analysis [12–14] and numerical simulation [15–18]. Up to date, most of the reported research is focused on the formation, evolution and mechanism of ice jams. According to field observations in natural rivers, many scholars believe that the development of an ice cover or jam will generally go through following stages, ice pans (floe) floating period, freeze-up period, ice-covered/jammed period and river break-up period.

During the ice-pans floating period, the surface layer of flowing water lost too much heat due to the decrease in air temperature [19]. On the other side, the turbulent mixing process of flowing water causes the heat exchange between the water surface layer and the main water body. The heat loss from flowing water in a river is carried out rapidly, especially in the main channel. Thus, with the decrease in the temperature, the production of frazil ice increases quickly and float on water surface. As a consequence, ice pans on water surface will be gradually developed and transported downstream.

During the period of river freeze-up, an ice cover is generally developed at cross section with relatively slow velocity, provided the coverage of ice pans on water surface is high enough [20]. Once an initial ice cover is formed on water surface, the incoming floating ice pans/floes will be either stopped in front of this initial ice cover if the flow velocity is low; or submerged under this initial ice cover if the flow velocity is high enough, and consequently an initial ice jam will be developed. The process of an ice jam formation and evolution is accompanied by the continuous advancement of the ice jam head toward upstream and the ice jam toe toward downstream. During the period of development of an ice jam, the ice jam thickness varies depending on the flow velocity under the ice jam and the amount of incoming ice pans/floes from upstream. When the flow condition in a river does not change much, the ice cover or ice jam formed in this river normally becomes stable during an ice-covered period.

Once a river is covered by an ice cover or ice jam, the distribution of flow velocity under an ice cover is completely different from that under an open flow condition. The presence of an ice cover/jam results in the complexity of the river hydraulics and sometimes significantly affects the hydrodynamic characteristics of a river [21], due to the increase in the wetted perimeter of a flow, and thus the resistance to the flow. Affected by the roughness coefficient of both ice cover/jam and riverbed, the maximum flow velocity under an ice-covered jammed condition is located between the channel bed and ice cover instead of at water surface under an open flow condition when the resistance caused by the air is neglected [22,23]. Furthermore, the increase in the resistance to the flow due to the appearance of an ice cover on water surface results in the decrease in flow velocity. For the same flow discharge, the presence of an ice cover/jam on water surface will cause a significantly increase in water level comparing to that under an open flow condition [24].

With the increase in temperature in spring, ice cover/jam in a river will be broken up. During the river breakup process, a lot of ice floes with different size will be generated. These ice floes may be congested at narrow sections of a river and form ice jams again [25]. The river ice jam during a river breakup period was termed by some scholars as the breakup ice jam, which is normally accompanied by ice flooding disaster.

The development and evolution of an ice jam continues during the entire ice-covered period. The migration process of vast amount of frazil ice particles at the bottom of an ice jam is similar to motion of sediment particles on a riverbed (as shown in Figure 1), which is commonly called as bed-load movement or "sand wave" [26]. Considering the similarity of sediment movement on a riverbed to that of frazil ice particles at the bottom of an ice jam, the movement of frazil ice particles should also have a "cover"-load layer (similar to the "bed-load" for sediment motion on a riverbed) and a suspended layer of frazil ice particles (similar to the "suspended load" of sediment over a riverbed). So, frazil ice particles at the bottom of an ice jam should be also delivered to downstream as the waved-shaped migration which is similar to the process of sand waves on a riverbed when certain conditions are met, that can be named as "ice wave" [27].

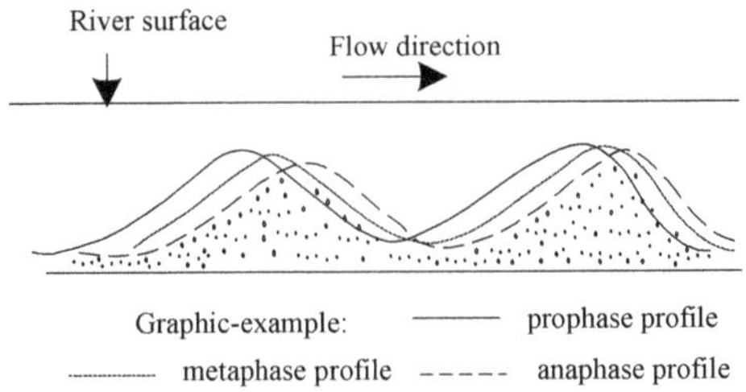

Figure 1. Sketch of the migration of sand waves.

The migration of ice waves at the bottom of an ice jam or cover is a special form of transportation of frazil ice particles. The migration of an ice wave can be described by various characteristic parameters such as wave crest, wave trough, wavelength, and migration speed (or wave speed). In this paper, Equations for describing ice jam thickness and flow condition during the migration of ice waves have been obtained. The criteria for the occurrence of an ice wave phenomenon have been proposed, and the equation for calculating the ice wavelength has been derived. By using equations mentioned above, calculation results of ice wave thickness at both wave crest and wave trough as well as wavelength are verified by results of laboratory experiments. The factors influencing the migration speed of ice waves have been also analyzed.

2. Materials and Methods

2.1. Laboratory Experiment Setup

Laboratory experiments have been carried out in a flume at Hefei University of technology. As shown in Figure 2, the flume for this experimental study is 26.68 m long, 0.4 m width. The smooth bottom of the flume is made of concrete with the bottom slope of zero. In total, 22 cross sections (CS) for measurements have been setup. The spacing distance between adjacent cross sections is 1.2 m. The model frazil ice particles are made of polyethylene material, which is also commonly used to model frazil ice particles in relevant research work in the world. The mass density of model frazil ice particles is 0.917 g/cm^3, which is nearly the same as that of natural ice. The model frazil ice particles have an ellipsoid shape. These particles are about 3.5 mm long with the length-to-thickness ratio of about 1.7. The static friction angle of ice particles is about 45. An ice feeder which is located between CS-4 and CS-5 was used to discharge ice particles to the channel with designated rates. One model ice cover made of Styro foam panel which is 1 m long and 0.4 m wide was placed between CS-20 and CS-21 in the downstream. This model ice cover was used to initiate an initial ice jam at the end of the flume.

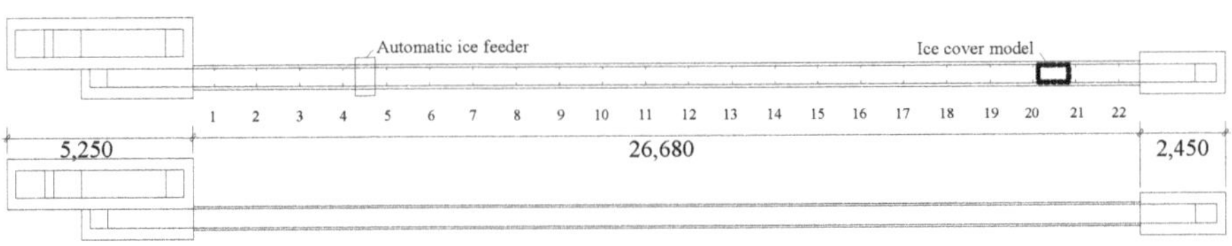

Figure 2. Layout plan of the flume for experiments.

The initial hydraulic condition for each experimental run (including approaching flow velocity, flow depth and ice discharge from the ice feeder) at CS-4 is used as the control condition. After ensuring that the flow discharge is constant and uniform flow is developed in the flume for each experiment run, the ice feeder was turned on to discharge model frazil ice particles into the flume. For each experimental run, the ice discharge rate is kept as a constant. During each experimental run, both water level and ice jam thickness at all cross sections were measured. The water level was measured by an external pressure measuring tube, and the ice jam thickness was read using a scale.

One can observe from experiments that, after model ice particles were discharged into the flume at CS-4, they will float on water surface and delivered to downstream but will be stopped and accumulated in front of model ice cover at the end of the flume. Gradually, an initial ice jam will develop and advance to the upstream section. During the development of an initial ice jam, if all ice particles are submerged at the leading edge of the ice jam, this initial ice jam can no longer advance to the upstream. The accumulation process of ice particles under an ice jam continues until it reaches an equilibrium state along the entire channel. The present experimental study belongs to a conceptual study, and does not consider the geometric similarity or geometric scale.

The real-time data such as pressure head and ice jam thickness at each cross section are recorded every 30 min. Each experiment lasted 20 h to make sure the evolution of the ice jam achieves an equilibrium state.

2.2. Mechanical Analysis of an Equilibrium Ice Jam

An ice jam is resulted from the accumulation of a large amount of frazil ice particles and ice floes under an ice cover. During the development and evolution of an ice jam, frazil ice particles are normally transported at the bottom of the ice jam in the form of "cover-load" (similar to the "bed-load" on a riverbed). Similar to the movement of sediment particles on a rivered, the motion of frazil ice particles at the bottom surface of an ice cover/jam also has a cover-load particle layer and a suspended particle layer. In the present conceptual study, the 1 m long Styro foam panel was placed on water surface at the end of the flume to initiate the formation of an initial ice jam along the channel. This model ice cover (Styro foam panel) did not affect the development of ice waves during the experiments.

In a laboratory flume, the movement of frazil ice particles at the bottom surface of an ice cover clearly demonstrates a phenomenon similar to the sand wave phenomenon on a riverbed, which can be called as "ice wave phenomenon" (as shown in Figure 3). Different from the sand wave movement on a riverbed, the migration of ice waves is more complex and more easily changeable due to the less difference in mass density between water and ice comparing to that between sand and water. Additionally, the changes in the incoming ice discharge also affect the appearance of ice waves.

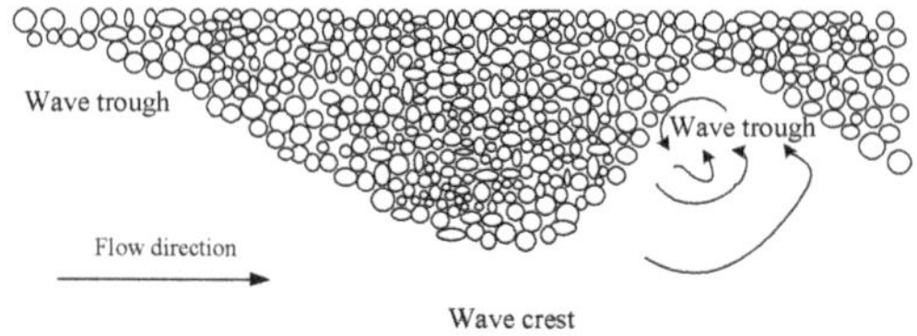

Figure 3. Schematic diagram of ice wave phenomenon.

During the evolution process of an ice jam, the cross-sectional area for flow changes depending on flow velocity and incoming ice discharge. When the cross-sectional area for flow becomes smaller (i.e., at the cross section where wave crest appears), the flow intensity will be increases, leading to an increase in the erosion process of ice particles at the wave crest and thus a decrease in the jam thickness. When the cross-sectional area for flow becomes larger, it will lead to less erosion activity of ice particles. More ice particles will be deposited in the zone of the wave trough. When the flow velocity under the ice jam reaches a certain value, ice waves at the bottom of an ice jam begins to migrate downstream. In the meantime, flow velocity varies accordingly during the migration process of ice waves due to changes in the cross-sectional area for flow.

With the increase in the length of an ice jam, its internal force will be also increased. Obviously, if the internal strength of an ice jam is less than the stress caused by the external force, the ice jam will partially collapse. As a consequence, the ice jam thickness downstream may be increased due to increased ice particles caused by the partially collapse of an ice jam in the upstream section until the ice jam can withstand a higher new stress. As pointed out by Beltaos [28], the theory for a wide river jam was first proposed by Pariset et al. [29]. By applying the limiting-stability relationship used for soil mechanics, the one-dimensional mechanical equilibrium differential Equation for an ice jam in river can be expressed as follows

$$\frac{d}{dx}(\overline{\sigma}_x t) + \frac{2k_0 k_1}{B}(\overline{\sigma}_x t) = \rho_i g t J + \tau_0 - \frac{2tC_i}{B} \quad (1)$$

According to the definition of Michel [30], σ_x is the normal stresses (including hydrostatic pressure) in x directions; t is the ice jam thickness; B is the channel width; k_1 is the lateral thrust coefficient; C_i is the adhesion force of an ice jam; k_0 = tanφ, φ is the internal friction angle of an ice jam; ρ_i is the mass density of ice; τ_0 is the flow shear stress at the bottom of an ice jam; J is the hydraulic gradient. By applying the theory of mechanics and continuity equation, considering internal force and assuming various boundary conditions, Equation (1) has been solved by Beltaos [28].

Pariset et al. [29] assumed that for the cohesionless an ice jam ($C_i \approx 0$), a wide ice jam can occur under limiting conditions. From the viewpoint of mechanics, the internal resistance of an ice jam is equal to the sum of the external forces, also represents a complete mobilization of the passive resistance of an ice jam. The general expression for a wide ice jam proposed as follows:

$$\frac{BV^2}{C^2H^2} = \mu \frac{\frac{\rho_i}{\rho}\left(1 - \frac{\rho_i}{\rho}\right)\frac{t^2}{H^2}\left(1 - \frac{\rho_i}{\rho}\frac{t}{H}\right)^3}{1 + \frac{\rho_i}{\rho}\frac{t}{H}} \quad (2)$$

where, C is the Chézy coefficient which represents the composite friction coefficient of the channel bed and the ice jam bottom. Results of calculation based on measured data showed that the Chézy coefficient value for both experimental study and field observations varies, indicating the Chézy coefficient is a comprehensive coefficient reflecting the roughness of a channel. Under an ice-covered/jammed flow condition, the wetted perimeter for a flow in a laboratory flume is $2(h + B)$; and the wetted perimeter for a flow in a wide natural channel can be approximated as $2B$. As shown in Figure 4, H is the approaching flow depth and V is the approaching flow velocity in front of the ice jam and equal to the discharge divided by flow cross sectional area. According to Beltaos [28], coefficients k_0, k_1 and k_2 have following relationship: $\mu = k_0k_1k_2 = 1.28$, k_2 is the passive soil pressure coefficient.

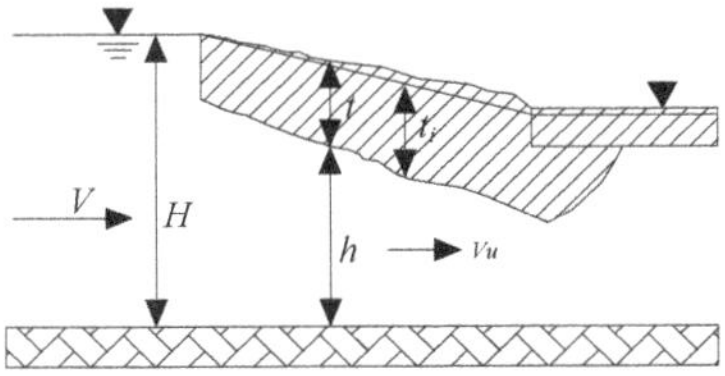

Figure 4. Schematic diagram of an ice jam.

According to Beltaos [28], to derive Equation (2), some assumptions have been introduced, such as the channel has been assumed as a prism channel with a rectangular cross-section. Convert the left end of Equation (2) to the flow Froude number through algebraic transformation, the following Equation can be obtained:

$$Fr = f\left(\frac{\mu C^2}{g}, \frac{H}{B}, \frac{t}{H}\right) = \sqrt{\frac{\mu C^2}{g}\left(\frac{H}{B}\right)\frac{\frac{\rho_i}{\rho}\left(1-\frac{\rho_i}{\rho}\right)\frac{t^2}{H^2}\left(1-\frac{\rho_i}{\rho}\frac{t}{H}\right)^3}{1+\frac{\rho_i}{\rho}\frac{t}{H}}} \quad (3)$$

According to the stable flow depth under the condition of equilibrium ice jam measured from four experimental runs, the value of H/B is 0.30, 0.40, 0.54 and 0.64, respectively. The influence curves of the flow Froude number Fr in front of the leading edge of an ice jam depending on the relative thickness of an ice jam t/H are showed in Figure 5 (Note: The Chézy coefficient C is taken as 14 which is an average value of laboratory results, as showed in Table 1). It can be seen from this figure that the upstream flow Froude number Fr has a limiting value (maximum) once the evolution of an ice jam achieves a stable state under the condition of a certain ratio of flow depth to the channel width (H/B).

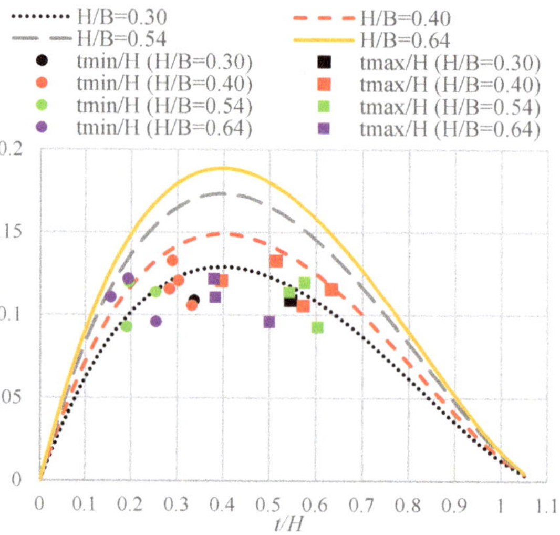

Figure 5. Dependence of flow Froude number Fr on the relative thickness of an ice jam t/H under condition of different H/B.

Table 1. Calculated Chézy coefficients for 11 experimental runs.

Initial Condition				Equilibrium Ice Jam					
Flow Depth (cm)	Flow Rate (L/s)	Ice Discharge (L/s)	Fr	Flow Depth (cm)	Average Ice Thickness (cm)	Water Head (cm)	Ice Jam Length (cm)	Hydraulic Gradient	Chézy Coefficient
10	5.61	0.01299	0.109	11.9	5.25	2.71	7.08	1.07×10^{-3}	14.10
15	9.00	0.02177	0.106	16.6	7.5	2.74	11.39	9.77×10^{-4}	14.83
15	9.60	0.00993	0.121	15.9	5.55	3.35	9.83	6.32×10^{-4}	16.22
15	10.81	0.0367	0.116	17.7	8.1	3.19	11.75	1.52×10^{-3}	13.17
15	12.02	0.03699	0.133	17.3	6.95	4.07	10.14	1.09×10^{-3}	15.45
20	12.02	0.03558	0.093	22.04	8.75	2.49	17.08	1.38×10^{-3}	9.45
20	14.43	0.03694	0.114	21.75	8.65	3.25	15.49	1.26×10^{-3}	12.10
20	15.20	0.04025	0.120	21.7	8.375	3.51	14.99	1.01×10^{-3}	13.94
25	16.00	0.01599	0.096	26	9.8	2.78	20.44	7.18×10^{-4}	12.95
25	18.01	0.01327	0.111	25.62	6.9	3.30	19.11	3.69×10^{-4}	16.36
25	20.00	0.02034	0.122	25.82	7.4	3.77	18.51	5.54×10^{-4}	15.20

One can see from Figure 5, if the upstream approaching flow Froude number Fr is below the limiting value (or maximum value), there are two relative thickness t/H ratios correspond to one Fr value. One can draw a horizontal line which is parallel to the X axis with the flow Froude number Fr less than the limiting value. The intersection on the left limb of the curve is the minimum value for the thickness of an ice jam, and the intersection on the right limb of the curve is the maximum value for the thickness of an ice jam. Thus, one can say from Pariset's result (Equation (3)) that the ice jam thickness during the mechanical evolution process varies dynamically between the minimum thickness and the maximum thickness. These two thicknesses are approximately corresponding to the thickness of ice wave trough and ice wave crest, respectively. One can say from this finding that when the measured flow Fr value is lower than the limiting Fr value, ice wave phenomenon would occur at the bottom of an ice jam.

Considering different ratios of the flow depth to the channel width (H/B), the limiting value of flow Froude number for an ice jam to achieve an equilibrium state is changed with the value of H/B. One can also see from Figure 5, with a certain channel width in the laboratory tests, the limiting flow Froude number for an ice jam to achieve an equilibrium state only depends on the flow depth, the deeper the flow depth, the larger the limiting flow Froude number for an ice jam to achieve an equilibrium state.

According to Figure 3, the following Figure 6 shows the schematic diagram of ice wave thickness and wavelength, where, $t_{\min}$ represents the thickness at wave trough and $t_{\max}$ represents the thickness at wave crest.

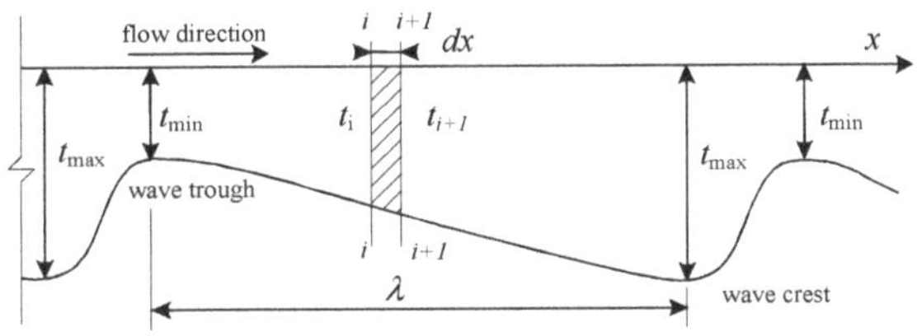

Figure 6. Schematic diagram of the thickness and length of an ice wave.

The periodic distance between the wave trough to the wave crest is the ice wavelength which is described by λ in Figure 6. The adhesion force Ci for a cohesionless ice jam in Equation (1) is relatively small and can be ignored, Equation (1) can be derived as follows:

$$\frac{df}{dx} + \frac{2k_0k_1f}{B} = \rho_i g t J + \tau_0 \tag{4}$$

where, $f = \sigma_x t$. As pointed out by Pariset et al. [29], the internal resistance f can be expressed as follows based on the soil mechanics:

$$f = tg^2\left(\frac{\pi}{4} + \frac{\varphi}{2}\right)\rho_i\left(1 - \frac{\rho_i}{\rho}\right)\frac{gt^2}{2} = k_2\rho_i\left(1 - \frac{\rho_i}{\rho}\right)\frac{gt^2}{2} \tag{5}$$

where, the coefficient k_2 can be determined by the theory of granular mechanics.

Assuming that the internal resistance f reaches the ultimate internal force at every cross section, taking the first derivative of Equation (5) into Equation (4), so:

$$dx = \frac{k_2\rho_i\left(1 - \frac{\rho_i}{\rho}\right)gtdt}{\rho_i g t J + \tau_0 - \frac{2k_0k_1f}{B}} \tag{6}$$

The above formula can be solved discretely by means of the different method. Along an ice wave, n cross sections have been divided with an equal spacing distance, that is, the equation for calculating the wavelength can be expressed as follows:

$$\lambda = \sum_{i=1}^{n-1}(x_{i+1} - x_i) = \sum_{i=1}^{n-1}\frac{k_2\rho_i\left(1 - \frac{\rho_i}{\rho}\right)gt_i(t_{i+1} - t_i)}{\rho_i g t_i J + \tau_0 - \frac{2k_0k_1f_i}{B}} \tag{7}$$

3. Results and Discussion

3.1. The Limiting Conditions for the Occurrence of Ice Wave Phenomenon

The formation, development, migration, and extinction of ice waves are accompanied by the continuous changes of various hydraulic factors in a river. Based on results of theoretical analysis expressed by Equation (3) and laboratory experiments, the limiting conditions for the occurrence of ice wave phenomenon have been studied.

The first derivative of the upstream flow Froude number to the ratio of the ice jam thickness to flow depth (t/H) could be obtained from Equation (3). Letting $d(Fr)/d(t/H) = 0$, one can obtain the limiting value for flow Froude number Fr under condition of different depth-width ratios (H/B). The calculated flow Froude number is the limiting value for the development of ice waves under an ice cover/jam. Results are presented in Figure 7.

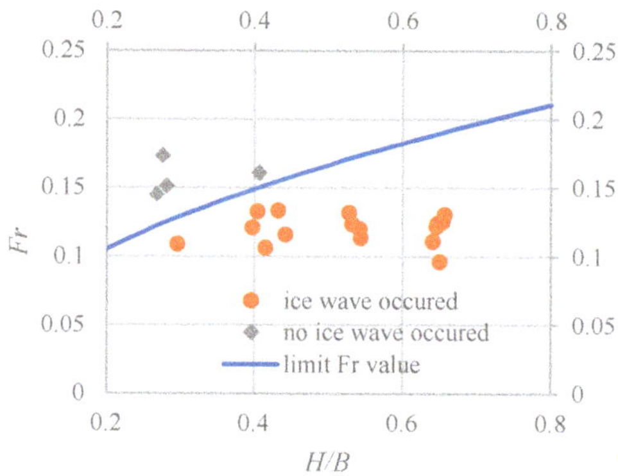

Figure 7. Calculated limiting flow Froude number Fr for developing an ice wave compared to those of laboratory experiments.

In total, 21 laboratory experimental runs have been conducted to assess the migration process of ice waves during the accumulation process of ice jams in a flume. By means of the regression analysis, the Chézy coefficient under the condition of an ice-covered flow ranged from 9 to 16 with an average value of 14 (As shown in Table 1, this value may only be applicable to this experimental study. The resistance effect of ice waves on water may increase the roughness of the wet boundary of the flume). In Table 2, the measured data of 21 experimental runs are summarized including flow depth, flow and ice discharge, and upstream approaching flow Froude number Fr. The occurrence of ice wave phenomenon is also recorded. One can see from Table 2 and Figure 7 that if the flow Froude number is located above the limiting Fr-curve in Figure 7, an ice wave phenomenon will not occur. Obviously, ice waves will present at the bottom of an ice jam when the flow Froude number is located below the limiting Fr-curve. It should be recommended that the limiting Fr-curve can be used to assess whether an ice wave occurs during the accumulation process of an ice jam.

Table 2. Hydraulic conditions for 21 experiment runs with and without presence of ice waves.

Initial Flow Depth (cm)	Flow Discharge (L/s)	Ice Discharge (L/s)	Flow Depth during Equilibrium Ice Jam (cm)	Fr during Equilibrium Ice Jam	Location of Ice Jam Head at CS	Occurrence of Ice Wave
10	5.61	0.01299	11.90	0.109	5	Yes
10	6.40	0.00625	10.75	0.145	5	No
10	6.40	0.0105	10.75	0.145	5	No
10	7.20	0.01054	11.31	0.151	18	No
10	8.00	0.0274	11.08	0.173	17	No
15	9.00	0.02177	16.60	0.106	5	Yes

Table 2. Cont.

Initial Flow Depth (cm)	Flow Discharge (L/s)	Ice Discharge (L/s)	Flow Depth during Equilibrium Ice Jam (cm)	Fr during Equilibrium Ice Jam	Location of Ice Jam Head at CS	Occurrence of Ice Wave
15	9.60	0.00993	15.90	0.121	5	Yes
15	10.81	0.00613	16.17	0.132	15	Yes
15	10.81	0.0367	17.70	0.116	5	Yes
15	12.02	0.03699	17.30	0.133	5	Yes
15	13.20	0.02365	16.26	0.161	14	No
20	14.43	0.03694	21.75	0.114	5	Yes
20	15.20	0.04025	21.70	0.120	5	Yes
20	15.20	0.00743	21.28	0.124	17	Yes
20	16.03	0.0259	21.12	0.132	15	Yes
25	16.00	0.01599	26.00	0.096	5	Yes
25	18.01	0.01327	25.62	0.111	6	Yes
25	20.00	0.02034	25.82	0.122	5	Yes
25	20.50	0.01298	25.88	0.124	16	Yes
25	21.00	0.01327	26.16	0.125	18	Yes
25	22.00	0.02579	26.27	0.130	19	Yes

3.2. Ice Jam Thickness of Wave Crest and Wave Trough

Considering fully developed ice jams (developed to the upstream cross Section 4) with the occurrence and migration of ice waves, 11 out of these 21 experimental runs have been selected to study the thickness of wave crest and wave trough. By using Equation (3), both the lower limiting value and the upper limiting value of the ice jam thickness have been calculated, as shown in Table 3.

Table 3. Calculated results of ice jam thickness at wave crest and trough compared to those of laboratory experiments.

Initial Condition			Equilibrium Ice Jam		Calculated Ice Jam Thickness (cm)		Measured Ice Jam Thickness (cm)			Diff. Btw. Calculated and Measured Results	
Flow Depth (cm)	Flow Rate (L/s)	Ice Discharge (L/s)	Fr	Flow Depth (cm)	Lower Limiting (t_{LL})	Upper Limiting (t_{UL})	At Wave Trough (t_{WT})	At Wave Crest (t_{WC})	Ratio: t_{WC}/t_{WT}	($t_{LL} - t_{WT}$) (cm)	($t_{UL} - t_{Wc}$) (cm)
10	5.61	0.01299	0.109	11.9	2.71	7.08	4.0	6.5	1.63	−1.29	0.58
15	9.00	0.02177	0.106	16.6	2.74	11.39	5.5	9.5	1.73	−2.76	1.89
15	9.60	0.00993	0.121	15.9	3.35	9.83	4.8	6.3	1.31	−1.45	3.53
15	10.81	0.0367	0.116	17.7	3.19	11.75	5.0	11.2	2.24	−1.81	0.55
15	12.02	0.03699	0.133	17.3	4.07	10.14	5.0	8.9	1.78	−0.93	1.24
20	12.02	0.03558	0.093	22.04	2.49	17.08	4.2	13.3	3.17	−1.71	3.78
20	14.43	0.03694	0.114	21.75	3.25	15.49	5.5	11.8	2.15	−2.25	3.69
20	15.20	0.04025	0.120	21.7	3.51	14.99	4.25	12.5	2.94	−0.74	2.49
25	16.00	0.01599	0.096	26	2.78	20.44	6.6	13.0	1.97	−3.82	7.44
25	18.01	0.01327	0.111	25.62	3.30	19.11	4.0	9.8	2.45	−0.7	9.31
25	20.00	0.02034	0.122	25.82	3.77	18.51	5.0	9.8	1.96	−1.23	8.71

Comparing the calculated upper limiting values and lower limiting values of the ice jam thickness to those of measurements in laboratory, it can be seen that the measured thickness of ice wave crest and ice wave trough are all within the range of calculated ice jam thickness from the lower limiting values to the upper limiting values (Figure 5). This means that the measured results of ice wave crest and ice wave trough in laboratory experiments should be affected by some factors, and will be attenuated compared to the limiting value of thickness of ice waves. The ratio of the thickness at wave crest to the thickness at wave trough ranges from 1.31. to 3.17. An ice jam which has a thickness within the range of calculated limiting values is obviously in the state of mechanical equilibrium.

In addition, the randomness and determinacy of the thickness of ice waves are verified from Table 3. Here, the randomness means that the thickness of ice waves at any designated cross section changes dynamically during the migration process of ice waves, which varies from the thickness of ice wave crest to that of ice wave trough. The determinacy of thickness of ice waves means that the thickness of ice wave crest and trough ranges from the lower limiting value to the upper limiting value obtained from Equation (3).

Under the same hydraulic condition, comparison has been conducted between the upper/lower limiting of ice jam thickness obtained from Equation (3) and the thickness at wave crest/trough measured from experiments. Results indicate that the measured thickness at ice wave crest from experiments approaches to the upper limiting value calculated from Equation (3), while the measured thickness at ice wave trough approaches to the lower limiting value calculated from Equation (3). As showed in Table 3, there is some difference between calculated results using Equation (3) and those of measurements from experiments in laboratory, this difference might be resulted from the approximately calculation of hydraulic radius which should be applicable to a wide and shallow river, namely, Equations (2) and (3) have been derived by assuming that the channel is a wide-and-shallow river. Therefore, the smaller the depth-width ratio H/B is, the higher accuracy of calculated results by using Equation (3) is.

3.3. *Flow Froude Number under the Ice Jam*

Besides the flow Froude number Fr, some scholars either use the flow Froude number under an ice jam (Fr_u) or ice jam Froude number (Fi) to study the ice jam evolution characteristics, expressed as follows:

$$Fr_u = \frac{V}{\sqrt{gh}} \quad or \quad Fi = \frac{V}{\sqrt{gt}} \tag{8}$$

Taking the flow Froude number (Fr_u) as example, results of these 11 experimental runs showed the relative thickness of ice wave (t/H) increases with the flow Froude number under an ice jam Fr_u. It means, the larger the flow Froude number under an ice jam, the thicker the ice jam. One can see from Figure 8, for a larger thickness of an ice jam at cross section where a wave crest is located, the flow Froude number under the wave crest is clearly more than that under the ice wave trough. With a large flow Froude number under the wave crest, frazil ice particles at the wave crest will be eroded. Thus, the flow cross sectional area at the wave crest gradually increased. Thus, ice waves will be gradually migrated downstream. Considering the ice jam Froude number (Fi) as the variable, the large the ice jam Froude number, the thinner the ice jam. For a larger thickness of an ice jam at the wave crest, the ice jam Froude number is clearly less than that at the ice wave trough. This finding is reasonable, since the flow depth under an ice jam is $h = H - (\rho_i/\rho)t$, namely the thicker the ice jam (at the wave crest section), the shallower the flow under the jam, where, H is the approaching flow depth in front of an ice jam; h is the flow depth under an ice jam, ρ and ρ_i are mass density of water and ice, respectively.

The ice discharge rate from upstream river reach has an important influence on the ice jam thickness. The higher the ice discharge is, the thicker the ice jam will be. As showed in Figure 8 and Table 3, for a dynamic equilibrium ice jam, the thickness of an ice jam at both wave crest and wave trough will be affected. Thus, the migration process of ice waves will be affected by the incoming ice discharge rate from upstream.

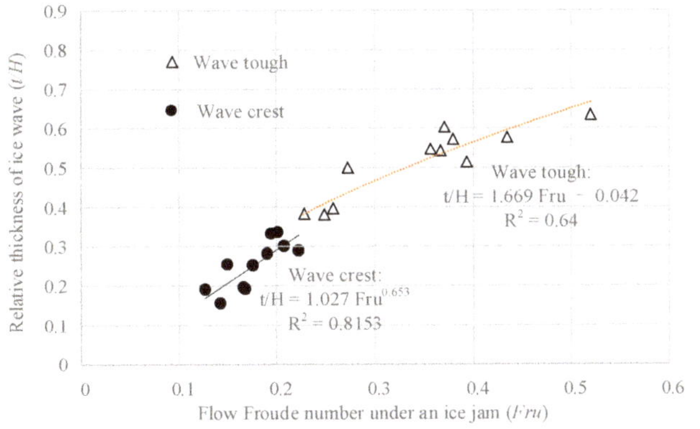

Figure 8. Relationship between the relative thickness of ice wave (at wave crest and wave trough) and flow Froude number under an ice jam based on laboratory experiments.

3.4. Wave Length and Migration Speed

Considering fully developed ice jams (covered the entire flume to the upstream cross Section 4) with the migration of ice waves, 11 out of those 22 experimental runs as mentioned above have been selected to study the wavelength and migration speed of ice waves. The wavelength has been determined using Equation (7), as shown in Table 4.

Table 4. Calculated ice wavelength compared to those of laboratory experiments and measured speed of ice wave migration.

Initial Condition			Equilibrium Ice Jam		Wave Length (cm)		Measured Results	
Flow Depth (cm)	**Flow Rate (L/s)**	**Ice Discharge (L/s)**	**Fr**	**Flow Depth (cm)**	**Calculated Result**	**Measured Result**	**Ice Wave Periodic Migration Time (s)**	**Ice Wave Migration Speed (m/s)**
10	5.61	0.01299	0.109	11.90	33.11	29	109	0.00266
15	9.00	0.02177	0.106	16.60	40.50	34	143	0.00238
15	9.60	0.00993	0.121	15.90	23.98	25	137	0.00182
15	10.81	0.03699	0.116	17.70	70.76	66	179	0.00369
15	12.02	0.03699	0.133	17.30	69.93	64	151	0.00424
20	12.02	0.03558	0.093	22.04	66.70	62	192	0.00323
20	14.43	0.03699	0.114	21.75	68.23	69	242	0.00285
20	15.20	0.04025	0.120	21.70	74.70	81	253	0.00320
25	16.00	0.01599	0.096	26.00	38.07	54	428	0.00126
25	18.01	0.01327	0.111	25.62	24.27	61	664	0.00092
25	20.00	0.02034	0.122	25.82	24.80	50	330	0.00152

It is found that when the flow depth is small (or H/B is small), compared to results of laboratory experiments, the difference between the calculated wavelength and that of measurement in laboratory experiment is very small. One can say that Equation (7) has certain accuracy. However, when the flow depth at the control section is 25 cm (or H/B is large), the calculated wavelength is smaller than the measured result from experiment, and there is a certain calculation error. This might be resulted from the approximated calculation of some integral expressions in the denominator of Equation (7). Further optimization of some parameters should be carried out to enhance the calculation accuracy.

Results of migration speed of ice wave from laboratory experiments are also presented in Table 3. The periodic time of ice wave migration in Table 3 refers to the time required for an ice wave to migrate a distance equal to the wavelength to downstream. This means, the ratio of the wavelength to the periodic time of ice wave migration is the migration speed of an ice wave.

Figure 9 shows the relationship between the migration speed of ice wave and the ice-water discharge ratio (Qi/Q) based on the 11 laboratory experimental runs. This linear relationship indicates that the migration speed of ice wave is mainly affected by the ice transport discharge (which is equal to the inlet ice discharge) and hydraulic condition of the experimental run. When the hydraulic condition does not change, the larger the ice discharge is, the faster the migration speed of ice waves during the evolution process of an ice jam is (see the test data of No.5 and No.6 in Table 4). When the ice discharge is constant, the smaller the flow rate is, the faster the migration speed of ice waves is (see the test data of No.4 and No.7 in Table 4).

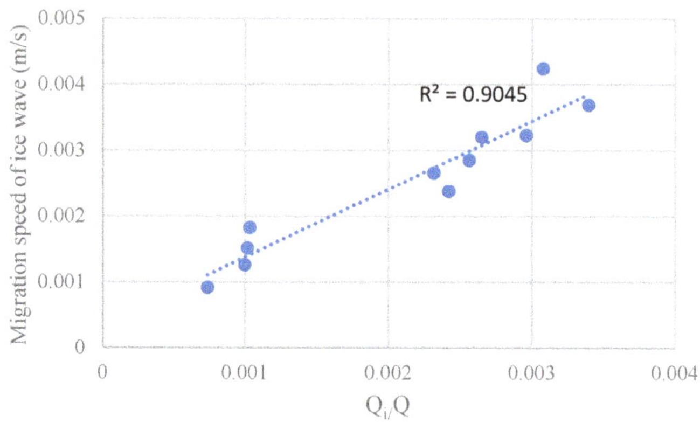

Figure 9. The relationship between the migration speed of ice wave and the ice-water discharge ratio (Q_i/Q) from laboratory experiments.

It is worth mentioning that the above-mentioned migration speed of ice waves actually refers to the average velocity for an ice wave advancing toward downstream. From laboratory experiments, it has been observed that the thickness of the "cover-load" layer along different cross section of an ice wave is not the same. The average thickness of the "cover-load" layer can be obtained based on the ice discharge, the average migration speed and the channel width. However, the calculated result only represents an average thickness of the "cover-load" layer, and cannot preciously express actual thickness of the cover-load layer at each cross section.

3.5. Case Studies

The ice jam accumulation and evolution process are very complicated due to various size of ice floes in natural rivers as well as the combined effects of thermal and hydraulic factors. Based on filed observations of ice jams in natural rivers, following five types of ice jam accumulation may occur in natural rivers during the ice-covered periods.

(a) Quasi-static ice jam: This kind of ice jam is also known as the continuous juxtaposition form. It is characterized by the continuous juxtaposition of a layer of ice floes with a specific thickness. The ice jam thickness completely depends on the initial ice thickness and the thermal effect during winter period, which is similar to but slightly different from the static ice cover in lakes.
(b) Frontal progression: The kind of accumulation refers to an ice jam whose thickness exceeds its initial thickness after the submerged ice floes accumulate at the front of an ice jam (or the leading edge of an ice jam). A relatively stable ice jam head in a river is essential for the development of this kind of ice jam.
(c) Packing ice jam: The front edge of an ice jam in rivers may advance and retreat during ice-jammed period, and when it advances upstream, local accumulation (packing) will often occur.

(d) Shoving ice jam: In natural rivers, when the cohesive force of an ice jam is not enough to support the positive stress at the front edge of an ice jam, a long and thin ice cover is often broken suddenly and leads to a thick accumulation body at the downstream. During the period of the initial stage, this accumulation body is normally loose without cohesion, but it may form a solid accumulation ice jam later due to the influence of thermal factors.

(e) Hanging dam: A hanging ice dam is referred to the development of an ice jam along a specific river section (such as the junction of rapid flow and slow flow), where the flow velocity is too high for ice floes to accumulate upstream. Under such a flow condition, the incoming ice floes will be entrained by water and accumulate under the bottom of the ice cover. The cross sectional area for flow under the ice accumulation decreases until the flow velocity is high enough to erode ice floes further to downstream.

In recent years, more hydraulic structures such as piers for bridges have been built in natural rivers. Many data of field measurements under the influence of natural conditions have been collected. Therefore, this study uses the valuable data of field observations collected in the St. Lawrence River, the Beauharnois Canal in the early 1950s and the La Grande River in the early 1970s. Data of field measurements provide excellent information including ice jam thicknesses at closely spaced cross sections, which measured sooner after the formation of ice cover/jams (so that there should be little impact from thermal aspect on the growth of the ice jam). Following field data are also available: cross sectional area of ice cover/jam, cross-sectional area for flow, flow discharge and river width. The flow velocity in front of ice cover was calculated using the flow discharge divided by total area for flow at time during the formation of an ice cover, and the velocity under the cover by using a similar calculation method. The ice cover/jam thickness is determined by total cross-sectional area of ice cover/jam divided by its surface width. The detailed data of field measurements are summarized in Michel's research [30].

It can be seen from those measurements data that the ratios of the flow depth to the channel width (H/B) of natural rivers are far less than those of laboratory experiments, the H/B values for the St. Lawrence River and Beauharnois Canal range from 0.002 to 0.015, and the H/B values for the La Grande River are from 0.005 to 0.087. Since it is difficult to accurately distinguish the frontal progression and packing ice jam in natural rivers, these two types of ice jam are considered as one category. Therefore, ice jams in natural rivers can be mainly divided into following four categories: quasi-static ice jam, frontal and packing ice jam, shoving accumulation, and ice dam. It should be noted that the ice dam phenomenon does not happen often in the above mentioned three field sites (only occurred in the La Grande River during winter periods from 1977 to 1978.

In addition, according to Michel [30], the Chézy coefficient C used for these field sites is ranged from 24 to 40. In the present study, the C value has taken the larger value. The value of μ is ranged from 1.28 to1.3; H/B is the value of the actual measurement data, and the t value associated with the data point in the diagram is the actual measurement. The limiting Fr- curve for the natural rivers is drawn according to Equation (3). The calculated limiting flow Froude number Fr for natural rivers using Equation (3) is compared to those of field observations for different types of ice jam accumulation in the St. Lawrence River, Beauharnois Canal and the La Grande River, as showed in Figures 10 and 11.

It can be seen from Figures 10 and 11 that when either quasi-static or frontal/packing ice jams occur in natural rivers, all data points of field measurements are located either above or near the limiting Fr-curve. On the contrary, when ice jams in natural rivers appear as either shoving accumulation or ice dam, all data points of field measurements are located below the limiting Fr-curve. From the perspective of the mechanical mechanism of the development of an ice jam, hydraulic factors generally play a leading role in the accumulation process of ice jams that classified as either quasi-static or frontal/packing ice jams. The thickness of both quasi-static and frontal/packing ice jams is relatively small, and the value of flow Froude number corresponding to the open flow condition has a relatively large value. However, for ice jams appear as either the shoving accumulation or ice dam, they generally reach an equilibrium state when the internal force of an ice jam is equal to the external force. The accumulation process for both shoving ice jam and ice dam is dominated by mechanical factors, and their thickness is relatively large. The value of the flow Froude number corresponding to the open flow is small (especially for the ice dam, the thickness often reaches a maximum value, with a minimum *Fr* value in front of the leading edge of an ice dam).

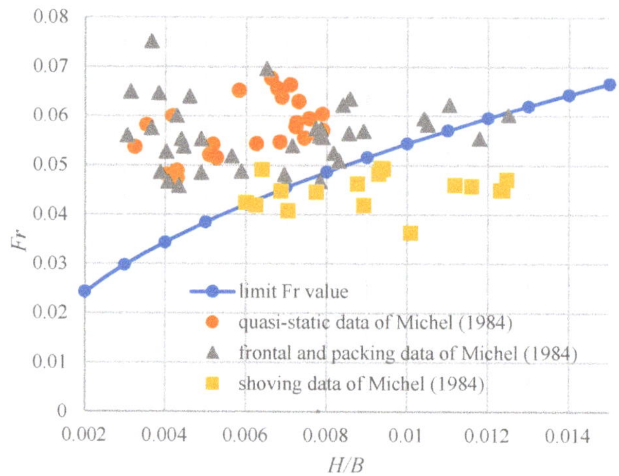

Figure 10. Calculated limiting flow Froude number *Fr* for natural rivers compared to those of field observations of the St. Lawrence River and Beauharnois Canal.

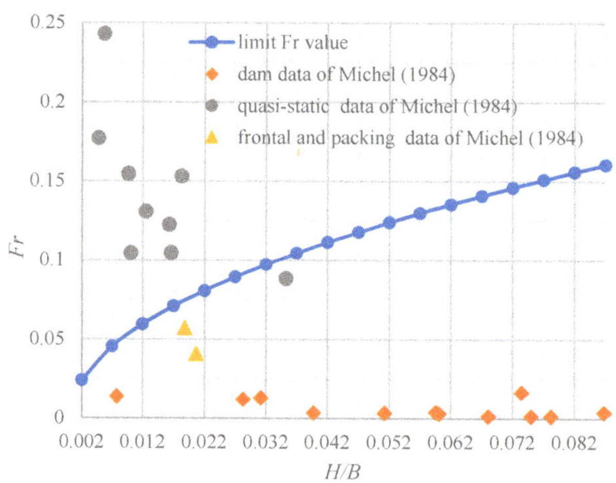

Figure 11. Calculated limiting flow Froude number *Fr* for natural rivers compared to those of field observations of the La Grande River.

The evolution of a river ice jam in a natural river is a process of gradual transition from flow with a high *Fr* value to that with a low *Fr* value. The mechanical action gradually dominates the development/accumulation of an ice jam, and the impact of the hydraulic action gradually decreases. From Figures 10 and 11, one can see results of field observations compared to those of calculation. Note: the limiting *Fr* value mentioned in this study represents the maximum *Fr* value in front of the leading edge of an ice jam for an equilibrium ice jam under the mechanical action, which means that the ice jam accumulation can reach the mechanical equilibrium state only when the *Fr* value under the open flow condition is less than or equal to the limiting *Fr* value. Figure 10 shows that almost all measured data of the two rivers during different winter periods are located near the limiting *Fr*-curve. One may conclude that the ice jam evolution achieves an optimal state when the value of flow Froude number *Fr* in front of the leading edge of an ice jam approaches the limiting *Fr*-curve.

As mentioned above, results of experiments showed that the limiting *Fr*-curve can be used to judge whether the ice wave phenomenon occurred. The main difference between the condition for laboratory experiments and the condition in natural rivers is whether there is a continuous supply of ice from the upstream. It should be noted that the formation of an ice dam phenomenon in a natural river generally requires the continuous incoming ice supply from the upstream. It can be concluded that when the ice jam has reached the mechanical equilibrium state, the incoming ice particles/floes floating on water surface would lead to the decrease in flow Froude number in the open section upstream of the ice jam and slowly move toward the front of the ice jam. However, under the ice jam, the flow velocity is faster than that in the open flow section since the cross section is partially occupied by ice. As a consequent, the incoming ice will be entrained by flowing water and submerged at the leading edge of ice jam and the accumulate under ice jam. Thus, the thickness of the ice jam will increase and the ice wave phenomenon may be in present under the bottom of the ice jam. Obviously, the incoming ice discharge affect the magnitude of flow Froude number. Under extreme condition with vast amount of incoming ice from upstream, the ice dam phenomenon may appear in a natural river with a sharply decrease in the cross sectional area for flow with high flow velocity under and ice jam in addition a significant increase in flow depth in front of the ice dam. Thus, the ice accumulate under the ice dam will be transported to the downstream. In such a case, the flow Froude number *Fr* in front of the leading edge of an ice dam decreases dramatically and approaches to zero (as shown in Figure 12). Thus, the ice dam could be considered as a special case for the ice wave phenomenon.

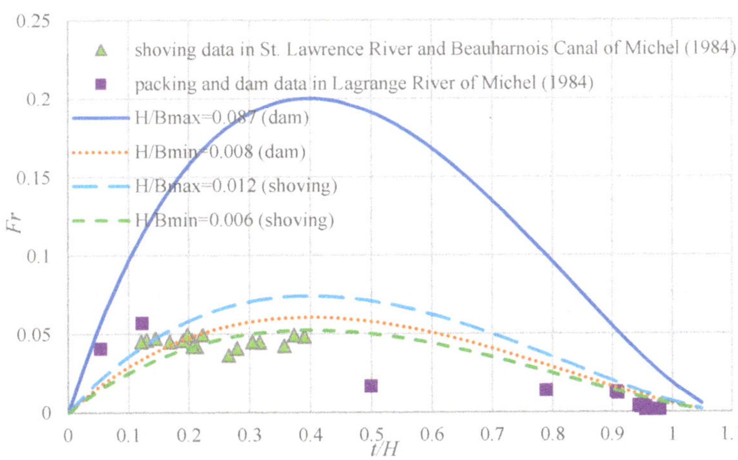

Figure 12. Variations of flow Froude number in front of the leading edge of ice jams with the relative thickness of an ice jam t/H—calculated results (curves) compared to those of field observations of the St. Lawrence River, Beauharnois Canal and La Grande River.

Figure 12 shows the variation of flow Froude number in front of the leading edge of ice jams with the relative thickness of an ice jam t/H for the shoving ice accumulation in the St. Lawrence River and the Beauharnois Canal (17 measured data in total), packing ice jam and ice dam in the La Grande River (14 measured data in total), compared to those of calculation using Equation (3) (curves). It can be seen from this figure that all data points of field measurements are within the range of the upper and lower limits of the ice thickness.

For the same flow Froude number in front of the leading edge of an ice jam, there will be some different relative thickness of an ice jam t/H in the natural rivers. Interestingly, when the relative thickness of an ice jam $t/H < 0.4$, all data points of field observations were for packing or shoving ice jams; while when $t/H > 0.4$, all data points of field observations are corresponding to the ice dam events. This means the relative thickness of an ice jam (t/H) of 0.4 is the limiting value for assessing whether an ice jam in a river belongs to an ice dam. This finding agrees well with result of Sui et al. (2002) [8] who carried out study regarding river ice accumulation in the Hequ Reach of the Yellow River in China. Based on many data of field observations along this 70 km-long river reach for six winters, Sui et al. found that the turning point of t/H is about 0.4 from an ice jam to an ice dam [8].

4. Conclusions

The ice wave phenomenon will occur under certain conditions during the evolution process of an ice jam in a river. Through the mechanical analysis of wide jams theory, the relationship between the flow Froude number in front of an ice jam and the ratio of ice jam thickness to flow depth (t/H) as well as the ratio of flow depth to channel width (H/B) has been discussed. Results showed that the ice jam thickness during the process of dynamic evolution varies dynamically between the minimum jam thickness and the maximum jam thickness, depending on the flow Froude number.

When the approaching flow Froude number reaches a certain value, ice waves appear at the bottom of the ice jam and begin to migrate downstream. Otherwise, ice waves cannot be developed during the evolution process of an ice jam. One curve has been developed for assessing if ice waves will occur at the bottom surface of an ice jam. If the value of approaching flow Froude number is located above the limiting Fr-curve, the ice wave phenomenon will not occur. Ice waves will present provided the value of the approaching flow Froude number is located below the limiting Fr-curve.

Comparing results calculated for both upper and lower limiting values of the ice jam thickness to those of laboratory experiments, the measured jam thicknesses at the crest and trough of ice waves are all within the calculated range from the lower limiting values to the upper limiting values of ice jam thickness. The measured thicknesses at ice wave crest from experiments approaches to the upper limiting value calculated from Equation (3); while the measured thicknesses at ice wave trough approaches to the lower limiting value calculated from Equation (3).

The relative thickness of ice wave (t/H) increases with the flow Froude number under an ice jam Fr_u. Namely, the large the flow Froude number under an ice jam, the thicker the ice jam. This means, for a larger thickness of an ice jam at cross section where a wave crest is located, the flow Froude number under the wave crest is clearly more than that under the ice wave trough. With a large flow Froude number under the wave crest, frazil ice particles at the wave crest will be eroded. Thus, the flow cross sectional area at the wave crest gradually increased. This leads to a gradually migration of ice waves toward downstream.

An Equation for calculating the ice wavelength has been derived and verified by experimental data. The relationship between the ice wave migration speed and the ice-water flow ratio (Qi/Q) has been analyzed. Results showed that, the ice wave migration speed is mainly affected by the ice discharge and the hydraulic conditions of the flume. When the hydraulic condition does not change, the larger the ice discharge is, the faster the migration speed of ice waves during the evolution process of an ice jam is. When the ice discharge is constant, the smaller the flow rate is, the faster the migration speed of ice waves is.

Ice wave is a microscopic phenomenon during the evolution of ice jams, which is observed in experimental study. However, it has not been reported in the natural rivers due to limitation in field observations. The continuous incoming ice particles during the laboratory experiments may also be one of the prerequisites for the development of ice waves. However, both the experimental study and prototype data reflect the periodic change of ice jam thickness when the hydraulic conditions are constant.

The ice jam evolution achieves an optimal state when the value of flow Froude number *Fr* in front of the leading edge of an ice jam approaches the limiting *Fr*-curve. When the ice jam has reached the mechanical equilibrium state, the incoming ice particles/floes floating on water surface would lead to the decrease in flow Froude number in the open section upstream of the ice jam and slowly move toward the front of the ice jam. The incoming ice will be entrained by flowing water and submerged at the leading edge of ice jam and the accumulate under ice jam. Thus, the thickness of the ice jam will increase and the ice wave phenomenon may be in present under the bottom of the ice jam. Interesting results of case studies about ice accumulation in natural rivers show that the relative thickness of an ice jam (t/H) of 0.4 is the limiting value for assessing whether an ice jam in a river belongs to an ice dam. When the relative thickness of an ice jam $t/H < 0.4$, all data points of field observations were for packing or shoving ice jams; while when $t/H > 0.4$, all data points of field observations are corresponding to the ice dam events. This finding agrees well with result of Sui et al. (2002) [8] who carried out study regarding river ice accumulation in the Hequ Reach of the Yellow River in China.

Author Contributions: P.C.: data curation, formal analysis, investigation, methodology, experiments, calculation, software, writing—original draft; J.S.: formal analysis, investigation, methodology, calculation, validation; writing reviewing; G.C.: data curation, investigation, methodology, validation; T.C.: data curation, investigation, methodology, validation. All authors have read and agreed to the published version of the manuscript.

Funding: This research received no external funding.

Data Availability Statement: Data is contained within the article and supplementary material.

Acknowledgments: Thanks to all the pioneer scholars in the field of river ice research for their contributions, especially for the prototype observation results provided by Michel B.

Conflicts of Interest: The authors declare no conflict of interest.

References

1. Xiao, Y.; Pavelsky, T.M.; Allen, G.H. The past and future of global river ice. *Nature* **2020**, *577*, 69–73.
2. Sui, J.; Karney, B.; Fang, D. Variation in water level under ice-jammed condition—Field investigation and experimental study. *Hydrol. Res.* **2005**, *36*, 65–84. [CrossRef]
3. Beltaos, S.; Burrell, B.C. Hydrotechnical Advances in Canadian River-Ice Science and Engineering during the Past 35 Years. *Can. J. Civ. Eng.* **2015**, *42*, 583–591. [CrossRef]
4. Sui, J.; Hicks, F.; Menounos, B. Observations of riverbed scour under a developing hanging ice dam. *Can. J. Civ. Eng.* **2006**, *33*, 214–218. [CrossRef]
5. White, K.D. Review of prediction methods for breakup ice jams. *Can. J. Civ. Eng.* **2003**, *30*, 89–100. [CrossRef]
6. Lindenschmidt, K.E.; Das, A.; Rokaya, P.; Chu, T. Ice jam flood risk assessment and mapping. *Hydrol. Process.* **2016**, *30*, 3754–3769. [CrossRef]
7. Sui, J.; Wang, J.; Balachandar, R. Accumulation of frazil ice along a river bend. *Can. J. Civ. Eng.* **2008**, *35*, 158–169. [CrossRef]
8. Sui, J.; Karney, B.W.; Sun, Z.; Wang, D. Field Investigation of Frazil Jam Evolution: A Case Study. *J. Hydraul. Eng.* **2002**, *128*, 781–787. [CrossRef]
9. Beltaos, S.; Carter, T. Field studies of ice breakup and jamming in lower Peace River, Canada. *Cold Reg. Sci. Technol.* **2009**, *56*, 102–114. [CrossRef]
10. Healy, D.; Hicks, F.E. Experimental study of ice jam formation dynamics. *J. Cold Reg. Eng.* **2006**, *20*, 117–139. [CrossRef]
11. Wang, J.; Hua, J.; Chen, P.; Sui, J.; Wu, P.; Whitcombe, T. Initiation of ice jam in front of bridge piers—An experimental study. *J. Hydrodyn.* **2019**, *31*, 117–123. [CrossRef]
12. Daly, S.F.; Axelson, K.D. Stability of floating and submerged blocks. *J. Hydraul. Res.* **1990**, *28*, 737–752. [CrossRef]
13. Wang, J.; Sui, J.Y.; Zhang, H.Y.; Chen, P.P.; Hirshfield, F. Mechanisms of ice accumulation in a river bend—An experimental study. *Int. J. Sediment Res.* **2012**, *27*, 521–537. [CrossRef]
14. Hicks, F.E.; Healy, D. Determining winter discharge based on hydraulic modeling. *Can. J. Civ. Eng.* **2003**, *30*, 101–112. [CrossRef]
15. Wang, J.; Chen, P.; Sui, J. Progress in studies on ice accumulation in river bends. *J. Hydrodyn.* **2011**, *23*, 737–744. [CrossRef]
16. Beltaos, S. Progress in the study and management of river ice jams. *Cold Reg. Sci. Technol.* **2008**, *51*, 2–19. [CrossRef]
17. Beltaos, S. Numerical prediction of ice-jam profiles in lower Athabasca River. *Can. J. Civ. Eng.* **2019**, *46*, 722–731. [CrossRef]
18. Wang, J.; Sui, J.; Chen, P. Numerical simulations of ice accumulation under ice cover along a river bend. *Int. J. Environ. Sci. Technol.* **2009**, *6*, 1–12. [CrossRef]

19. Sola, D.; Scott, K.A. Efficient Shallow Network for River Ice Segmentation. *Remote Sens.* **2022**, *14*, 2378. [CrossRef]
20. Altena, B.; Kääb, A. Quantifying river ice movement through a combination of European satellite monitoring services. *Int. J. Appl. Earth Obs. Geoinf.* **2021**, *98*, 102315. [CrossRef]
21. Sui, J.; Faruque, M.A.A.; Balachandar, R. Local scour caused by submerged square jets under ice cover. *ASCE J. Hydraul. Eng.* **2009**, *135*, 316–319. [CrossRef]
22. Sui, J.; Wang, J.; He, Y.; Krol, F. Velocity profiles and incipient motion of frazil particles under ice cover. *Int. J. Sediment Res.* **2010**, *25*, 39–51. [CrossRef]
23. Jafari, R.; Sui, J. Velocity field and turbulence structure around spur dikes with different angles of orientation under ice covered flow conditions. *Water* **2021**, *13*, 1844. [CrossRef]
24. Namaee, M.; Sui, J. Velocity profiles and turbulence intensities around side-by-side bridge piers under ice-covered flow condition. *J. Hydrol. Hydromech.* **2020**, *68*, 70–82. [CrossRef]
25. Cheng, T.; Wang, J.; Chen, P.; Sui, J. Simulation of ice accumulation around bridge piers during river breakup periods using a discrete element model. *J. Hydrodyn.* **2022**, *34*, 94–105. [CrossRef]
26. Shen, H.T.; Wang, D.S. Under cover transport and accumulation of frazil granules. *J. Hydraul. Eng.* **1995**, *121*, 184–195. [CrossRef]
27. Wang, J.; Wu, Y.; Sui, J.; Karney, B. Formation and movement of ice accumulation waves under ice cover—An experimental study. *J. Hydrol. Hydromech.* **2019**, *67*, 171–178. [CrossRef]
28. Beltaos, S. *River Ice Jams;* Water Resources Publications: Littleton, CO, USA, 1995.
29. Pariset, E.; Hausser, R.; Gagnon, A. Formation of ice covers and ice jams in rivers. *J. Hydraul. Div.* **1966**, *92*, 1–24. [CrossRef]
30. Michel, B. Comparison of field data with theories on ice cover progression in large rivers. *Can. J. Civ. Eng.* **1984**, *11*, 798–814. [CrossRef]

water

MDPI

Article

Impact of Local Scour around a Bridge Pier on Migration of Waved-Shape Accumulation of Ice Particles under an Ice Cover

Zhixing Hou [1], Jun Wang [1,*], Jueyi Sui [2,*], Feihu Song [1] and Zhicong Li [1]

1 School of Civil and Hydraulic Engineering, Department of Municipal Engineering, Hefei University of Technology, Hefei 230009, China; houhfut@163.com (Z.H.); songfeihu2016@163.com (F.S.); lzc1641222981@163.com (Z.L.)
2 School of Engineering, Environmental Engineering Program, University of Northern British Columbia, Prince George, BC V2N 4Z9, Canada
* Correspondence: junwanghfut@126.com (J.W.); jueyi.sui@unbc.ca (J.S.); Tel.: +86-137-0551-0008 (J.W.); +1-250-960-6399 (J.S.)

Abstract: The migration of a waved-shape accumulation of ice particles under an ice cover (referred to as "ice wave" in this study) is a phenomenon of transport of ice particles during an ice accumulation process in rivers. The migration of an ice wave will affect the pier scour. On the other hand, the local scour at the pier will affect the migration of ice waves. The interaction between the migration of ice waves and local scour around a pier is a very complicated process since not only the channel bed deforms, but also the ice jam develops simultaneously. By conducting a series of flume experiments, the interaction between the local scour around bridge piers and the migration of ice waves was studied. By applying both continuity and momentum equations, an empirical equation has been derived for predicting the thickness of ice waves around the pier. The impacts of the scour hole on the thickness of ice waves around the pier have been studied. The thickness of the wave crest and the migration speed of ice waves have been investigated. Similar to a scour hole in a sand bed, an "ice scour hole" appeared at the bottom of the ice jam around the pier. The existence of the "ice scour hole" affects the development of ice waves. A formula for calculating ice transport capacity has been obtained. Results calculated using the derived formula are in good agreement with those of laboratory experiments.

Keywords: bridge piers; ice jam; local scour; transverse thickness distribution; ice wave crest; waved-shape accumulation of ice particles (ice wave)

Citation: Hou, Z.; Wang, J.; Sui, J.; Song, F.; Li, Z. Impact of Local Scour around a Bridge Pier on Migration of Waved-Shape Accumulation of Ice Particles under an Ice Cover. *Water* **2022**, *14*, 2193. https://doi.org/10.3390/w14142193

Academic Editor: Karl-Erich Lindenschmidt

Received: 22 May 2022
Accepted: 8 July 2022
Published: 11 July 2022

1. Introduction

The presence of ice jams in rivers leads to a decrease in the flow capacity and could potentially result in flooding and collapse of hydraulic structures [1]. When an ice cover is formed in a river, frazil ice particles from upstream are entrained by water and submerged under the ice cover and then accumulate under the ice cover. A special phenomenon called ice jam is formed along a river reach depending on the hydraulic and climate conditions [2,3]. During the formation process of an ice jam, a waved-shape accumulation of ice particles (defined as the "ice wave" in this study) has been noticed [4]. The migration of an ice wave is a movement of a group of ice particles under an ice cover. This process should be similar to the phenomenon of sand waves on a riverbed surface during sediment movement and bedload motion. When a bridge pier is present in a river, the pier affects the transport capacity of ice particles and floes, the thickness of ice jams, and the water level around it. In the meantime, the appearance of an ice jam results in intense local scour around piers [5]. The presence of scour holes around bridge piers changes the flow structure and influences the migration of the ice wave process around bridge piers.

The evolution of an ice jam is affected by hydraulic, thermal, and boundary conditions. The mechanism of ice jam evolution in a channel has been investigated by researchers, either based on field measurements in natural rivers, flume experiments in laboratories, or numerical simulations. Beltaos et al. [6–8] made long-term observations on ice jam thickness during the spring breakup period of the Matapedia River. Theoretical analysis on the internal strength characteristics of ice jam was also conducted. Based on field observations of ice jams in the Yellow River, which is 70-km long, Sui et al. [9,10] studied the mechanism of ice jam evolution and the riverbed deformation. Results indicated that variations of the ice jam thickness, water levels, and the changes of riverbed deformation depend on flow Froude number. Some numerical simulations have been conducted to study ice jam evolution and ice transport. Knack and Shen [11] modified the Rouse formula for calculating the suspended sediment transport discharge. Yang et al. [12] proposed an equation to describe the formation of an ice jam and simulated the evolution process of an ice jam in the Songhua River. Mao et al. [13] and Wang et al. [14] built some synthetic and dynamic mathematic models for simulating river ice jam. Szydłowski and Kolerski [15] used the DynaRICE model to simulate river ice hydrodynamics in the vicinity of bridge piers. Results indicated that the new structure in a channel has a notable effect on flow velocity and ice transport capacity. Instead of the impact of ice jam on bridge piers, some researchers carried out research work regarding the interaction of debris and bridge piers. Istrati et al. [16] investigated the tsunami-borne debris loading on bridges to mention as an example here.

The formation of a stable ice jam is the sequence of the balance between the drag force caused by flowing water, the partial force of the gravity of the ice jam along the flow direction, and the supporting force of the channel bank to the ice jam [1]. By considering the effect of the secondary vortex and neglecting the cohesive force between ice particles, equations for describing the changes of ice jam thickness have been derived [17].

The flow condition in a river changes significantly in the presence of bridge piers and then affects the evolution process of an ice jam. Beltaos et al. [18] pointed out that when the resultant force of the drag force and gravity force is greater than the resistance caused by bridge piers, it is difficult to develop an ice jam. Wang et al. [19] studied the stability of an ice jam around a pier and developed an equation for describing an ice jam under an equilibrium condition. By conducting a series of flume experiments, Wang et al. [20,21] further studied the stability of an ice jam around a bridge pier and proposed the criteria for the formation of an ice jam around piers. The influence of bridge piers on the evolution of an ice jam was also investigated.

The presence of an ice cover/jam in a river results in a shift of the maximum velocity from the water surface to the point between the ice cover/jam and channel bed, depending on the roughness of the ice cover/jam and channel bed. This means, the maximum flow velocity under an ice cover/jam is shifted closer to the channel bed, comparing to that under an open flow condition. As a consequence, shear stresses near the channel bed under an ice-covered or ice-jammed flow condition increase and more sediment will be eroded comparing to that under an open flow condition [22]. Carr and Tuthill [23] studied the scour hole at the Melvin Price Lock and claimed that a large scour hole developed with a smaller discharge under an ice-covered flow condition. To date, some important experimental studies on local scour around bridge piers under ice-covered conditions have been conducted [24–32]. Results of those studies showed that the depth of scour holes under an ice cover was clearly deeper than that under an open flow condition. The maximum scour depth increased with the increase in the roughness of an ice cover. Wang et al. [5] analyzed the local scour under an ice-jammed flow condition through laboratory experiments. Results of these experimental studies show that the size of a scour hole is affected by the dynamic process of the accumulation of an ice jam, namely, the variation of the thickness of an ice jam around bridge piers influences the local scour process. It is found that the maximum scour depth under an ice jam was much larger than that under both an open flow and sheet ice-covered flow conditions.

At present, the existing studies are mainly focused on the influence of bridge piers on the evolution process of an ice jam. Very few studies have been carried out by considering the impacts of the scouring process at piers on the evolution of an ice jam. Very few data about systematic observations of the migration of ice waves under ice jams are available [33]. The theory and mechanism of the migration of an ice wave has not been explored. A scour hole with deeper depth and larger scour area will be formed under an ice-jammed flow condition [33]. On the other hand, the existence of a scour hole should affect the migration process of the ice wave and ice transport process. In the present paper, based on laboratory experiments, the interaction between the local scour at a bridge pier and the migration process of ice waves around the pier is studied. The migration speed of an ice wave and ice transport process at the pier was investigated, and a formula for determining the ice transport capacity was derived.

2. Methods and Materials

Laboratory experiments were carried out in a flume, as shown in Figure 1. The flume was 26.68 m long and 0.4 m wide. With an equal spacing distance of 1.2 m, 22 cross sections (CS) for measurements along the flume were setup. A pressure transducer was installed at each measurement CS. The wall of the flume was smooth, tempered glass. The bottom of the flume between CS-2 and CS-22 was covered by a layer of sand to form a sand bed. The initial thickness of the sand bed was 10 cm. The inhomogeneity coefficient (η) of sand is 1.61 mm with a median grain size (d_{50}) of 0.713. Before each experimental run, the entire sand bed was leveled with the slope of 0. To initiate the formation of an ice jam, a Styrofoam panel, which was 1.0 m long and 0.39 m wide, was placed on the water surface at CS-20. The CS-4 was chosen as the control cross section. At the beginning of each experimental run, the average approaching flow velocity (V_0) and flow depth (H_0) at CS-4 were used as the initial hydraulic condition. In this experimental study, the model ice particles were made of polyethylene material with a mass density of 0.918 g/cm^3, which was nearly identical to that of natural ice of 0.917 g/cm^3. The shape of the model ice particles was flat-ellipsoid with the longest diameter of 3.5 mm. Model ice particles were discharged from an ice hopper that was located over the flume between CS-2 and CS-3. The discharge rate of ice particles from the ice hopper could be adjusted as required. A cylinder with a diameter of D = 2 cm was used to model a bridge pier and installed in the center of the flume at CS-16.

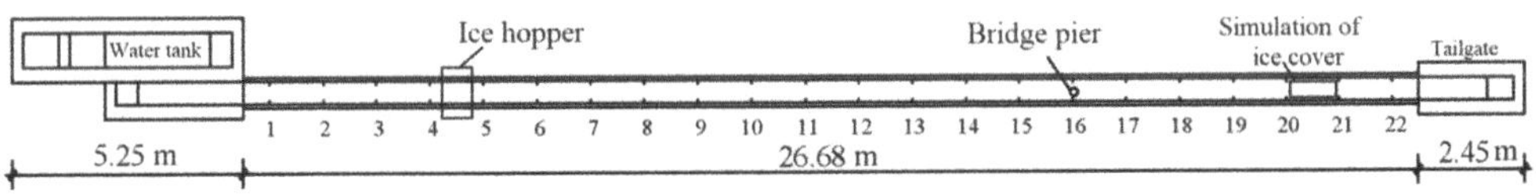

Figure 1. The setup of the experimental flume in the laboratory.

Results of the experiments showed that if the flow intensity is too high, the frazil ice particles are unable to accumulate for the development of an ice jam and the ice wave phenomenon does not occur. In this study, the flow Froude number was used as an important factor for describing the flow intensity relevant for the accumulation process of an ice jam. Thus, when the flow Froude number is more than the critical Froude number for the development of an ice jam, frazil ice particles cannot accumulate under an ice cover, but will be washed out and transported to the downstream section of the channel, referred as to the "wash-load" [34]. To investigate the impact of the scour hole on the migration of ice waves, flow conditions for all experimental runs were controlled with the flow Froude number less than its critical value. Under such a flow condition, ice waves could be generated in each experimental run. In the present study, to compare the result of ice wave migration with the presence of scour hole at the pier to that without a local scour, experiments were conducted with and without the presence of a local scour around the pier. The flow conditions for 16 experimental runs are summarized in Table 1. In the present study, all scour experiments used clear water scour with an experiment duration of 12 h to ensure that both local scour and accumulation of an ice jam achieved equilibrium conditions as long as the water level and scour depth at the pier did not change. The selections of proportion of relevant parameters in this experimental study are conformed to the General Code for Design of Highway Bridges and Culverts (JTG D60-2004) [35].

Table 1. Experimental conditions.

Experimental Run #	v_0 (m/s)	H_0 (m)	D (cm)	Q_i (L/s)	d_{50} (mm)	H_S (m)
A1	0.17	0.25	2	0.0205	0.713	0.0319
A2	0.15	0.25	2	0.0205	0.713	0.0286
A3	0.17	0.20	2	0.0205	0.713	0.0343
A4	0.15	0.20	2	0.0205	0.713	0.0323
A5	0.18	0.25	2	0.018	0.713	0.0288
A6	0.18	0.25	2	0.022	0.713	0.0327
A7	0.18	0.25	2	0.026	0.713	0.0350
A8	0.18	0.25	2	0.030	0.713	0.0365
A9	0.18	0.20	2	0.018	0.713	0.0308
A10	0.18	0.20	2	0.022	0.713	0.0339
A11	0.18	0.20	2	0.026	0.713	0.0384
A12	0.18	0.20	2	0.030	0.713	0.0391
B1	0.17	0.25	2	0.0205	/	0
B2	0.15	0.25	2	0.0205	/	0
B3	0.17	0.20	2	0.0205	/	0
B4	0.15	0.20	2	0.0205	/	0

Note: v_0 is the average approaching velocity at CS-4; H_0 is the approaching flow depth at CS-4; D is the pier diameter; Q_i is the discharge rate of ice particles released from the ice hopper; d_{50} is the median grain size of sand particles; H_S is the maximum scour depths. B1–B4 represent experiments without local scour (without bed sand on the flume bottom).

3. Interaction of Local Scour and Ice Wave Migration

Wang et al. [5] pointed out that the development of an ice jam can be divided into following three stages: the initial development of an ice jam from downstream toward upstream of a channel (stage 1); the formation of an initial ice jam along the entire channel (stage 2); and the thickening process of the initial ice jam accompanied by the migration process of ice waves (stage 3). During the third stage, the incoming ice particles from the upstream section will be entrained by flowing water and submerged and accumulated under the initial ice jam and gradually transport downstream. As can be seen in Figure 2, ice waves may appear during this process, depending on flow Froude number.

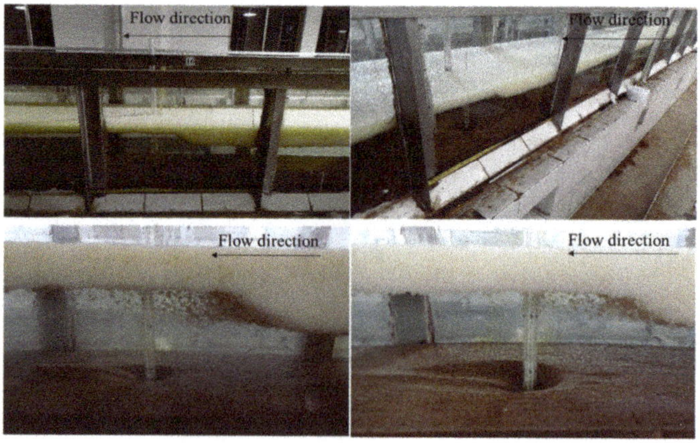

Figure 2. Ice wave migration in experiments.

As shown in Figure 3, during the migration process of an ice wave, along the flow direction, points "A", "B", and "C" represent the "wave starting point", "wave crest", and "wave trough" of an ice wave, respectively. The A-B section is the stoss side of an ice wave, and the B,C section is the lee side of an ice wave [34]. The appearance of ice waves under an ice jam creates extra resistance to the flowing water. Due to variations in the thickness of the ice jam along the A-C channel section, the velocity field under an ice wave (along the channel section A-C, as shown in Figure 3) becomes complicated. The flow depth under point "B" is the least, implying the maximum flow velocity comparing to that at another place. Thus, ice particles at the crest of an ice wave will be eroded and delivered downstream. In the third stage, the local scour process around a bridge pier interacts with the ice jam evolution. It has been observed that both the scour hole and ice wave morphology does not change significantly or reach a dynamic equilibrium. After a scour hole around a pier reaches an equilibrium condition, the scour hole has the deepest depth and largest range and thus it should have the greatest influence on the migration of an ice wave.

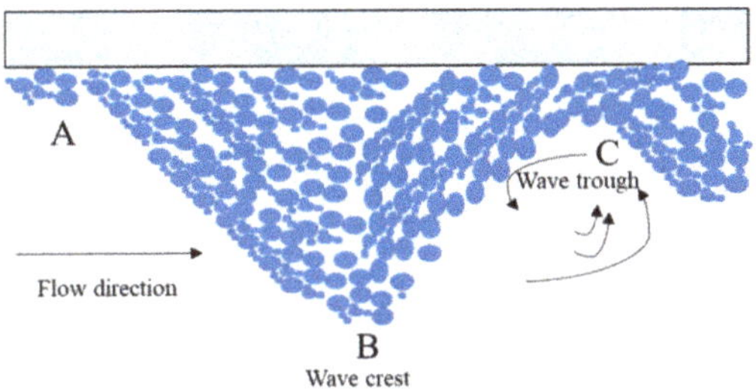

Figure 3. Diagram of an ice wave migration.

Although the migration process of an ice wave is similar to that of sand waves formed by sediment on riverbeds, the drag force, buoyancy force, and gravity of ice particles under an ice jam are different from those of sediment particles in a riverbed. The sediment particles in a riverbed are subjected to more gravity than the buoyancy force, while ice particles at the bottom of an ice jam are subjected to larger buoyancy force than gravity. An ice particle is more easily disturbed by flowing water than a sediment particle, since the difference between water mass density and ice mass density is less than that between sand mass density and water mass density. With a local scour at the pier, the presence of a scour hole further changes flow conditions around the pier. Depending on the location of ice particles and the kinetic energy of flowing water, the moving speeds of ice particles are different. Thus, different ice particles will be deposited at different places. It should be noted that the features of ice waves around the pier are the main objective in this study.

Regarding the influence of the thickness of an ice wave on the depth of the scour hole, according to Wang et al. [34], the thicker the ice jam, the deeper the scour hole, and thus, the higher risk to the pier. Under the action of an ice wave, a channel reach from the upstream ice jam head (or wave front or free water surface) to the pier is considered the study "control volume". Forces acting on this control volume are shown in Figure 4.

The continuity equation for flow water passing this control volume under an ice wave can be written as follows:

$$\int_0^B H(x)v_{w0}dx = \int_0^B (H_{w1}(x) + H_{s1}(x))v_{w1}dx \quad (1)$$

where B is flume width; x is the distance from the left flume wall (perpendicular to the flume wall); dx is the width of each unit profile; $H(x)$ is the flow depth in front of an ice wave at a distance of x from the left flume wall; $v_{w0}(x)$ is the depth-average flow velocity in front of the ice wave at a distance of x from the left flume wall; $H_{w1}(x)$ is the flow depth above the initial sand bed at the pier at a distance of x from the left flume wall; $v_{w1}(x)$ is the depth-average flow velocity at the pier under the ice jam at a distance of x from the left flume wall; $H_{s1}(x)$ is the depth of the scour hole at a distance of x from the left flume wall; and $H_{s1}(x) = 0$ in a sand bed without a scour, as shown in Figure 4.

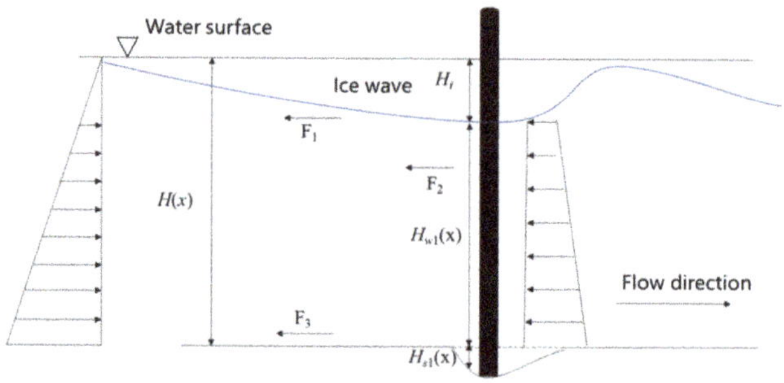

Figure 4. Forces acting on the control volume of flowing water under an ice wave around the bridge pier.

The flow velocity distribution along the transverse direction (perpendicular to the flow direction or flume wall) can be approximately expressed as follows [36]:

$$v_{w0}(x) = \left(\frac{d}{B/2}\right)^{0.11} Kv_{m0} \quad (2)$$

$$v_{w1}(x) = \left(\frac{d}{B/2}\right)^{0.11} Kv_{m1} \quad (3)$$

where d is the distance from this profile to the left or right-side flume wall; v_{m0} is the depth-average flow velocity in the center of the flume in front of the ice wave, v_{m1} is the depth-average flow velocity in the center of the flume at the pier (note, the bridge pier is located at the center of the flume); K is the transverse distribution parameter of flow velocity.

As shown in Figure 5, the width of the channel cross section at the pier (transverse direction of the flume) was divided into n units. The width of each unit can be expressed as:

$$\sum_{1}^{n} Hv_{w0}b = \sum_{1}^{n} (H_{w1} + H_{s1})v_{w1}b \quad (4)$$

where b is the width of each unit, and the sum of width of all units is B; H is the flow depth in front of an ice wave in this transverse unit; H_{w1}, H_{s1} are flow depth under an ice jam (above the initial sand bed) and the scour hole depth in this transverse unit, respectively, as shown in Figure 5.

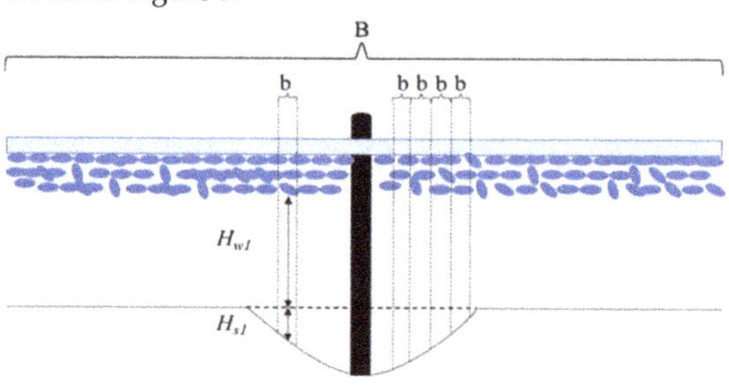

Figure 5. Channel cross-section at the pier (where, b is the width of each unit, and B is the sum of width of all units).

In each transverse unit, the continuity equation near the pier along the flow direction can be written as:

$$Hv_{w0} = (H_{w1} + H_{s1})v_{w1} \tag{5}$$

Considering the influence of the local scour around the pier, flowing water under an ice wave is taken as the control volume, and the momentum equation along the flow direction is expressed as:

$$\sum F = \rho_w Q_w (v_{w1} - v_{w0}) \tag{6}$$

where ΣF is the total force in the direction of flow; ρ_w is the mass density of water; Q_w is the discharge of flow.

The difference of pressure force between the upstream control surface (cross section) and downstream control surface of flow under the ice wave is:

$$F_P = \frac{1}{2}\rho_w g B H_i^2 \tag{7}$$

where F_P is the difference between the pressure force at upstream control surface (at the ice wave front) and that at downstream cross section (at the pier); H_i is the thickness of ice wave crest; g is gravitational acceleration.

The drag force resulted from an ice wave to the control volume can be expressed as:

$$F_1 = \frac{1}{2}C_i H_i \rho_w B {v_{w0}}^2 \tag{8}$$

where F_1 is the drag force caused by an ice wave; C_i is the drag coefficient, which ranges from 5.5 to 12.5 and increases with the increase of the ice discharge rate Q_i.

The resistance force caused by the pier when flow passes the pier is:

$$F_2 = \frac{1}{2}C_D \rho_w D H {v_{w0}}^2 \tag{9}$$

where F_2 is the resistance force caused by the pier, and other transverse units without the pier are $F_2 = 0$; C_D is the resistance coefficient, which is related to the Reynolds number; D is the pier diameter.

The resistance to the control volume caused by the channel bed and side walls of the flume can be expressed as:

$$F_3 = C_d \rho_w (B + 2H) L {n_c}^2 {v_{w0}}^2 \tag{10}$$

where F_3 is the resistance caused by the channel bed and side wall of the flume; C_d is the drag coefficient; n_c is composite roughness coefficient of the flume; L is the length of the control volume between the upstream and downstream control surfaces.

Combining Equations (5)–(10), the following equation is derived:

$$\begin{aligned}\rho_w H B v_{w0}\left(\tfrac{Hv_{w0}}{H-H_{s1}-H_i} - v_{w0}\right) = \tfrac{1}{2}\rho_w g B H_i^2 - \tfrac{1}{2}C_i H_i \rho_w B {v_{w0}}^2 - \tfrac{1}{2}C_D \rho_w D H {v_{w0}}^2 \\ - C_d \rho_w (B + 2H) L {n_c}^2 {v_{w0}}^2\end{aligned} \tag{11}$$

It can be concluded that the thickness of the ice wave crest under the influence of the local scour process around the pier can be expressed as follows:

$$\begin{aligned}H_i = -\tfrac{b}{3a} + \sqrt[3]{-\tfrac{27a^2d-9abc+2b^3}{54a^3} + \sqrt{\left(\tfrac{27a^2d-9abc+2b^3}{54a^3}\right)^2 + \left(\tfrac{3ac-b^2}{9a^2}\right)^3}} + \\ \sqrt[3]{-\tfrac{27a^2d-9abc+2b^3}{54a^3} - \sqrt{\left(\tfrac{27a^2d-9abc+2b^3}{54a^3}\right)^2 + \left(\tfrac{3ac-b^2}{9a^2}\right)^3}}\end{aligned} \tag{12}$$

where $a = -\frac{1}{2}gB$; $b = \frac{1}{2}gBH - \frac{1}{2}gBH_{s1} + \frac{1}{2}C_i B {v_{w0}}^2$;
$c = -BH{v_{w0}}^2 - \frac{1}{2}C_i {v_{w0}}^2 H + \frac{1}{2}C_i B {v_{w0}}^2 H_{s1} + \frac{1}{2}C_D D H {v_{w0}}^2 + C_d(B + 2H)L{n_c}^2{v_{w0}}^2$;

$d = -BHH_{s1}v_{w0}^2 - \frac{1}{2}C_D DH^2 v_{w0}^2 + \frac{1}{2}C_D DHH_{s1}v_{w0}^2 - C_d(B+2H)HLn_c^2 v_{w0}^2 + C_d(B+2H)H_{s1}Ln_c^2 v_{w0}^2$.

Equation (12) is simplified as follows:

$$H_i = A + \sqrt[3]{B + \sqrt{B^2 + C^3}} + \sqrt[3]{B - \sqrt{B^2 + C^3}} \tag{13}$$

where $A = -\frac{b}{3a}$; $B = -\frac{27a^2d - 9abc + 2b^3}{54a^3}$; $C = \frac{3ac - b^2}{9a^2}$.

The appearance of an ice wave under an ice cover leads to changes in the thickness of an ice jam, since both the crests and troughs of ice waves appear alternately along an ice jam from upstream to downstream. In fact, an equilibrium ice jam belongs to a dynamic equilibrium. The equilibrium thickness of an ice jam reported in the present study corresponds to the average thickness of wave crest and wave trough. Results showed that the ratio of the wave crest thickness to the average thickness of an ice jam is about 2–3, and this ratio increases with the increase in the ice discharge rate Q_i. Because ice wave affects the local scour depth and the mechanical load on the pier, an ice wave with a thicker crest should be vulnerable to failure of the bridge. By substituting the experimental data into the above formula, the thickness of an ice wave can be obtained. At the cross section where the pier is located, the origin of the transverse coordinates is located at the center of the flume (the pier center) is set. The transverse distribution of thickness of ice waves at this cross section is shown in Figure 6. The calculated results using the above formula agree well with results of experiments.

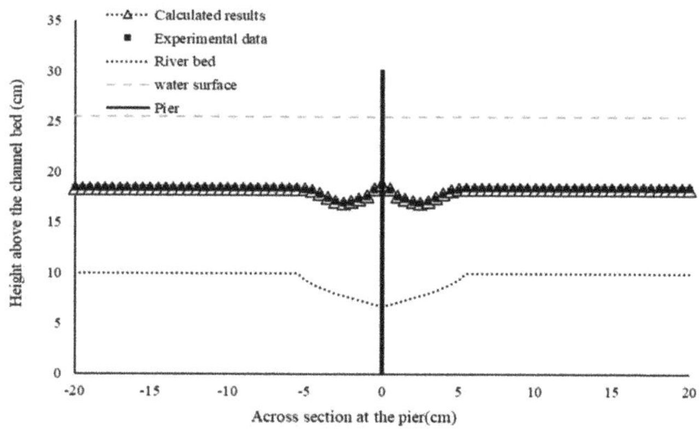

Figure 6. Transverse distribution of ice wave thickness at the cross section where the pier is located.

The distribution of thickness of an ice wave in the transverse direction has the following features:

(1) As shown in Figure 6, the closer to the central axis of the flume, the smaller the thickness of an ice wave and the deeper the scour hole. Thus, the cross-sectional area for flow at the pier is larger. Around the pier, it is difficult for ice particles to accumulate here because of the existence of the downflow and vortices around the pier. Results of experiments showed that, similar to a scour hole in a sand bed, an "ice scour hole" appeared at the bottom of the ice jam around the pier. The appearance of the "ice scour hole" is resulted from the presence of the pier. The existence of the "ice scour hole" affects the development of ice waves. Interestingly, the thickness of the wave crest at this cross section (where the pier is located) is less than that at other cross sections without the presence of a pier.

(2) With the increase in the distance from the flume center toward the flume wall, the thickness of the wave crest increases first and then decreases. In the zone around the pier, a scour hole in the sand bed developed, and the wave crest thickness increases

slightly because the presence of a scour hole leads to the increase in the cross-sectional area for flow. Further away from the flume center (or the pier), for zones without the presence of scour holes in the sand bed, the wave crest thickness reaches a constant toward the flume side wall.

To analyze the impacts of the local scour in the channel bed on the migration of an ice wave under different experiment conditions, the ice wave thickness of each transverse unit was averaged to obtain the average ice wave crest thickness at the cross section where the pier presents and was used as the ice wave crest thickness for this cross section. The comparison between the calculated results and those of experiments is shown in Table 2. In Table 2, the wave crest thickness with the presence of the local scour in the bed is calculated by Equation (13), and the wave crest thickness without the presence of the local scour is calculated by the formula proposed by Wang et al. [34]. As shown in Table 2, the calculated results agree well with those of experiments. Clearly, under the same hydraulic condition, the appearance of a scour hole around the pier leads to an increase in the thickness of the ice wave crest at the pier.

Table 2. Comparison of the calculated ice wave crest thickness to that of experiments.

Experiment Run #	Calculated Value (m)	Experimental Value (m)	Experiment Run #	Calculated Value (m)	Experimental Value (m)
A1	0.082	0.073	B1	0.068	0.070
A2	0.082	0.079	B2	0.053	0.077
A3	0.065	0.056	B3	0.059	0.051
A4	0.065	0.068	B4	0.053	0.065

The flow velocity under an ice wave around the pier is also one of important factors affecting ice transport capacity. Results indicated that the local scour at the pier affects the migration speed of an ice wave passing the pier. The migration of an ice wave is a complex process. During the evolution process of an ice jam, an ice jam may reach a dynamic equilibrium state due to the presence of ice waves along the ice-jammed channel. The wave height and wavelength of each ice wave vary constantly. The simplified schematic diagram of an ice wave is given in Figure 7.

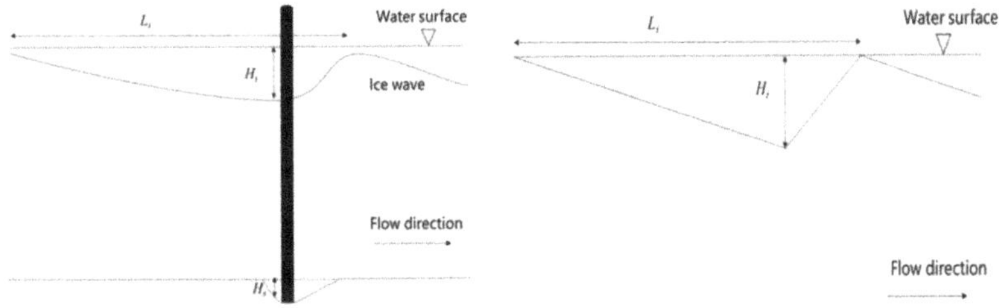

Figure 7. Simplified schematic diagram of an ice wave.

The average height of an ice wave crest was obtained from measured values of an ice wave crest after both the ice jam and local scour process at the pier achieved an equilibrium state, and similarly for obtaining the average length of ice wave. The average migration speed of an ice wave was calculated based on the ice discharge rate during this period. The formula for calculating these factors is expressed as follows:

$$\overline{H_i} = \sum_{1}^{n} H_i/n \tag{14}$$

$$\overline{L_i} = \sum_{1}^{n} L_i/n \tag{15}$$

$$\overline{V_i} = \frac{1}{2}\overline{H_i L_i B} \quad (16)$$

where $\overline{H_i}$ is the average height of ice wave crest; $\overline{L_i}$ is the average length of ice wave; $\overline{V_i}$ is the average ice wave volume; L_i is the length of the ice wave.

Then, the migration speed of an ice wave can be calculated as follows:

$$v_i = \frac{\sum_{1}^{T} Q_i}{\overline{V_i}} \frac{L_i}{T} \quad (17)$$

where v_i is the migration speed of an ice wave; T is the duration of the selected period for certain time for collecting data.

The migration speed of an ice wave in each experiment was calculated, as shown in Table 3. Compared with the experiment conditions in Table 1, the migration speed of an ice wave with the presence of a local scour is lower than that without a local scour in the bed. Additionally, the faster the flow velocity under an ice jam is, the higher the migration speed of an ice wave is. Under the same flow condition, the presence of a local scour slows down the migration of an ice wave around the pier.

Table 3. The calculated migration speed of ice wave.

Number	H_s (m)	v_i (m/s)	Number	H_s (m)	v_i (m/s)
A1	0.0319	0.00142	B1	0	0.00147
A2	0.0286	0.00128	B2	0	0.00134
A3	0.0343	0.00184	B3	0	0.00200
A4	0.0323	0.00152	B4	0	0.00159

With the presence of a local scour, considering the effects of various influencing factors on the migration speed of an ice wave, the following formula can be used to express the migration speed of an ice wave:

$$v_i = f(v, H, g, n_i, n_b, \rho_i, \rho_w, B, d_i, H_{w1}, Q_w, Q_i, L_i, T, J) \quad (18)$$

where n_i is the roughness coefficient of the ice jam; n_b is the composite roughness coefficient of the channel bed and flume wall; ρ_i is the density of ice; d_i is the particle grain size of ice; T is the temperature; J is the water surface slope.

In this experimental study, the median grain sizes of ice and sediment particle are consistent. Considering other major variables influencing the migration speed of an ice wave, the following equation for calculating the migration speed of an ice wave under an ice jam was derived:

$$\frac{v_i}{v} = 8.165\left(\frac{v}{\sqrt{gH}}\right)^{1.424}\left(\frac{H_{s1}+H_{w1}}{H}\right)^{-0.009}\left(\frac{Q_i}{Q_w}\right)^{0.542} \quad (19)$$

As showed in the above equation, the migration speed of an ice wave is affected by the flow Froude number, the sum of the scour depth and the water depth under the ice jam where the pier presents, and the ice discharge rate. This equation indicates that the higher the flow Froude number is, the faster the ice waves migrate; the greater the ratio of the water depth under an ice jam, including the scour hole depth to total water depth, the smaller the migration speed of an ice wave; the greater the ice discharge, the greater the migration speed of an ice wave.

The ice transport capacity through the migration speed of an ice wave with the presence of a scour hole refers to the ice transport capacity per unit time at the pier. After the ice jam achieved a dynamic equilibrium state, although the migration process of ice waves along the ice-jammed channel continues, the mean value of the ice transport capacity per unit time tends to be stable. After an ice jam achieved a dynamic equilibrium, the ice transport capacity at the pier was equal to the incoming ice discharged from the ice hopper. By applying the mass conservation equation to the process of ice jam accumulation, the ice transport capacity without the presence of a local scour at the pier was given as follows [34]:

$$\frac{\partial Q_T}{\partial x} + (1-p)B\frac{\partial y_i}{\partial t} = 0 \tag{20}$$

where Q_T is the ice transport capacity; p is the porosity of an ice jam in this study, $p = 0.4$; y_i is the thickness of an ice jam at any cross section; and t is time.

The migration speed of an ice wave can be expressed as:

$$\frac{\partial y_i}{\partial t} + v_i\frac{\partial y_i}{\partial x} = 0 \tag{21}$$

The following equation can be obtained by combining Equations (20) and (21):

$$\frac{\partial Q_T}{\partial x} - (1-p)Bv_i\frac{\partial y_i}{\partial x} = 0 \tag{22}$$

At any location of x_0 and at any instant time of t_0, Equation (22) can be expressed as:

$$Q_T(y, t_0) - Q_T(h_0, t_0) = \int_{h_0}^{y} (1-p)Bv_i dy = (1-p)Bv_i(y - h_0) \tag{23}$$

where h_0 is the ice jam thickness at the cross section of the ice wave trough.

By ignoring the ice transport capacity below the trough of an ice wave, the ice wave transport capacity without a presence of a local scour at the pier can be expressed by the following formula:

$$Q_T = \alpha(1-p)BH_iv_i \tag{24}$$

where α is the correction coefficient of ice wave bottom morphology. The value ranges from 0.6 to 0.8.

Substituting the ice wave crest thickness and the migration speed of an ice wave by considering the influence of a local scour at the pier into the above equation, the ice transport capacity at the pier with the presence of a local scour can be derived as:

$$\begin{aligned} Q_T &= 8.165\alpha(1-p)Bv(A + \sqrt[3]{B + \sqrt{B^2 + C^3}} + \sqrt[3]{B - \sqrt{B^2 + C^3}})(\frac{v}{\sqrt{gH}})^{1.424} \\ &\quad (\frac{H_{s1} + H_{w1}}{H})^{-0.009}(\frac{Q_i}{Q_w})^{0.542} \end{aligned} \tag{25}$$

The ice transport capacity by means of the migration process of ice waves with a presence of a scour hole under different hydraulic conditions (such as ice discharge rate, flow velocity, and water depth) can be determined by Equation (25). It can be seen from this equation that the ice transport capacity at the pier decreases with the increase in both the scour hole depth and the flow depth under an ice jam. This is because the presence of a local scour increases the cross sectional area of flow at the pier and the migration speed of an ice wave at the pier slows down. As shown in Figure 8, based on data collected from experiments, the ice transport capacities calculated using the above equation with the presence of a local scour at the pier agree well with those of laboratory experiments.

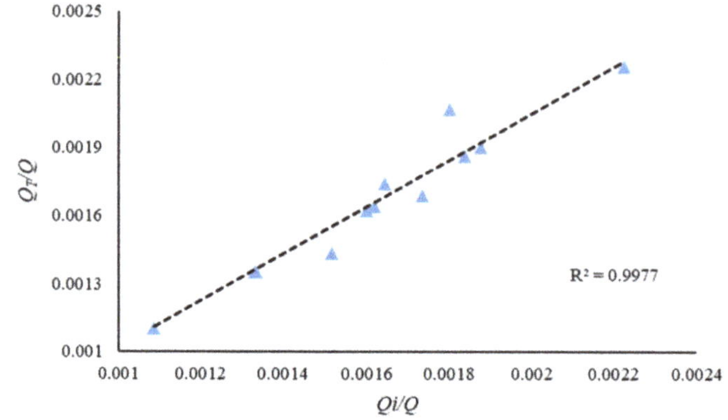

Figure 8. Calculated ice transport capacities compared to those measured from experiments.

4. Conclusions

In the present study, to analyze the influence of a scour hole on the migration of ice waves, laboratory experiments have been carried out under different flow and boundary conditions. The following results have been drawn:

1. By combining the continuity equation and momentum equation, the equation for calculating the thickness of ice wave crest under the influence of a local scour was derived. Results showed that the ice wave thickness near the central axis of the flume is smaller. This "ice scour hole" phenomenon at the bottom of an ice jam near the pier is similar to the scour hole at the pier in the sand bed. With the increase in the distance from the flume center, the thickness of the wave crest increases first and then decreases. Further away from the central axis of the flume, for zones without the presence of scour holes in the sand bed, the wave crest thickness reaches a constant toward the flume walls. The calculated average ice wave thickness of the whole section is in good agreement with result of experiments regardless of whether or not a local scour is present in channel bed. Under the same hydraulic condition, the presence of a local scour at a pier lead to an increase in the thickness of the wave crest.
2. Under the same hydraulic condition, the appearance of a scour hole slows down the migration speed of an ice wave at the pier. With the presence of a local scour at the pier, the dimensionless relation for determining the migration speed of an ice wave was obtained based on experimental data. Results showed that the larger the ice discharge rate and flow Froude number is, the higher the migration speed of an ice wave is. The greater the ratio of the water depth under an ice jam, including the scour hole depth to total water depth, the smaller the migration speed of an ice wave.
3. The ice transport capacity with the presence of a scour hole was analyzed. The ice transport capacity at the pier decreases with the increase in both the scour hole depth and the flow depth under an ice jam. The developed equation can be used to determine the ice transport capacity by means of the migration speed of ice waves considering a local scour process at a bridge pier. The calculated results agree well with those of experiments in the laboratory.
4. The interaction between an ice jam and local scour at a pier is very complicated and has hardly been conducted. This study, based on laboratory experiments, belongs to conceptual research instead of research based on field prototype data from a natural river. More field observation data are needed to verify results obtained from laboratory experiments. Considering the influence of different pier types, river bends, and side wall effects, relevant experiments need to be further carried out.

Author Contributions: Conceptualization, J.W. and J.S.; methodology, Z.H.; investigation, Z.H. and J.S.; software, Z.H.; validation Z.H.; data curation, Z.H.; formal analysis, Z.H.; writing-original draft preparation, Z.H.; writing—review and editing, J.S.; funding acquisition, J.W.; laboratory works, Z.H., F.S. and Z.L. All authors have read and agreed to the published version of the manuscript.

Funding: This research was supported by the National Natural Science Foundation of China (grant nos. 51879065). The authors are grateful for the financial support.

Institutional Review Board Statement: Not applicable.

Informed Consent Statement: Not applicable.

Data Availability Statement: The data are available.

Conflicts of Interest: The authors declare no conflict of interest.

References

1. Beltaos, S. River Ice Jams: Theory, Case Studies, and Applications. *J. Hydraul. Eng.* **1983**, *109*, 1338–1359. [CrossRef]
2. Sui, J.; Karney, B.W.; Fang, D. Variation in water level under ice-jammed condition–Field investigation and experimental study. *Hydrol. Res.* **2005**, *36*, 65–84. [CrossRef]
3. Sui, J.; Wang, J.; Balachandar, R.; Sun, Z.; Wang, D. Accumulation of frazil ice along a river bend. *Can. J. Civ. Eng.* **2008**, *35*, 158–169. [CrossRef]
4. Sui, J.; Wang, D.; Karney, B.W. Suspended sediment concentration and deformation of riverbed in a frazil jammed reach. *Can. J. Civ. Eng.* **2000**, *27*, 1120–1129. [CrossRef]
5. Wang, J.; Hou, Z.X.; Sun, H.J.; Fang, B.H.; Sui, J.; Karney, B. Local scour around a bridge pier under ice-jammed flow condition—An experimental study. *J. Hydrol. Hydromech.* **2021**, *69*, 275–287. [CrossRef]
6. Beltaos, S.; Burrell, B.C. Field measurements of ice-jam-release surges. *Can. J. Civ. Eng.* **2005**, *32*, 699–711. [CrossRef]
7. Beltaos, S. Internal strength properties of river ice jams. *Cold Reg. Sci. Technol.* **2010**, *62*, 83–91. [CrossRef]
8. Beltaos, S.; Burrell, B.C. Ice-jam model testing: Matapedia River case studies, 1994 and 1995. *Cold Reg. Sci. Technol.* **2010**, *60*, 29–39. [CrossRef]
9. Sui, J.; Karney, B.W.; Sun, Z.; Wang, D. Field investigation of frazil jam evolution: A case study. *J. Hydraul. Eng.* **2002**, *128*, 781–787. [CrossRef]
10. Sui, J.; Hicks, F.; Menounos, B. Observations of riverbed scour under a developing hanging ice dam. *Can. J. Civ. Eng.* **2006**, *33*, 214–218. [CrossRef]
11. Knack, I.; Shen, H.T. Sediment transport in ice-covered channels. *Int. J. Sediment Res.* **2015**, *30*, 5. [CrossRef]
12. Yang, K.L.; Liu, Z.P.; Li, G.F.; Chen, C.J.; Liu, C.J.; Hu, H.D. Simulation of ice jams in river channels. *Water. Resour. Hydr. Eng.* **2002**, *33*, 40–47.
13. Mao, Z.Y.; WU, J.J.; Zhang, L.; Zhang, R.T. Numerical simulation of river ice jam. *Adv. Water Sci.* **2003**, *14*, 700–705.
14. Wang, J.; He, L.; Chen, P.P.; Sui, J. Numerical simulation of mechanical breakup of river ice-cover. *J. Hydrodyn.* **2013**, *25*, 415–421. [CrossRef]
15. Szydłowski, M.; Kolerski, T. Numerical modeling of water and ice dynamics for analysis of flow around the Kiezmark Bridge piers. In *Free Surface Flows and Transport Processes*; Springer: Cham, Switzerland, 2018; pp. 465–476.
16. Istrati, D.; Hasanpour, A.; Buckle, I.G. Numerical investigation of tsunami-borne debris damming loads on a coastalbridge. In Proceedings of the 17th World Conference on Earthquake Engineering (17WCEE), Sendai, Japan, 13–18 September 2020.
17. Wang, J.; Shi, F.Y.; Chen, P.P.; Wu, P.; Sui, J. Simulations of ice jam thickness distribution in the transverse direction. *J. Hydrodyn.* **2014**, *26*, 840–847. [CrossRef]
18. Beltaos, S.; Miller, L.; Brian, B.C.; Sullivan, D. Hydraulic effects of ice breakup on bridges. *Can. J. Civ. Eng.* **2007**, *34*, 539–548. [CrossRef]
19. Wang, J.; Shi, F.Y.; Chen, P.P.; Wu, P.; Sui, J. Impact of bridge pier on the stability of ice jam. *J. Hydrodyn.* **2015**, *27*, 865–871. [CrossRef]
20. Wang, J.; Hua, J.; Sui, J.; Wu, P.; Lu, T.; Chen, P.P. The impact of bridge pier on ice jam evolution—An experimental study. *J. Hydrol. Hydromech.* **2016**, *64*, 75–82. [CrossRef]
21. Wang, J.; Hua, J.; Chen, P.P.; Sui, J.; Wu, P.; Whitcombe, T. Initiation of ice jam in front of bridge piers—An experimental study. *J. Hydrodyn.* **2019**, *31*, 117–123. [CrossRef]
22. Wang, J.; Sui, J.; Karney, B.W. Incipient motion of non-cohesive sediment under ice cover—An experimental study. *J. Hydrodyn.* **2008**, *20*, 117–124. [CrossRef]
23. Carr, M.L.; Tuthill, M.A. Modeling of Scour-Inducing Ice Effects at Melvin Price Lock and Dam. *J. Hydraul. Eng.* **2012**, *138*, 85–92. [CrossRef]
24. Batuca, D.; Dargahi, B. Some experimental results on local scour around cylindrical piers for open and covered flow. In Proceedings of the 3rd International Symposium on River Sedimentation, Jackson, MS, USA, 31 March–4 April 1986.

25. Ackermann, N.L.; Shen, H.T.; Olsson, P. Local scour around circular piers under ice covers. Ice in the Environment. In Proceedings of the 16th IAHR International Symposium on Ice, Dunedin, New Zealand, 2–6 December 2002.
26. Hains, D.B. An Experimental Study of Ice Effects on Scour at Bridge Piers. Ph.D. Thesis, Lehigh University, Bethlehem, PA, USA, 2004.
27. Wu, P.; Hirshfield, F.; Sui, J.; Wang, J.; Chen, P.P. Impacts of ice cover on local scour around semi-circular bridge abutment. *J. Hydrodyn.* **2014**, *26*, 10–18. [CrossRef]
28. Namaee, M.R.; Sui, J. Impact of armour layer on the depth of scour hole around side-by-side bridge piers under ice-covered flow condition. *J. Hydrol. Hydromech.* **2019**, *67*, 240–251. [CrossRef]
29. Namaee, M.R.; Sui, J. Local scour around two side-by-side cylindrical bridge piers under ice-covered conditions. *Int. J. Sediment Res.* **2019**, *34*, 355–367. [CrossRef]
30. Namaee, M.R.; Sui, J. Velocity profiles and turbulence intensities around side-by-side bridge piers under ice-covered flow condition. *J. Hydrol. Hydromech.* **2020**, *68*, 70–82. [CrossRef]
31. Valela, C.; Sirianni, D.; Nistor, I.; Rennie, C.D.; Almansour, H. Bridge pier scour under ice cover. *Water* **2021**, *13*, 536. [CrossRef]
32. Jafari, R.; Sui, J. Velocity field and turbulence structure around spur dikes with different angles of orientation under ice covered flow conditions. *Water* **2021**, *13*, 1844. [CrossRef]
33. Hu, H.; Wang, J.; Cheng, T.; Hou, Z.; Sui, J. Channel Bed Deformation and Ice Jam Evolution around Bridge Piers. *Water* **2022**, *14*, 1766. [CrossRef]
34. Wang, J.; Wu, Y.; Sui, J.; Karney, B. Formation and movement of ice accumulation waves under ice cover—An experimental study. *J. Hydrol. Hydromech.* **2019**, *67*, 171–178. [CrossRef]
35. *JTG D60–2004*; General Code for Design of Highway Bridges and Culverts. Ministry of Transportation of China (MTC): Beijing, China, 2004.
36. Yan, J.; Wang, E.P.; Sun, D.P.; Dong, Z.H. Experimental study on distribution properties of velocity in rectangular open channel. *Eng. J. Wuhan Univ.* **2005**, *38*, 59–64.

water

MDPI

Article

Channel Bed Deformation and Ice Jam Evolution around Bridge Piers

Haotian Hu [1], Jun Wang [1,*], Tiejie Cheng [1], Zhixing Hou [1] and Jueyi Sui [2,*]

1 School of Civil and Hydraulic Engineering, Hefei University of Technology, Hefei 230009, China; 18756718559@163.com (H.H.); hfut_chengtj@126.com (T.C.); houhfut@163.com (Z.H.)
2 School of Engineering, University of Northern British Columbia, Prince George, BC V2N 4Z9, Canada
* Correspondence: junwanghfut@126.com (J.W.); jueyi.sui@unbc.ca (J.S.); Tel.: +86-13705510008 (J.W.); +1-250-960-6399 (J.S.)

Abstract: The interaction between the evolution of an ice jam and the local scour at bridge piers becomes much more complicated due to the evolution of both the channel bed and ice jam. Thus, research work regarding this topic has been hardly conducted. In the present study, experiments under different flow conditions with three different pier shapes were carried out. Through laboratory experiments, the development of scour holes around bridge piers under open flow, ice-covered, and ice-jammed flow conditions was compared. The results show that under the same hydraulic condition and with the same ice discharge rate (Q_i/Q), the development of an initial ice jam with a local scour around bridge piers along the entire flume takes a relatively short time. However, it takes a longer time for an ice jam to achieve an equilibrium state. With the presence of a local scour at bridge piers, after an ice jam reaches an equilibrium state, the ice jam thickness, water level, and water depth for flow are relatively larger compared to that without a local scour at the pier. The equilibrium ice jam thickness around the pier is negatively correlated with the initial flow Froude number. When the development of an initial ice jam is dominated by a mechanical thickening process, the rate of the development of a scour hole around a pier is faster. On the other hand, when the development of an initial ice jam is dominated by a hydraulic thickening process, the development of a scour hole around a pier can be treated as a scour process under an ice-covered flow condition. An equation was developed to determine the scour depth around a pier under an ice-jammed flow condition by considering related factors such as the flow Froude number, ice jam thickness, and ice discharge rate. The results of this research can provide a reference for bridge design and safety protection, as well as the interaction mechanism of local scour and ice jam evolution.

Keywords: bridge pier; local scour; ice jam evolution; ice jam thickness; scour hole

Citation: Hu, H.; Wang, J.; Cheng, T.; Hou, Z.; Sui, J. Channel Bed Deformation and Ice Jam Evolution around Bridge Piers. *Water* **2022**, *14*, 1766. https://doi.org/10.3390/w14111766

Academic Editor: Giuseppe Oliveto

Received: 19 May 2022
Accepted: 27 May 2022
Published: 31 May 2022

1. Introduction

Ice jams in rivers are a common ice phenomenon in cold regions in winter that affect both hydraulic conditions and boundary conditions of watercourses. Sometimes, under the same flow condition, the presence of an ice cover in a river leads to a higher water level, which can induce ice flood disasters. The presence of bridge piers in a river narrows the flow cross section, changes the flow state of water around the piers, and forms a horseshoe vortex, which often results in a local scour around bridge piers. As reported by researchers, the existence of bridge piers in rivers increases the possibility of the formation of ice jams [1,2], and the formation and development of an ice jam aggravate the local scour of bridge piers [3]. Therefore, the mutual influence of flow conditions, local scour, and ice jam evolution makes relevant problems more complex.

Recently, some research has been conducted to study the hydraulic interaction between bridge piers and ice jams. The results of laboratory experiments showed that the existence of bridge piers results in an increase in the ice transport capacity near bridge piers and a decrease in the ice jam thickness [2]. Wang et al. [4,5] carried out an experimental study on ice jam blockage at bridge piers, which showed that ice blockage was affected by the ice cube size, spanning distance between piers, and flow conditions. Based on experiments conducted in a curved flume in a laboratory, it was found that the bridge pier located at the apex of the river bend had less influence on ice jam formation than the one installed in a straight section of a river (the junction of two bends) [6]. The results of laboratory experiments regarding the variation of water level around piers with different shapes under an ice jam showed that the increment of water level in the present of a cylindrical pier was the smallest, while it was the largest for a rectangular pier, and the regression formula for calculating the increase in water level caused by an equilibrated ice jam was given [7]. It was found that the ice accumulation process around bridge piers depends on the pier size, ice discharge rate, flow discharge, and Froude number. For a flow with a smaller Froude number and a larger ice discharge rate, with the presence of a smaller pier in the channel, the evolution process of an ice jam is similar to that without the presence of a pier [8]. There are very few studies on numerical simulation. Cheng et al. [9] used a discrete element model to simulate the ice floe accumulating process around bridge piers. In addition, Yang et al. [10] used a material point method; Istrati et al. [11] used a finite element method; Salciarini et al. [12] used a discrete element method, and Hasanpour et al. [13] used a smoothed particle hydrodynamics method to study tsunami-induced debris impacts on bridge structures, respectively, which has a certain reference value for the numerical simulation of hydraulic action between bridge piers and ice jams. Parola et al. [14] conducted experimental studies to develop a method for predicting debris forces on piers and superstructures. Malavasi et al. [15] studied the hydrodynamic interactions between an open flow and a bridge deck. The results of this work will provide a theoretical reference for the study of the interaction between ice floes and bridge decks under extreme hydraulic conditions.

Some researchers have done research work regarding the local scour process around circular piers under both open flow and ice-covered (instead of ice-jammed) flow conditions. In an open flow, to improve the applicability of the equation developed by Melville and Coleman in laboratory and field data, Pandey et al. [16] proposed new K-factors for a nonuniform gravel bed. Memar et al. [17] studied the influence of the size and position of circular collars from the sediment bed on serial bridge piers. Zhao et al. [18] derived one formula for calculating scour depth around a skew bridge in a curved road by solving the deflection angle of water flow and the width of the water crossing section. The existence of an ice cover changes the flow structure [19–22] and affects the development of a local scour. Compared to the adjacent position of the collar on the bed, placing the collars below the bed would reduce the scouring speed around a pier. Bacuta and Dargahi [23], Ackermann and Shen [24], and Hains and Zabilansky [25,26] studied the local scour at cylindrical bridge piers under ice-covered flow conditions and found that compared to open flow conditions, the local scour depth is greater under an ice-covered flow condition. Munteanu [27] performed experiments under four different boundary conditions (open flow, totally ice-covered, two sides partially ice-covered and one side partially ice-covered) and pointed out that the local scour process around the bridge pier was the most intense under two sides in a partial ice-covered condition, and the maximum scour depth was about 55% higher than that under an open flow condition. Christopher et al. [28] investigated the scour problem around bridge piers under different ice cover thicknesses and found that the scour depth increased with the increase in ice cover thickness. Wu et al. [29–31] investigated the local scour process under ice-covered flow conditions based on laboratory experiments and reported that the maximum scour depth increased with the increase in cover roughness. They also established empirical formulas for calculating the scour depth under an ice-covered flow condition. Wang et al. [32] analyzed the relationship between the flow intensity and dimensionless scour depth, as well as the difference in the rate of change of scour depth under an ice-covered flow condition compared to that under an open flow condition and developed an empirical equation for determining the change of scour depth with the time under an ice-covered condition. Namaee and Sui [33–37] studied the influence of ice covers with different roughness on the local scour process around side-by-side piers. They found that compared with the smooth-covered flow condition, the near-bed flow velocity was higher, and the maximum flow velocity position was closer to the channel bed under the rough cover. The maximum scour depth around piers increased with the decrease in particle size and the increase in cover roughness and densimetric Froude number. The smaller the pier size and the larger the pier spacing distance, the weaker the horseshoe vortex around the pier and the shallower the scour hole. The grain size of the armour layer obviously affects the depth of the scour hole.

The existing research studies either focus on the influence of bridge piers on the stability and evolution of ice jams or the local scour process around piers under an ice-covered (instead of ice-jammed) flow condition. However, no research work has been reported regarding the local scour process around bridge piers during the evolution process of an ice jam considering the thickness of an ice jam and the time for achieving the equilibrium condition. In the present experiment study, the influence of a local scour at bridge piers on the evolution of ice jams was studied. The development of scour holes under different evolution modes of the ice jam was investigated. The results of the present study can provide a basis for bridge design and safety protection in practical engineering.

2. Materials and Methods

2.1. Experimental Setup

Laboratory experiments were carried out in a flume. The plan layout of the flume is shown in Figure 1. The length of the flume is 26.68 m with a width of 0.4 m and a depth of 1.3 m. Along this flume from upstream to downstream, 22 cross sections (CS) for measurement with an equal spacing distance of 1.2 m were set up. Between the upstream CS-4 and CS-5, an ice hopper was setup for discharging model ice particles into the flowing water. A Styrofoam panel that was 0.6 m long and 0.4 m wide was placed on the water surface between CS-20 and CS-21 for assisting with the initiation of an initial ice jam in front of the foam panel in the flume, corresponding to the phenomenon of an initial ice jam head formation in natural rivers.

Polyethylene particles were used to model ice particles. The mass density of the polyethylene particles was 0.918 g/cm^3, which is nearly equal to the mass density of ice of 0.917 g/cm^3 in rivers. Polyethylene particles has a flat ellipsoid shape with the longest diameter of 3.5 mm (as shown in Figure 1c). Between CS-2 and CS-22, a sand bed was prepared to model a sand bed in rivers. The initial thickness of the sand bed in the flume was 10 cm. The median particle size (d_{50}) of the three sands used in the experiments was 0.49 mm, 0.713 mm, and 1.04 mm, respectively. The inhomogeneity coefficient of these three sands was 2.0, 1.61, and 2.68, respectively. The mass density was 1.423 g/cm^3. Three different shapes of model piers were used, namely cylindrical, rectangle, and round end-shaped piers, as shown in Figure 1. The model pier was installed in the flume center at CS-16. Cross section 3 was the control section for the initial flow depth (H_0) and initial velocity (V_0) and the ice discharge rate (Q_i) from the ice hopper. In engineering design, the ratio of the pier width to a single span between adjacent piers is 0.04~0.16 [38]. Considering the flume width of 40 cm, the pier width (D) was 2 cm. In addition, this research work is a conceptual study, which is not targeted at a specific engineering prototype, but mainly investigates the changes of local scours and median particle sizes of sediment on the speed of ice jam evolution and scour hole development.

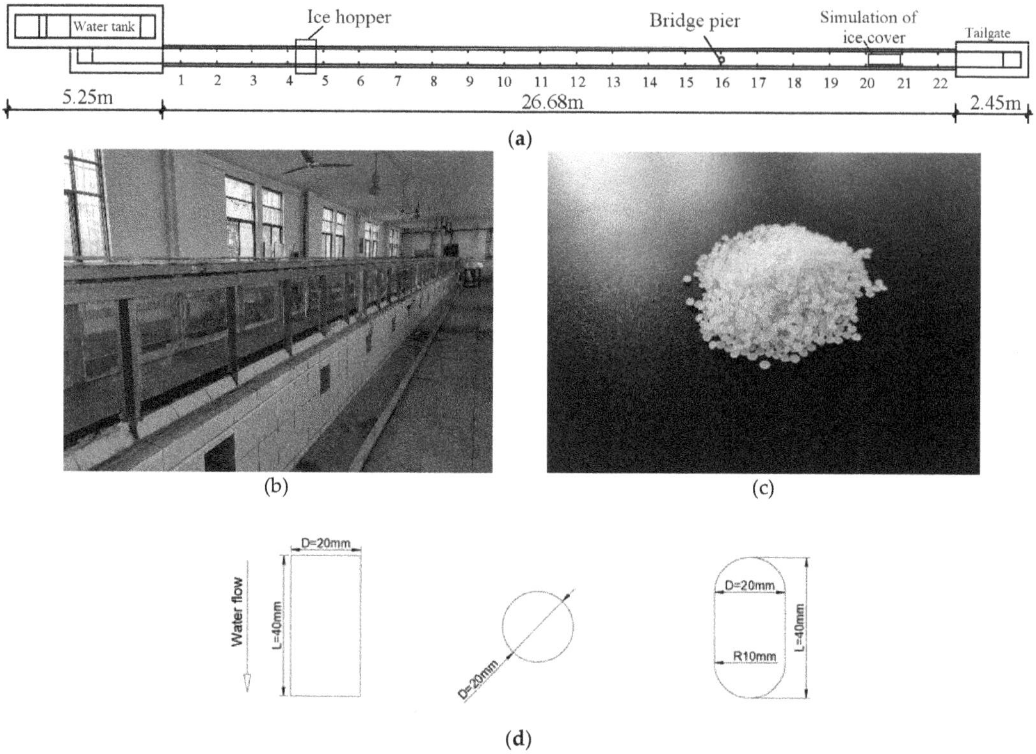

Figure 1. (**a**,**b**) Flume layout for experiment; (**c**) polyethylene particles; (**d**) cross section of model piers.

2.2. Experimental Process

(1) Firstly, the sand scraper was used to level the sediment bed along the study channel so that the flow pattern at each cross section was approximately the same.

(2) The flume was filled with water slowly to prevent sediment particles in the channel bed from the initiation of motion until the water level in the flume reached the target value. The flow discharge of each experiment running through the triangular thin-walled weir with an accuracy of 0.1 mm was kept as constant. The water level in the flume was controlled by adjusting the tailgate of the flume.

(3) Styrofoam panels were placed on the water surface between CS-20 and CS-21 to simulate the initiating point for an ice jam. A four-channel ultrasonic current meter with an accuracy of 0.01 m/s was arranged at the cross section where the pier was located to measure flow velocities around the model pier, which was placed in the center at CS-16.

(4) Ice particles were filled in a barrel with a volume of 7.40 L. The mass of each barrel of ice particles was 6.786 kg. The time required for each barrel of ice particles to enter the flume from the ice hopper was calculated. At this point, the test preparation work was completed. Note: the ice discharge rate was controlled as the target value during each experiment run.

(5) At CS-16, where the piers were located, the depth and length of scour holes were measured every 10 min for the first 1 h and then every 30 min. The thickness of the ice jam around this cross section was measured using a ruler with an accuracy of 1 mm, every 10 min in the previous hour, and then every 30 min (Note: when the ice wave phenomenon did not occur at this cross section, the thickness of the

ice jam was the average of the largest and smallest values along this cross section; when an ice wave phenomenon appeared at this cross section, the thickness of the ice jam at this cross section was the average of the thickness of the wave crest and wave trough). The water level was measured at the same frequency as that for the ice jam thickness, and both were performed simultaneously until the ice jam and local scour achieved an equilibrium state. Experiments showed that the time for the scour hole to achieve an equilibrium state under an ice-jammed flow condition was related to the hydraulics and ice flow conditions and ranged from 8 h to 11 h. However, all experiments lasted 24 h to ensure that the local scour under an ice jam condition achieved an equilibrium state.

(6) Subsequently, we stopped adding ice particles in the flume, raised the tailgate, and gradually reduced the flow discharge to increase the water level so that ice particles under an ice jam in the flume stopped moving. To obtain a precious bathymetry of the scour hole, laboratory measurements were carefully carried out in order to not disturb the scour hole and ice jam due to the addition of probes. A point gage and a spot measuring instrument with an accuracy of 0.1 mm were used for measuring the bathymetry of the deformation of the sand bed around the pier. For each experimental run, data were collected at 26 measurement points around the bridge pier. Some variables used in this study were defined and are shown in Figure 2.

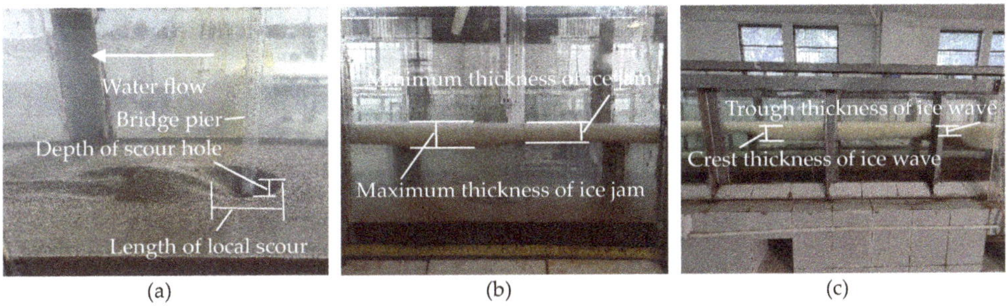

Figure 2. Scour hole at the pier under an ice jam and associated variables. ((**a**): scour hole with point measurement; (**b**,**c**): points for measuring ice jam thickness).

Each experimental run was considered to achieve an equilibrium state if the following conditions were met: (1) Water levels at all cross section did not change; (2) The thickness of the ice jam at any cross section did not change, namely, the amount of ice particles discharged from the ice hopper located upstream equalled that of the output at the downstream end of the ice jam; (3) The scour hole also reached an equilibrium condition, namely, not only the depth and shape of the scour hole around the pier did not change, but also the height and shape of the deposition dune downstream of the scour hole did not change. Whenever an experimental run satisfied these conditions, this experimental run was considered to reach the equilibrium state.

The evolution of an ice jam is divided into three stages: (1) The formation of the initial ice jam phase: the ice jam develops from the downstream Styrofoam panel to upstream with the continuous constant incoming ice particles from the ice hopper (as shown in Figure 3a). (2) The ice jam thickening stage: with the continuous incoming ice particles from the ice hopper, the ice jam gains thickness from upstream to downstream since the incoming ice particles are entrained by the flowing current and submerged at the head of an ice jam and then gradually delivered downstream. This process continues before the ice jam reaches an equilibrium state (as shown in Figure 3b). (3) The ice jam equilibrium stage: the amount of ice particles coming from upstream is equal to the output amount of ice particles (the amount of ice particles collected from the pond located at the downstream of the flume) from the toe of the ice jam body; however, the appearance of waves of ice accumulation migrate from upstream to downstream, and as a consequence, the ice jam becomes thicker whenever the crest of a wave of ice accumulation passes and thinner whenever the trough of such as wave passes (as shown in Figure 3c). In the present study, to assess the local scour around a pier, various experiments with local scour (as shown in Figure 4) and without local scour (namely, the fixed-bed experiments) were conducted, as summarized in Table 1.

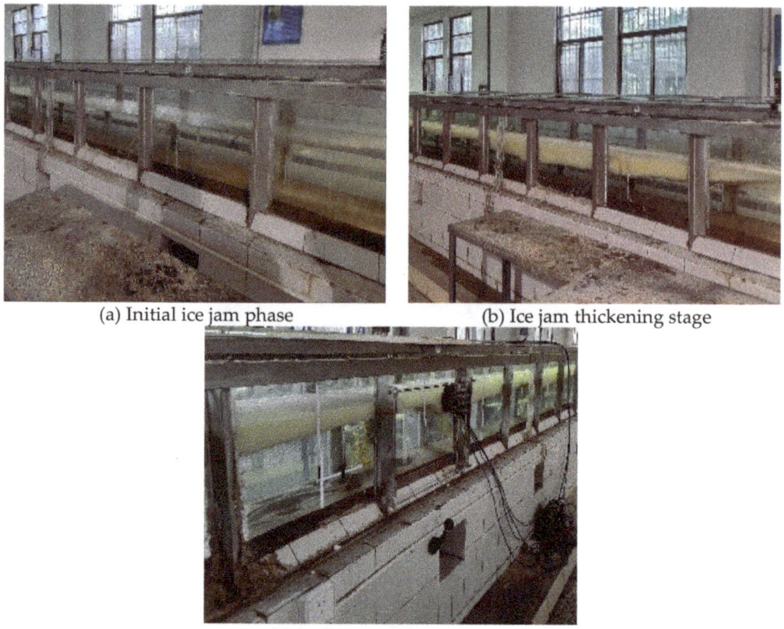

(a) Initial ice jam phase (b) Ice jam thickening stage

(c) Ice jam equilibrium stage

Figure 3. Development of an ice jam at different stages.

Figure 4. Local scour hole around a pier.

Table 1. Summary of all experimental runs.

Number	Average Approaching Velocity V_0 (m/s)	Approaching Flow Depth H_0 (m)	Ice Flow Rate Q_i (L/s)	Grain Size of Sand Bed d_{50} (mm)	With or Without Local Scour	Bridge Pier Shape
A1	0.17	0.25	0.0205	0.713	With	cylindrical
A2	0.15	0.25	0.0205	0.713	With	cylindrical
A3	0.17	0.20	0.0205	0.713	With	cylindrical
A4	0.15	0.20	0.0205	0.713	With	cylindrical
A5	0.16	0.20	0.0205	0.713	With	round end-shaped
A6	0.18	0.25	0.0205	0.713	With	round end-shaped
A7	0.17	0.20	0.0205	0.713	With	round end-shaped
A8	0.15	0.15	0.0205	0.713	With	round end-shaped
A9	0.17	0.25	0.0205	0.713	With	rectangle
A10	0.15	0.25	0.0205	0.713	With	rectangle
A11	0.15	0.20	0.0205	0.713	With	rectangle
A12	0.15	0.15	0.0205	0.713	With	rectangle
A13	0.17	0.25	0.0205	0.490	With	cylindrical
A14	0.15	0.20	0.0205	0.490	With	cylindrical
A15	0.17	0.25	0.0205	1.040	With	cylindrical
A16	0.15	0.20	0.0205	1.040	With	cylindrical
B1	0.17	0.25	0.0205	-	Without	cylindrical
B2	0.15	0.25	0.0205	-	Without	cylindrical
B3	0.17	0.20	0.0205	-	Without	cylindrical
B4	0.15	0.20	0.0205	-	Without	cylindrical

Pier shape factor K_ζ refers to a table for calculating the pier shape factor and pier width in the appendix of the Code for Hydrological Specifications for Survey and Design of Highway Engineering (JTG C30-2015). The pier shape factors of different pier types involved in this study are shown in Table 2.

Table 2. The pier shape factors corresponding to different pier shape.

Pier Shape	Pier Shape Factor K_ζ (The Oblique Angle with Water Flow is 0°)
cylindrical	1
rectangle	1.2
round end-shaped	0.98

3. Interaction of Local Scour and Ice Jam Evolution

3.1. Impact of Local Scour on Ice Jam Evolution

Figure 5 shows the vertical distribution of flow velocity at the cross section where the bridge pier is located, with and without a scour in the vicinity of the pier. One can see in Figure 5 that the difference in velocity distribution before and after an initial ice jam reaches the cross section (CS-16) where the pier is located (as shown in Figure 6b). It should be noted here that when an initial ice jam does not reach CS-16, it means that the head of an initial ice jam is located between CS-17 and CS-16 or downstream of CS-16 (as shown in Figure 6a), while the ice jam that passes CS-16 corresponds to the fact that the head of an initial ice jam is located upstream of CS-15 (as shown in Figure 6c).

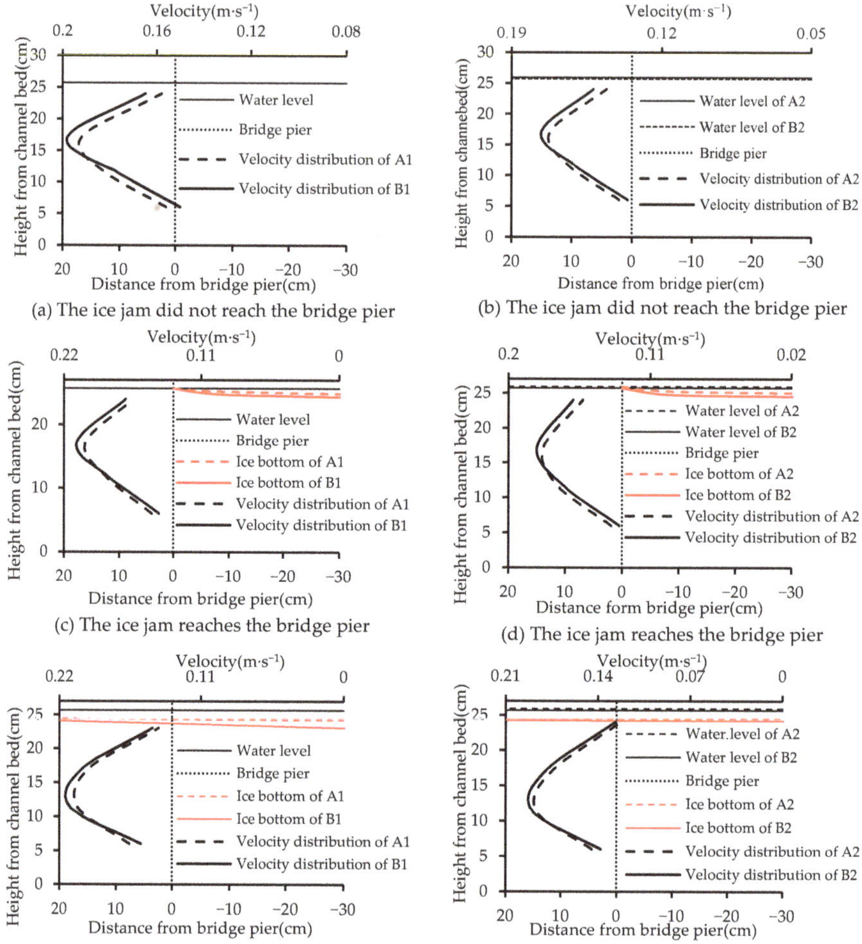

(a) The ice jam did not reach the bridge pier
(b) The ice jam did not reach the bridge pier
(c) The ice jam reaches the bridge pier
(d) The ice jam reaches the bridge pier
(e) The ice jam passes through the bridge pier
(f) The ice jam passes through the bridge pier

Figure 5. Velocity distributions with and without a pier scour for 3 cases: before the head of an initial ice jam reaching the pier (**a**,**b**); the head of an initial ice jam reaching the pier (**c**,**d**); after the head of an initial ice jam passing the pier (**e**,**f**).

(a) The ice jam did not reach the bridge pier
(b) The ice jam reaches the bridge pier
(c) The ice jam passes through reach the bridge pier

Figure 6. Pier scour for 3 cases: before the head of an initial ice jam reaching the pier (**a**); the head of an initial ice jam reaching the pier (**b**); after the head of an initial ice jam passing the pier (**c**).

As shown in Figure 5, before the head of an initial ice jam reaches and passes the pier, the flow velocity near the water surface or at the bottom of the initial ice jam at CS-16 is smaller. Also, the flow velocity near the channel bed is relative higher with the presence of a pier scour, compared to that without a local scour (fixed-bed condition), which makes the initial ice jam develop faster around the pier. When the head of an initial ice jam passes the pier (CS-16), the development of the initial ice jam is mainly influenced by the flow condition (velocity and depth) and the scour conditions (with or without a scour). As seen from Figure 7, in the presence of a pier scour, water levels upstream of the pier (CS-16) are greater after the initial ice jam passes CS-16; thus, the initial ice jam develops more rapidly toward the upstream section of the channel.

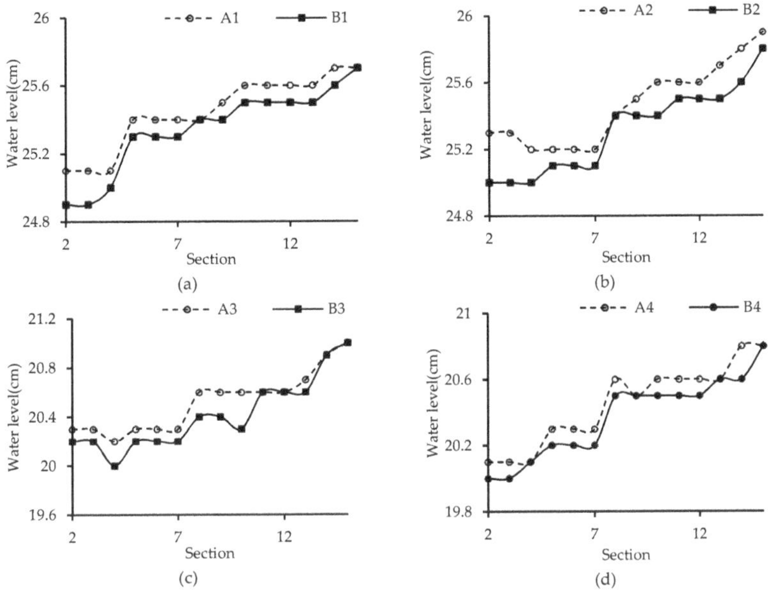

Figure 7. Comparison of water levels in the upstream cross section when the initial ice jam passes the pier (CS-16) with a local scour and those without a local scour. ((**a**): $H_0 = 0.25$ m, $V_0 = 0.17$ m/s; (**b**): $H_0 = 0.25$ m, $V_0 = 0.15$ m/s; (**c**): $H_0 = 0.20$ m, $V_0 = 0.17$ m/s; (**d**): $H_0 = 0.20$ m, $V_0 = 0.15$ m/s).

The comparison of the time required to complete the formation of an initial ice jam along the entire channel with and without a local scour is shown in Table 3. Compared with the results without a local scour process around the pier, the time required to complete the formation of an initial ice jam along the channel is shorter with the presence of a local scour around the pier. Because of the local scour process around the pier, the flow velocity near the pier is smaller, which leads to a more rapid development of the initial ice jam toward upstream. Meanwhile, with the propagation of the head of an initial ice jam toward upstream, the water levels upstream of the initial ice jam head are higher, and thus the flow velocity is lower, and it takes a shorter time to complete the formation of an initial ice jam along the channel.

Table 3. The time required to complete the development of an initial ice jam phase with and without local scour conditions.

Bridge Pier Shape	V_0 (m/s)	H_0 (m)	Q_i (L/s)		Time Required to Complete the Initial Ice Jam Phase (min)	
			With Local Scour	Without Local Scour	With Local Scour	Without Local Scour
cylindrical	0.17	0.25	0.0205	0.0205	138	195
cylindrical	0.15	0.25	0.0205	0.0205	80	107
cylindrical	0.17	0.20	0.0205	0.0205	269	300
cylindrical	0.15	0.20	0.0205	0.0205	100	118
round end-shaped	0.17	0.20	0.0205	0.0224	93	129
round end-shaped	0.16	0.20	0.0205	0.0212	87	175

The flow depths under an equilibrium ice jam with a local scour were compared to those without a local scour, as presented in Figure 8. At each designated cross section, under the same hydraulic condition and ice discharge rate, the flow depth under an equilibrium ice jam without a local scour process ("B") is smaller than that with a local scour process ("A"), and thus, the average flow velocity is larger, and the ice transport capacity is stronger. This means that the existence of a local scour should consume part of the energy of the flow under an ice jam, which is not conducive to the transportation of ice particles to the downstream section.

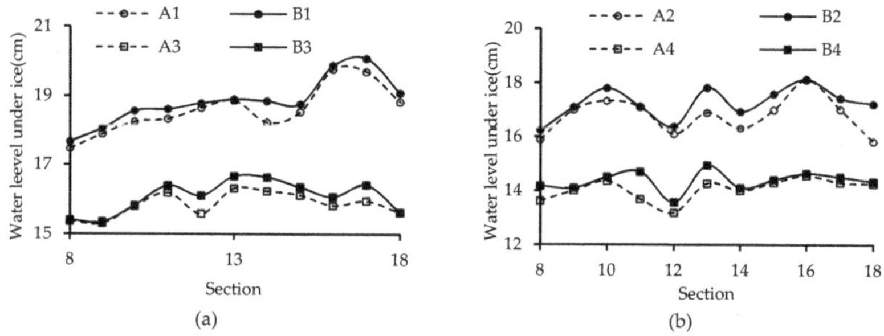

Figure 8. Comparison of water depth under an equilibrium ice jam with a local scour around the pier to those without a local scour. ((**a**): $H_0 = 0.25$ m, $V_0 = 0.17$ m/s and $H_0 = 0.20$ m, $V_0 = 0.17$ m/s; (**b**): $H_0 = 0.25$ m, $V_0 = 0.15$ m/s and $H_0 = 0.20$ m, $V_0 = 0.15$ m/s).

Figure 9 shows the comparison of water levels during the state of an equilibrium ice jam with a local scour to those without a local scour around a pier. It can be seen from Figure 9 that for an equilibrium ice jam, the water level at each cross section with a local scour is greater than that without a local scour around the pier. At the same time, it can be seen that in general, with the presence of a local scour at the pier, the increase in the water level along the channel reaches between CS-2 and CS-4 is the highest. This phenomenon results from the thickening stage of an ice jam, namely, the thicker the ice jam, the higher the upstream water level and the larger the hydraulic slope, as reported by Sui et al. [39]. The numbers in Figure 9 show that when the water level for an equilibrium ice jam without the presence of a local scour is used as the benchmark value, one can clearly see the rate of increase in the water level for an equilibrium ice jam with the presence of a local scour condition. The rate of increase in water level at the cross section where the pier is located is also larger because the presence of a local scour and the pier resulted in an increase in water level compared with that at the other cross sections. On the other hand, one can see from Figure 10 that the thickness of an equilibrium ice jam at each cross section with a local scour is greater than that without a local scour around a pier. Because of the existence of a local scour, the water flow will lose part of the energy through the bridge pier section, which reduces the ice transport capacity of the water flow near the bridge pier section, and during the evolution of the ice jam, the ice particles near the pier are transported downstream less, and the thickness of the cross-section ice jam increases. While the thickness of the ice jam near the pier thickens, the upstream water level increases, thereby reducing the water flow ice transport capacity of the upstream section, and there is more ice particle accumulation in the upstream section.

Table 4 shows the comparison of the time required for an ice jam to reach an equilibrium condition with a local scour process around the pier to that without a local scour. As shown in Table 4, it takes a longer time for the ice jam to reach an equilibrium jam stage with the presence of a local scour than that without a local scour at the pier. As mentioned above, the presence of a local scour consumes the flow energy near the pier. Because of the presence of a local scour at the pier, the water depth of the flow cross section increased. During the thickening stage of an ice jam, the ice transport capacity around the pier is weakened. Thus, there are more ice particles accumulated at the bottom of the ice jam near the pier; as a consequence, it causes an increase in the thickness of the ice jam and a reduction in the water depth under the ice jam. This will affect the transport capacity of ice particles. This local scour process and ice jam evolution interact with each other, and thus, it takes a longer time for an ice jam development to achieve an equilibrated stage with the presence of a local scour at the pier.

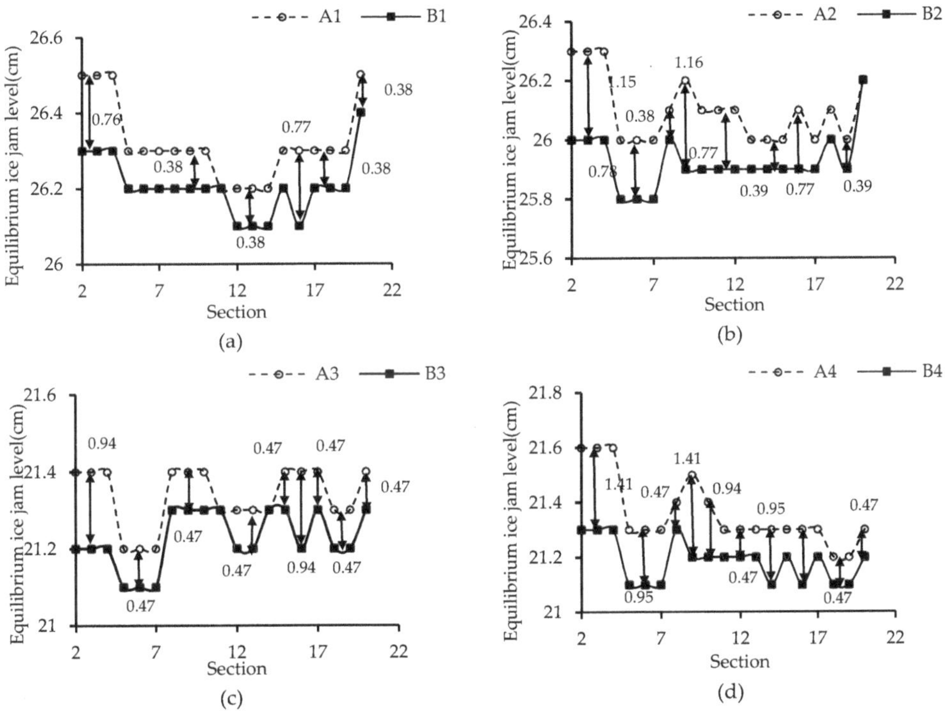

Figure 9. Comparison of water levels during an equilibrium ice jam with a local scour around the pier to those without a local scour. The numbers in Figure 9 show that when the water level for an equilibrium ice jam without the presence of a local scour is used as the benchmark value, the rate of increase in water level for an equilibrium ice jam with the presence of a local scour condition is in %. ((**a**): $H_0 = 0.25$ m, $V_0 = 0.17$ m/s; (**b**): $H_0 = 0.25$ m, $V_0 = 0.15$ m/s; (**c**): $H_0 = 0.20$ m, $V_0 = 0.17$ m/s; (**d**): $H_0 = 0.20$ m, $V_0 = 0.15$ m/s).

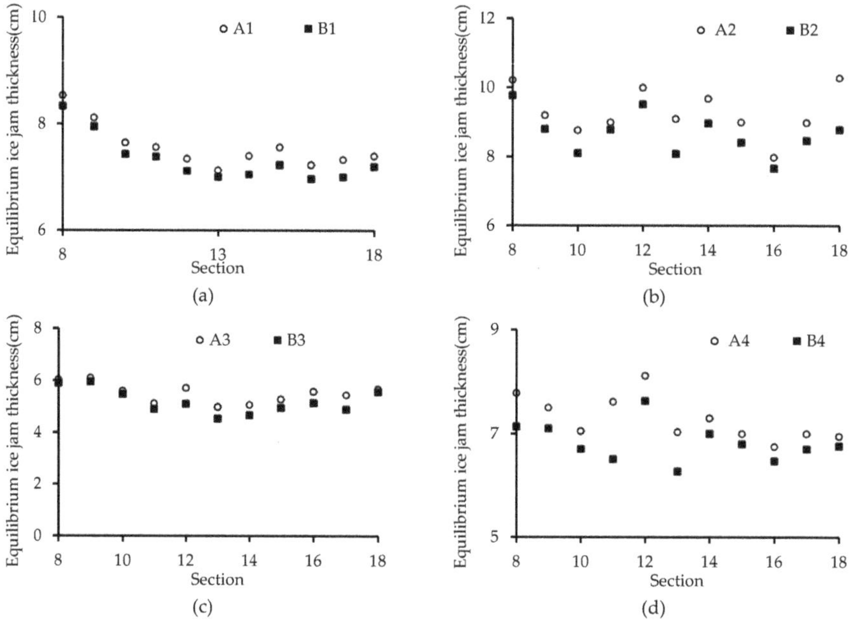

Figure 10. Comparison of the thickness of an equilibrium ice jam with the presence of a local scour at the pier to that without a local scour. ((**a**): $H_0 = 0.25$ m, $V_0 = 0.17$ m/s; (**b**): $H_0 = 0.25$ m, $V_0 = 0.15$ m/s; (**c**): $H_0 = 0.20$ m, $V_0 = 0.17$ m/s; (**d**): $H_0 = 0.20$ m, $V_0 = 0.15$ m/s).

Table 4. Comparison of the time required for an ice jam to reach an equilibrium condition with and without a local scour around the pier.

Bridge Pier Shape	V_0 (m/s)	H_0 (m)	Q_i (L/s)		Time Required to Complete the Initial Ice Jam Phase (min)	
			With Local Scour	Without Local Scour	With Local Scour	Without Local Scour
cylindrical	0.17	0.25	0.0205	0.0205	8.5	7.0
cylindrical	0.15	0.25	0.0205	0.0205	9.0	8.0
cylindrical	0.17	0.20	0.0205	0.0205	8.0	7.0
cylindrical	0.15	0.20	0.0205	0.0205	8.5	7.5
round end-shaped	0.17	0.20	0.0205	0.0224	7.5	5.5
round end-shaped	0.16	0.20	0.0205	0.0212	8.5	6.5

As seen in Figure 11, for the same ice discharge rate (Q_i), the ratio of jam thickness (T_i) to flow depth under an ice jam (h) T_i/h decreases with the increase in the flow Froude number F_r regardless of whether a local scour appears around the pier. An increase in the F_r means that the ice transport capacity of the flow under the ice jam increases, and more ice particles are transported downstream during the evolution of the ice jam (as shown in Figure 12). Thus, the thickness of the ice jam becomes relatively small (as shown in Figure 13). It was also found that with the same F_r, the T_i/h is larger when there is a local scour around the pier. Because of the presence of a local scour, the flow velocity at the cross section where the pier is located decreases. Therefore, the transport capacity of the ice decreases; the ice jam evolution process slows down, and fewer ice particles are delivered to the downstream of the pier. Thus, the thickness of the ice jam increases, resulting in a greater value of T_i/h in the equilibrium state. During the thickening process of an ice jam near the pier, the upstream water level increases. Thus, the ice transfer capacity in the upstream section is also reduced. More ice particles accumulate in the upstream section. The flow velocity under the ice jam increases and causes the local scour upstream of the pier. Thus, the T_i/h value in the upstream cross section is greater (as shown in Figure 14). In Figure 12, F_r is the flow Froude number of the initial flow at the cross section where the pier is located.

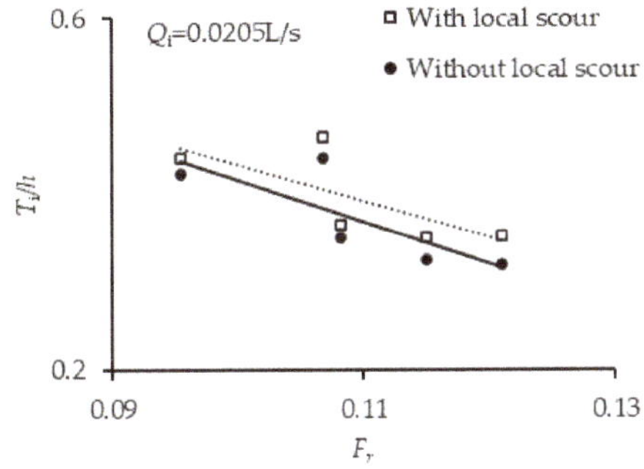

Figure 11. Relationship between Ti/h and F_r with and without a local scour.

Figure 12. Transportation of ice particles.

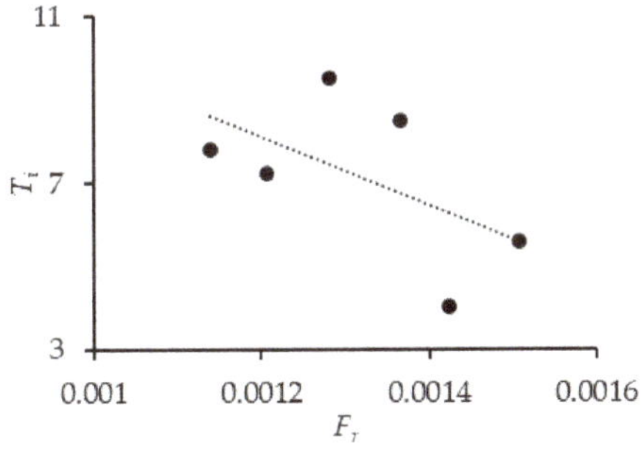

Figure 13. Relationship between Ti/h and F_r with the presence of a local scour.

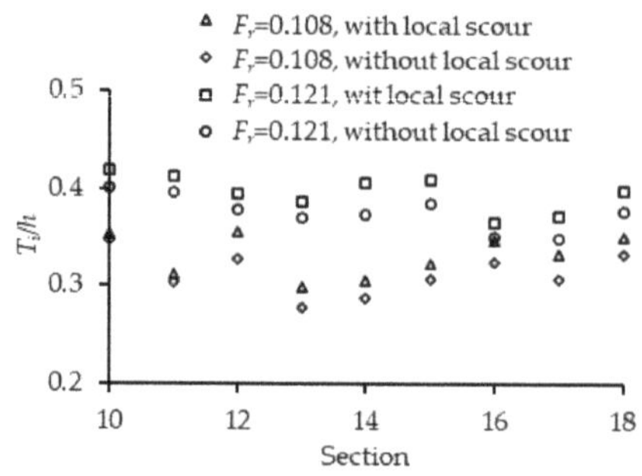

Figure 14. The ratio Ti/h at different cross sections with and without a local scour with the same F_r.

3.2. *Effect of Ice Jam Evolution on Local Scour*

The results show that compared to the formation of a scour hole under an open flow condition, the formation of a scour hole under an ice-jammed flow condition can be classified into the following two different categories based on the thickening process of the ice jam:

1. The bed shear stress changes dramatically after an ice cover forms on the water surface compared to that of open flow conditions. Thus, in the presence of an ice cover, the turbulence intensities in a scour hole increase, and the sediment transport will be affected significantly. As pointed out by Wu et al. [29–31], the rougher the ice cover, the closer the maximum velocity to the sand bed and thus, the greater the Reynolds shear stress inside the scour hole. As a consequence, the scour hole at a pier becomes deeper. When the development of an ice jam is dominated by a mechanical thickening process (an ice cover is unable to withstand the external forces acting on it and is broken up and extruded to form an ice jam) [40], the development process of a scour hole is similar to that under an open flow condition. In such a case, as the ice jam develops upstream, the thickness of the ice jam increases. When the initial ice jam reaches the pier, the thickness of the initial ice jam grows rapidly, and the flow depth decreases. Thus, the velocity under the initial ice jam at this section increases and results in an intensified scouring process around the pier with a rapid increase in the scour depth. The scouring process at the pier gradually decreases since the increase in the jam thickness gradually slows down and gradually approaches the equilibrium ice jam phase, as shown in Figure 15 (In the figure, L_i is the length of a scour hole, and h_i is the depth of a scour hole at the pier, respectively).
2. When the formation of an initial ice jam is dominated by a hydraulic thickening process (ice floes and frazil ice particles are entrained by flowing water and submerged at the front of the ice cover and accumulate under the ice cover) [40], with fewer ice particles entrained and accumulating under an ice cover/initial ice jam, the thickness of the initial ice jam is much smaller and grows more slowly compared to that dominated by a mechanical thickening process. Thus, the water depth under the ice jam changes less. The depth of the scour hole around a pier develops similarly to that under an ice-covered flow condition, and the depth of the scour hole around a pier increases slowly. During the thickening process of the ice jam (dominated by a hydraulic thickening process) along the channel, the bottom of the ice jam appears as a wave-shaped accumulation. It is noticed that when the crest of such a wave migrates from upstream to the cross section of the pier, the flow depth under the ice jam decreases, and the flow velocity increases. Under such a circumstance (migration of a wave of ice accumulation), the scouring process at the pier becomes intense. Gradually, the development of the ice jam approaches an equilibrium condition, and the scouring process around the pier slows down, as shown in Figure 16.

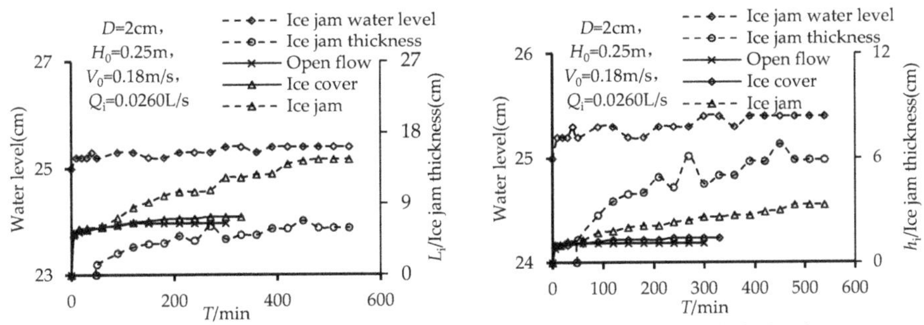

(a) Length of scour hole development (b) Depth of scour hole development

Figure 15. Comparison of the development of a scour hole around a pier under open flow, ice cover and ice jam conditions in Case 1.

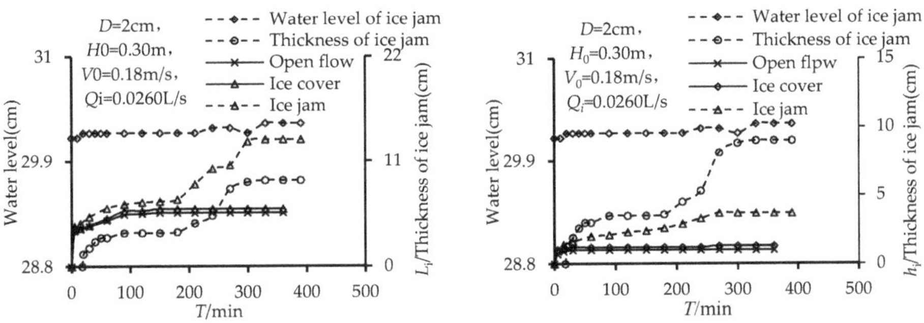

(a) Length of scour hole development (b) Depth of scour hole development

Figure 16. Comparison of the development of a scour hole around a pier under open flow, ice cover and ice jam conditions in Case 2.

4. Empirical Relationship between Scour Depth and Ice Jam Thickness

Wang et al. (2021) performed experiments to study the local scour at a pier under an ice-jammed flow condition [3]. In their study, the relationship between the equilibrium thickness of ice jams (T_i) and the maximum scour depth of scour holes at the pier (h_s) was derived as follows,

$$\frac{T_i}{H} = 1.751\left(\frac{h_s}{H}\right) - 0.016 \quad \text{with } R^2 = 0.491 \tag{1}$$

The dependence of the ratio of the maximum scour depth of scour holes to the flow depth (H) on the flow Froude number ($F_r = V/(gH)^{0.5}$) can be expressed as the following:

$$\frac{h_s}{H} = 2.014\frac{V}{\sqrt{gH}} - 0.146 \quad \text{with } R^2 = 0.951 \tag{2}$$

where V = average flow velocity and g = gravitational acceleration.

In the above study [3], the impacts of both pier size and particle size of bed material on the maximum depth of scour holes at piers under an ice-jammed flow condition were not explored. The effect of the ice discharge rate on the maximum depth of scour holes at piers under an ice-jammed flow condition was not investigated. In addition, the shape of the piers was not considered.

Namaee and Sui [33–37] carried out experiments in a large outdoor flume to study local scour around two side-by-side piers under ice-covered flow conditions by using three nonuniform sands. Considering factors influencing the depth of scour holes around the side-by-side piers, such as flow Froude number, ice cover bottom roughness, bed roughness, pier diameter, and pier spacing, the following equation for describing the maximum scour depth of scour holes was obtained:

$$\frac{y_{max}}{y_0} = 5.96\left(\frac{D_{50}}{y_0}\right)^{-0.070}\left(\frac{G}{D}\right)^{-0.256}\left(\frac{n_i}{n_b}\right)^{0.546}(F_r)^{1.677} \quad \text{with } R^2 = 0.900 \tag{3}$$

where y_{max} = maximum scour depth, D_{50} = median particle diameter, y_0 = approach flow depth, G = bridge spacing distance, D = pier diameter, n_i = ice cover roughness, and n_b = channel bed roughness.

In the above study [30], two types of model ice covers with different roughness coefficients were used. The evolution of an ice jam on the local scour process around piers was not studied.

Overall, the maximum depth of scour holes (h_s) in the vicinity of the pier under an ice jam depends on the flow intensity, pier size (D), particle size of the sand bed (d_b), ice particle size (d_i), thickness of an ice jam (T_i), ice discharge rate (Q_i), roughness of the channel bed (n_b), and ice cover (n_i), respectively. The maximum depth of a scour hole (h_s) can be expressed as the following relation:

$$h_S = f(T_i, H, V, Q_i, D, g, d_i, d_b, n_i, n_b) \tag{4}$$

From the dimensional analysis, the relative maximum depth of a scour hole (h_s/H) can be presented as follows,

$$\frac{h_S}{H_0} = f\left(\frac{T_i}{H_0}, \frac{V_0}{\sqrt{gH}}, \frac{Q_i}{Q_w}, \frac{D}{H_0}, \frac{d_i}{d_b}, \frac{n_i}{n_b}\right) \tag{5}$$

The dimensionless variable Froude number of the approaching flow is very important. In the current study, only one sand was used for bed material. For the model ice particle, only one size of model ice was used. Thus, only the dominant parameters affecting the maximum scouring depth around a pier under an ice jam were considered for developing an empirical relationship to describe the dependency of the maximum scour depth on the ice jam thickness, flow Froude number, and ice discharge rate. The ratio n_i/n_b was not considered in this study, assuming both n_i and n_b did not change much. In the present study, the pier coefficient was also considered since different pier shapes were used. Based on data collected from experiments, a regression analysis of the relative maximum scour depth (h_S/H) against other dimensionless variables including the relative ice jam thickness (T_i/H), flow Froude number (F_r), grain size of bed material (d_{50}/D), pier shape factor(K_ζ), and ice discharge rate (Q_i/Q) was performed. The following equation for calculating the maximum depth of a scour hole at a pier under an ice jam was derived:

$$\frac{h_s}{H_0} = 10.928\left(\frac{T_i}{H_0}\right)^{0.247}\left(\frac{V_0}{\sqrt{gH_0}}\right)^{1.483}\left(\frac{Q_i}{Q_w}\right)^{0.177}\left(\frac{d_{50}}{D}\right)^{-0.416}K_\zeta^{1.288} \quad \text{with } R^2 = 0.771 \tag{6}$$

where g = gravitational acceleration, Q = flow discharge, H = approaching flow depth, V = average approaching velocity, d_{50} = Grain size of sand bed, D = pier size, and K_ζ = pier shape factor.

Equation (3) implies that the relative maximum scour depth (h_S/H) of scour holes around the pier primarily depends on the relative ice jam thickness (T_i/H), Froude number ($Fr = V/(gH)^{0.5}$) of the approaching flow, and ice discharge rate (Q_i/Q). One can see from Equation (3) that the thicker the ice jam accumulation, the deeper the scour hole around the pier. The depth of a scour hole also increases with the flow Froude number. Interestingly, a larger ice discharge rate (Q_i/Q) also results in a deeper scour hole. This is because a larger ice discharge rate (Q_i/Q) leads to a thicker ice jam. On the other hand, the finer the bed material (d_{50}/D), the larger and the deeper the scour hole around the pier. For a bridge pier whose cross section has a rectangular shape, the pier shape factor is more than that of a cylindrical shape. Thus, the maximum scour depth around the rectangular-shaped pier is more than that around a cylindrical pier. As shown in Figure 17, the results of the calculation using Equation (3) agree well with those of the laboratory measurements. The h_s' in Figure 17 is the calculated maximum depth of a scour hole.

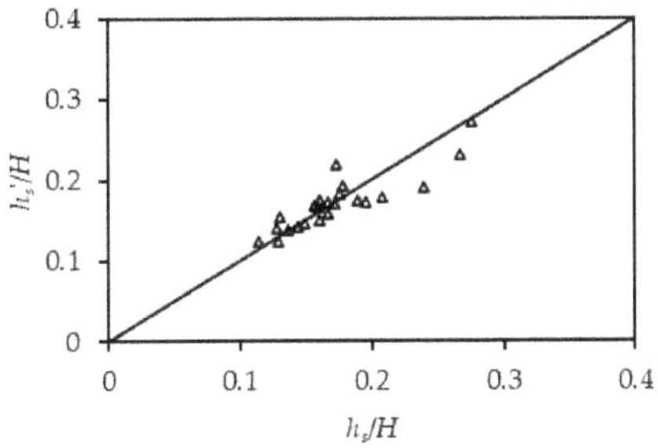

Figure 17. Comparison of the ratio of calculated maximum scour depths to H to those of the measurements.

5. Conclusions

In the present study, experiments in a flume were carried out to study the influence of the evolution of ice jams on the local scour around bridge piers. The following results were obtained:

1. During the formation of the initial ice jam around the piers with the local scour process, compared to the absence of the local scour around the pier, the depth of the water under the initial ice jam is greater with the local scour, implying that the flow under the initial ice jam has less capacity for transporting ice particles. Thus, during the thickening stage of an ice jam, more ice particles accumulate under the initial ice jam, and the thickness of the equilibrium ice jam is greater.
2. Compared to the results without a local scour, with the presence of a local scour around the pier, it needs more time for an ice jam to achieve an equilibrium condition; the upstream water level and jam thickness are clearly higher, and the rate of increase in the water level of equilibrium ice jams in upstream and pier sections is larger. In addition, the flow depth under an equilibrium ice jam with the scour case is smaller than that without a pier scour, implying that the presence of a local scour consumes part of the flow energy and is not conducive to transport ice particles.
3. Compared to the development of scour holes without the presence of an ice cover, the formation of scour holes under an ice jam can be classified into the following two different categories based on the thickening process of the ice jam: (a) When the development of an ice jam is dominated by a mechanical thickening process, as the ice jam develops upstream, the thickness of an ice jam increases. When the initial ice jam reaches the pier, the flow depth under the ice jam decreases. The development of a scour hole is accelerated. Experiments showed that the development of a scour hole under an ice-jammed flow condition is faster before an ice jam reaches an equilibrium state. (b) When the development of an initial ice jam is dominated by a hydraulic thickening process, during the thickening stage of an ice jam, when the ice wave crest reaches the pier, the thickness of an ice jam increases rapidly, and thus, the flow depth under the ice jam decreases and leads to a rapid development of the scour hole at the pier.
4. Regardless of whether a local scour appears or not, the T_i/h is negatively correlated with F_r. An increase in F_r results in an increase in the ice transport capacity. Therefore, more ice particles are transported downstream, and the ice jam thickness is relatively small. However, with the same flow Froude number F_r, the T_i/h value is larger in the presence of a local scour. The ice jam thickness near the piers increases, which leads to an upstream water level increase, and the ice transport capacity is weakened. Thus, the ice jam thickness in the upstream section of the pier increases

5. An equation for calculating the maximum depth of a scour hole around the pier under an ice-jammed flow condition is obtained. The results show that with the increase in the flow Froude number (F_r), ice flow (Q_i), and pier shape coefficient (K_ζ), the maximum depth of scour holes around the pier increases. The maximum depth of a scour hole under an ice-jammed flow condition increases with the decrease in the particle size of the sand bed. The calculation results using the proposed equation agree well with those of the measurements in the laboratory.
6. This paper mainly focuses on the interaction between local scour at a single pier and ice jam evolution under the flow condition whereby the leading edge of an initial ice jam can reach the upstream of the piers. The mechanical impact of ice particles on the pier and the interaction between piers (pier superstructure) and ice jams under extreme conditions were not investigated. The influence of different angles between the direction of the flow and the bridge pier on local scour at the pier was not assessed. In the event that the leading edge of an ice jam cannot reach the pier section, the local scour around the pier should be further studied.

Author Contributions: H.H.: Laboratory works, data curation, formal analysis, methodology, and writing—original draft preparation; J.W.: Conceptualization, laboratory supervision, methodology, funding acquisition, writing—original draft preparation; T.C.: Laboratory works, data curation, investigation; Z.H.: Laboratory works, data curation, investigation; J.S.: Conceptualization, formal analysis, methodology, writing—review and editing. All authors have read and agreed to the published version of the manuscript.

Funding: This research was supported by the National Natural Science Foundation of China (grant nos. 51879065). The authors are grateful for the financial support.

Institutional Review Board Statement: Not applicable.

Informed Consent Statement: Not applicable.

Data Availability Statement: The data are available in the case that it is required.

Conflicts of Interest: The authors declare no conflict of interest.

References

1. Sui, J.; Wang, J.; He, Y.; Krol, F. Velocity profiles and incipient motion of frazil particles under ice cover. *Int. J. Sediment Res.* **2010**, *25*, 39–51. [CrossRef]
2. Wang, J.; Wang, T.; Li, S.; Chen, P. Impacts of bridge pier on ice jam initiation and ice thickness in a curved channel—An experimental study. *J. Hydraul. Eng.* **2017**, *48*, 588–593.
3. Wang, J.; Hou, Z.; Sun, H.; Fang, B.; Sui, J.; Karney, B. Local scour around a bridge pier under ice-jammed flow condition—An experimental study. *J. Hydrol. Hydromech.* **2021**, *69*, 275–287. [CrossRef]
4. Wang, J.; Hua, J.; Chen, P.; Sui, J.; Wu, P.; Whitcombe, T. Initiation of ice jam in front of bridge piers—An experimental study. *J. Hydrodyn.* **2019**, *31*, 117–123. [CrossRef]
5. Wang, J.; Chen, P.; Qang, Q. lmpact of bridge piers on ice jam stage variation: An experimental study. *Adv. Water Sci.* **2015**, *26*, 867–873.
6. Urroz, G.E.; Schaefer, J.; Ettema, R. Bridge-pier location and ice conveyance in curved channels. *J. Cold Reg. Eng.* **1994**, *8*, 66–72. [CrossRef]
7. Wang, J.; Huang, N.; Wu, Y.; Cheng, T. Impacts of bridge-pier types on balancing ice jams: An experimental study. *Adv. Water Sci.* **2018**, *29*, 568–573.
8. Wang, J.; Hua, J.; Sui, J.; Wu, P.; Liu, T.; Chen, P. The impact of bridge pier on ice jam evolution an experimental study. *J. Hydrol. Hydromech.* **2016**, *64*, 75–82. [CrossRef]
9. Cheng, T.; Wang, J.; Chen, P.; Sui, J. Simulation of ice accumulation around bridge piers during river breakup periods using a discrete element model. *J. Hydrodyn.* **2022**, *34*, 94–105. [CrossRef]
10. Yang, W.C.; Shekhar, K.; Arduino, P.; Mackenzie-Helnweinb, P.; Millerb, G. Modeling tsunami induced debris impacts on bridge structures using the material point method. *Procedia Eng.* **2017**, *175*, 175–181. [CrossRef]
11. Istrati, D.; Hasanpour, A.; Buckle, I. Numerical Investigation of Tsunami-Borne Debris Damming Loads on a Coastal Bridge. In Proceedings of the 17 World Conference on Earthquake Engineering, Sendai, Japan, 27 September–2 October 2020; p. 27.
12. Salciarini, D.; Tamagnini, C.; Conversini, P. Discrete element modeling of debris-avalanche impact on earthfill barriers. *Phys. Chem. Earth Parts A/B/C* **2010**, *35*, 172–181. [CrossRef]

13. Hasanpour, A.; Istrati, D.; Buckle, I. Coupled SPH–FEM Modeling of Tsunami-Borne Large Debris Flow and Impact on Coastal Structures. *J. Mar. Sci. Eng.* **2021**, *9*, 1068. [CrossRef]
14. Parola, A.C.; Apelt, C.J.; Jempson, M.A. *Debris Forces on Highway Bridges*; No. 445; Transportation Research Board: Washington, DC, USA, 2000.
15. Malavasi, S.; Riva, M.; Vanali, M.; Larcan, E. Hydrodynamic forces on a submerged bridge. *WIT Trans. Built Environ.* **2001**, *56*. [CrossRef]
16. Pandey, M.; Oliveto, G.; Pu, J.H.; Sharma, P.K.; Ojha, C.S.P. Pier scour prediction in non-uniform gravel beds. *Water* **2020**, *12*, 1696. [CrossRef]
17. Memar, S.; Zounemat-Kermani, M.; Beheshti, A.; Rahimpour, M.; Cesare, G.D.; Schleiss, A.J. Influence of collars on reduction in scour depth at two piers in a tandem configuration. *Acta Geophys.* **2020**, *68*, 229–242. [CrossRef]
18. Zhao, J.; Zhang, Y.; Li, M. Experimental study of piers scour in complex boundary condition. *J. Sediment Res.* **2013**, *1*, 14–19.
19. Nyantekyi-Kwakye, B.; Essel, E.E.; Dow, K.; Clark, S.P.; Tachie, M.F. Hydraulic and turbulent flow characteristics beneath a simulated partial ice-cover. *J. Hydraul. Res.* **2021**, *59*, 392–403. [CrossRef]
20. Beltaos, S.; Burrell, B.C. Effects of River-Ice Breakup on Sediment Transport and Implications to Stream Environments: A Review. *Water* **2021**, *13*, 2541. [CrossRef]
21. Biron, P.M.; Buffin-Bélanger, T.; Martel, N. Three-dimensional turbulent structures at a medium-sized confluence with and without an ice cover. *Earth Surf. Process. Landf.* **2019**, *44*, 3042–3056. [CrossRef]
22. Fu, H.; Guo, X.; Kashani, A.H. Experimental study of real ice accumulation on channel hydraulics upstream of inverted siphons. *Cold Reg. Sci. Technol.* **2020**, *176*, 103087. [CrossRef]
23. Batuca, D.; Dargahi, B. Some experimental results on local scour around cylindrical piers for open and covered flow. In *Third International Symposium on River Sedimentation*; University of Mississippi: University, MS, USA, 1986; pp. 1095–1104.
24. Ackermann, N.L.; Shen, H.; Olsson, P. Local Scour Around Circular Piers Under Ice Covers. In Proceedings of the Ice in the Environment: 16th IAHR International Symposium on Ice, Dunedin, New Zealand, 2–6 December 2002; pp. 149–155.
25. Hains, D.; Zabilansky, L. *Laboratory Test of Scour under Ice: Data and Preliminary Results*; U.S. Army Engineer Research and Development Center, Cold Regions Research and Engineering Laboratory: Hanover, NH, USA, 2004; p. 182.
26. Hains, D.; Zabilansky, L. The Effects of river ice on scour and sediment transport. In Proceedings of the CRIPE 13th Workshop on the Hydraulics of Ice Covered Rivers, Hanover, NH, USA, 15–16 September 2005; pp. 102–107.
27. Munteanu, A. *Scouring around a Cylindrical Bridge Pier under Partially Ice-Covered Flow Condition*; University of Ottawa: Ottawa, ON, Canada, 2004.
28. Valela, C.; Sirianni, D.A.; Nistor, I.; Rennie, C.D.; Almansour, H. Bridge pier scour under ice cover. *Water* **2021**, *13*, 536. [CrossRef]
29. Wu, P.; Balachandar, R.; Sui, J. Local Scour around Bridge Piers under Ice-Covered Conditions. *J. Hydraul. Eng.* **2016**, *142*, 04015038. [CrossRef]
30. Wu, P.; Hirshfield, F.; Sui, J. Further studies of incipient motion and shear stress on local scour around bridge abutment under ice cover. *Can. J. Civ. Eng.* **2014**, *41*, 892–899. [CrossRef]
31. Wu, P.; Hirshfield, F.; Sui, J. Armour Layer Analysis of Local Scour around Bridge Abutments Under Ice Cover. *River Res. Appl.* **2015**, *31*, 736–746. [CrossRef]
32. Wang, J.; Li, Z.; Cheng, T.; Sui, J. Time-dependent local scour around bridge piers under ice cover-an experimental study. *Chin. J. Hydraul. Eng.* **2021**, *52*, 1174–1182.
33. Namaee, M.R.; Sui, J. Impact of armour layer on the depth of scour hole around side-by-side bridge piers under ice-covered flow condition. *J. Hydrol. Hydromech.* **2019**, *67*, 240–251. [CrossRef]
34. Namaee, M.R.; Sui, J. Effects of ice cover on the incipient motion of bed material and shear stress around side-by-side bridge piers. *Cold Reg. Sci. Technol.* **2019**, *165*, 102811. [CrossRef]
35. Namaee, M.R.; Sui, J. Local scour around two side-by-side cylindrical bridge piers under ice-covered conditions. *Int. J. Sediment Res.* **2019**, *34*, 355–367. [CrossRef]
36. Namaee, M.R.; Sui, J. Velocity profiles and turbulence intensities around side-by-side bridge piers under ice-covered flow condition. *J. Hydrol. Hydromech.* **2020**, *68*, 70–82. [CrossRef]
37. Namaee, M.R.; Sui, J.; Wu, Y.; Linklater, N. Three-dimensional numerical simulation of local scour around circular side-by-side bridge piers with ice cover. *Can. J. Civ. Eng.* **2021**, *48*, 1335–1353. [CrossRef]
38. CCCC High-Way Consultants Co, Ltd. *General Code for Design of Highway Bridges and Culverts: JTG D60—2015[S]*; China Communications Press: Beijing, China, 2015.
39. Sui, J.; Karney, B.; Fang, D. Variation in water level under ice-jammed condition–Field investigation and experimental study. *Hydrol. Res.* **2005**, *36*, 65–84. [CrossRef]
40. Ashton, G.D. *River and Lake Ice Engineering*; Water Resources Publication: Alexandria, VA, USA, 1986.

water

MDPI

Article

Characteristics of Turbulence in the Downstream Region of a Vegetation Patch

Masoud Kazem [1], Hossein Afzalimehr [1,*] and Jueyi Sui [2]

1 Faculty of Civil Engineering, Iran University of Science and Technology, Tehran 16846-13114, Iran; masoud_kazem@cmps2.iust.ac.ir
2 School of Engineering, University of Northern British Columbia, Prince George, BC V2N 4Z9, Canada; Jueyi.Sui@unbc.ca
* Correspondence: hafzali@iust.ac.ir; Tel.: +98-913-2175524

Abstract: In presence of vegetation patches in a channel bed, different flow–morphology interactions in the river will result. The investigation of the nature and intensity of these structures is a crucial part of the research works of river engineering. In this experimental study, the characteristics of turbulence in the non-developed region downstream of a vegetation patch suffering from a gradual fade have been investigated. The changes in turbulent structure were tracked in sequential patterns by reducing the patch size. The model vegetation was selected carefully to simulate the aquatic vegetation patches in natural rivers. Velocity profile, TKE (Turbulent Kinetic Energy), turbulent power spectra and quadrant analysis have been used to investigate the behavior and intensity of the turbulent structures. The results of the velocity profile and TKE indicate that there are three different flow layers in the region downstream of the vegetation patch, including the wake layer, mixing layer and shear layer. When the vegetation patch is wide enough ($D_v/D_c > 0.5$, termed as the patch width ratio, where D_v is the width of a vegetation patch and D_c is the width of the channel), highly intermittent anisotropic turbulent events appear in the mixing layer at the depth of z/H_v = 0.7~1.1 and distance of x/H_v = 8~12 (where x is streamwise distance from the patch edge, z is vertical distance from channel bed and H_v is the height of a vegetation patch). The results of quadrant analysis show that these structures are associated with the dominance of the outward interactions (Q1). Moreover, these structures accompany large coherent Reynolds shear stresses, anomalies in streamwise velocity, increases in the standard deviation of TKE and increases in intermittent Turbulent Kinetic Energy (TKE_i). The intensity and extents of these structures fade with the decrease in the size of a vegetation patch. On the other hand, as the size of the vegetation patch decreases, von Karman vortexes appear in the wake layer and form the dominant flow structures in the downstream region of a vegetation patch.

Keywords: coherent Reynolds shear stress; intermittent turbulence; mixing layer; quadrant analysis; vegetation patches

Citation: Kazem, M.; Afzalimehr, H.; Sui, J. Characteristics of Turbulence in the Downstream Region of a Vegetation Patch. *Water* **2021**, *13*, 3468. https://doi.org/10.3390/w13233468

Academic Editor: Achim A. Beylich

Received: 18 November 2021
Accepted: 1 December 2021
Published: 6 December 2021

1. Introduction

Submerged vegetation patches are fundamental participants of aquatic ecosystems, providing important habitats for fauna, including fishes [1–3]. As the reduction in the aquatic vegetation is a global concern, researchers have conducted studies to investigate the concept of a large-scale environmental issue [4–6]. However, while most of the current hydraulic-orientated research is focused on the flow characteristics in the canopy region of fully vegetated channels, finite dimensional vegetation patches with no fully developed region downstream of vegetation patches are very common in natural rivers. These sparse vegetation patches have a substantial effect on flow structures depending on the dimensions of the vegetation patch [7–9]. The effect of vegetation patch dimensions on flow structures has significant consequences for the ecosystem through various mechanisms. For example, an intermediate coverage (10–40%) of vegetation may promote high species richness due to the conditions provided by the optimal behavior of flow [10–12]. In addition, flow structures around a vegetation patch have a significant impact on the sediment transport, and consequently on the topography of the river bed [13–15]. Overall, the interaction between flow and vegetation in natural channels has attracted a lot of attention from hydraulic researchers. To mention only a few, the most recent research works have been conducted by researchers [16–18], in which various aspects of this interaction have been investigated.

Given the importance of the research topic, this present study aims to assess the effects of different sizes of vegetation patch on the flow characteristics in the downstream region of vegetation patches. The sizes of vegetation patches range from full coverage across the entire channel to a narrow vegetation patch located in the middle of the channel.

2. Flow Structures behind Vegetation Patches

Compared to a rough bed without a vegetation patch, the presence of a vegetation patch in a channel results in changes in the flow structures and momentum transfer. Therefore, flow resistance and turbulence characteristics are expected to differ in the downstream regions of vegetation patches [19]. To assess the characteristics of turbulent flow, the following methodologies are commonly used: spectral analysis of velocity time series, determination of Reynolds stresses, and Turbulent Kinetic Energy (TKE), as well as quadrant analysis [20–22]. The analysis of the coherent Reynolds shear stress is another method that is used to determine coherent turbulent structures. However, this approach has hardly been used in hydraulic-oriented works due to the difficulties in calculating the phase velocity.

In order to make clear the terminologies used in the above-mentioned approaches, an introduction is presented in this section alongside the literature review. In the classic terminology of turbulences, for a sufficiently long duration of velocimetry (T), and for the velocity component of $u_i(t)$, the average velocity ($\overline{u}_i$) is defined as:

$$\overline{u}_i = \lim_{T\to\infty}\frac{1}{T}\int_0^T u_i(t)dt \quad (1)$$

Considering $i = 1, 2, 3$ as the x, y and z directions in the Cartesian coordinate system, $\overline{u}, \overline{v}$ and $\overline{w}$ are defined as the averages of velocities in the x, y, and z directions, respectively. Consequently, the velocity fluctuations are defined by:

$$\begin{aligned} u' &= u(t) - \overline{u} \\ v' &= v(t) - \overline{v} \\ w' &= w(t) - \overline{w} \end{aligned} \quad (2)$$

Then, the Turbulent Kinetic Energy (TKE) is also calculated as:

$$\text{TKE} = 0.5\left(\overline{u'^2} + \overline{v'^2} + \overline{w'^2}\right) \quad (3)$$

In general, as was reported by researchers, a sharp increase in TKE is linked to the presence of vortexes [23–26]. In partially covered channels with emergent vegetation patches, where the flow is almost 2D, the flow structure in the downstream region of vegetation patches is dominated by the horizontal von Karman vortexes developed by instabilities in the trailing edges of the vegetation patch [27,28]. The distance between the trailing edge of the barrier and the formation of von Karman vortexes is known as L_{kv}. According to reported research based on TKE analysis, for a relatively dense vegetation patch with a diameter of D_v, in which the leakage velocity from the dense vegetation patch is negligible, $L_{kv} \approx 2.5D_v$ [27,28]. On the other hand, the 3D flow structures in the downstream region of a submerged vegetation patch are more complicated. In addition to the constant presence of a vertical recirculation zone beyond the vegetation patch, the presence and magnitude of the von Karman vortexes in the horizontal plane depends on the geometry of the vegetation patch [28–30]. For instance, the presence of von Karman vortexes at a depth of $H_v/H > 0.55\sim0.7$, where H_v is the height of the vegetation patch and H is the flow depth [28,30]. On the other hand, results indicated that the dominant structure of a wide submerged path occurs in the vertical plane, as vortexes forming a vertical recirculation zone [28,29]. The distance between the trailing edge of the barrier and the forming of the vertical vortex is known as L_{kv}. For a solid submerged barrier, there is no gap between the trailing edge of the barrier and the forming vortex, and the length of wake behind the barrier is zero ($L_{vv} = 0$) [31]. However, for a porous vegetation patch, the velocity of flow passing through the vegetation patch may delay the formation of the recirculating region ($L_{vv} \neq 0$). The reported value of L_{vv} ranges between $1H_v$ and $5H_v$ depending on the porosity and diameter of the vegetation patch [28,29]. As reported by Liu et al. (2018), the value tends to be about $1H_v$ in a relatively wide vegetation patch, and increases up to $5H_v$ in vegetation patches with a low blockage ratio [28]. These results indicate that L_{vv} and L_{kv} are affected by the vegetation patch dimensions of h and D, respectively. In addition, the bleed velocity that penetrates into the wake region can also increase these values considerably [23,31].

In addition to the above-mentioned classic variables, coherent turbulent variables have become fundamental terms of flow analysis in recent years [32–34]. According to the triple decomposition approach, the instantaneous velocity can be written as [35]:

$$u(x,t) = \overline{u}(x) + \widetilde{u}_c(x,t) + u_r(x,t) \tag{4}$$

where $u(x,t)$ is the instantaneous velocity, u_r is the deflection of velocity (known as u' in the classic approach), and $\overline{u}(x)$ is the classic average of velocity during time period T. In addition to the classic averaging method, the periodic phase average can be defined as:

$$\langle u(x,t)\rangle = \lim_{N\to\infty}\frac{1}{T_P}\sum_1^N u_i(x,t+iT_P) \tag{5}$$

where T_p is the period of occurrence of a coherent structure and is equal to $1/f_d$, where f_d is the dominant frequency of the occurrence. Generally, when a coherent event occurs in a particular region of flow, f_d will reach an obvious peak in the power density spectrum of the related time series in the same region. For a time series with a length of T, N is the number of cycles with a period of T_p, which can be calculated as $N = T/T_p$. Following these definitions, the coherent velocity deflection is:

$$\widetilde{u}_c(x,t) = \langle u(x,t)\rangle - \overline{u}(x) \tag{6}$$

Considering the same approach for other components of velocity, the coherent Reynolds shear stress can be written as:

$$\tau_c = -\rho\langle\widetilde{u}_c\widetilde{w}_c\rangle \tag{7}$$

The total coherent and non-coherent Reynolds shear stress is defined as follows:

$$\tau_r = -\rho\overline{\overline{u}_r\overline{w}_r} \tag{8}$$

Although it is very rare to implement τ_c in studies of river hydraulics with the presence of vegetation patches, a few research works have been conducted to study the variation in τ_r in river hydraulics in the presence of vegetation patches. For instance, in the downstream region of a submerged vegetation patch, there is a high gradient pattern of Reynolds shear stress in the vertical plane, which increases toward the surface of the flow [36]. This phenomenon is induced by the high-speed flow that passes above the submerged vegetation patch and causes a considerable velocity gradient when compared with the low velocity flow leaking through the vegetation patch. The region with a high velocity gradient is known as the shear layer. This region of flow has particular turbulence characteristics that will be studied in this paper.

As another method of analysis, the decomposition of bursting events (known as quadrant analysis in 2D space and octant analysis in 3D space) is also widely applied to determine the dominant turbulent events in the presence of both emerged and submerged vegetation patches [22,37–39]. The occurrence probability of event k is calculated as the normalized occurrence frequency, f_k, for a particular class of events related to different classes of events [40]:

$$f_k = \frac{n_k}{N} \text{ with } N = \sum_1^4 n_k \; k = 1,2,3,\ 4 \tag{9}$$

where n_k is the number of events belonging to class k, and N is the total number of events.

For the quadrant analysis, a "hole" region has been proposed in the majority of previous studies. In a "hole" region, the event must be filtered and not be considered. In octant analysis, this threshold can be written as:

$$|u'(t)w'(t)| > C_H\left|\overline{u'w'}\right| \tag{10}$$

For the threshold parameter C_H, the low-intensity events below a certain limit were omitted, which were scaled by the average of the velocity fluctuations. The high value of C_H implies the selection of the strongest events, but the total number of instantaneous $[u'(t)\ w'(t)]$ decreases so much that the contribution region in each quadrant becomes meaningless. For quadrant analysis, the threshold level $C_H = 1$ is suggested for use to reach a good compromise between the clear identification of the events and the preservation of a number of instantaneous events of each class [40,41]. In this paper, C_H is set as 1. The classes of events are also defined as:

- Q1—occurrence of the outward interaction when $u' > 0$ and $w' > 0$;
- Q2—occurrence of the ejection when $u' < 0$ and $w' > 0$;
- Q3—occurrence of the inward interaction when $u' < 0$ and $w' < 0$;
- Q4—occurrence of the sweep when $u' > 0$ and $w' < 0$.

To date, most of the reported research has applied quadrant analysis to study coherent flow structures within or above a vegetation patch. However, there are a few published studies in which quadrant analysis has been conducted to assess the coherent flow structures in the downstream region of a vegetation patch. Okamoto and Nezu (2013) point out that in the region immediately behind the submerged vegetation patch, there is a small margin of ejection dominancy around the top edge [7]. Similarly, based on data collected from experiments performed on dryland vegetation in a wind tunnel, Mayaud et al. (2016) revealed that there were elevated frequencies of Q2 (ejection) and Q4 (sweep) events in the immediate toe of the vegetation patch [42]. In contrast, it is reported the dominance of outward and inward interactions in the shear layer induced by the flow passing above the vegetation patch, which is a notable characteristic of the flow in the downstream region of a submerged vegetation patch in a more distant region [38]. It must be mentioned that the relation between quadrant occurrences and velocity structures is an interesting research topic that has received attention from researchers recently. Wang et al. (2019) established interconnections between the classic definition of vortex groups and quadrant occurrences, which are used in the next section of this paper [43].

In conclusion, the introduced techniques have been successfully used in the hydrodynamics analysis in channels with the presence of vegetation patches. However, the reported studies on the hydrodynamics of channels with submerged vegetation patches are limited to those with the flow either above the patches or immediately downstream of the patches. For example, the flow structures was traced up to a distance of $8H_v$ from the toe of the patch towards the downstream [36]. Liu et al. (2018) studied the flow structures to a distance of just 5~$6D_v$ from the patch toe in the downstream direction [28]. The quadrant analysis has been conducted in the field; here, the distance from the patch toe was even more limited, to ~$3D_v$ [38]. Although this distance from the patch toe is enough to address the characteristics of patch-induced turbulence in the near field, it is not enough to cover the more distant events of turbulence. For example, the investigation of the characteristics of the shear layer generated above the patch requires an extended range of velocimetry in the downstream direction. In addition, there is a need to combine the above-mentioned methods to evaluate the turbulent characteristics of flow behind the submerged vegetation patches. This requires a particular emphasis on the implementation of τ_c in the turbulence analysis, which has hardly been studied in recent works. In order to fill the above-mentioned gaps in the previous research, in the present study, a combined approach to turbulent analysis is used to investigate the turbulent characteristics of flow in the region downstream of a vegetation patch with an extended range up to $17H_v$.

3. Experimental Setup

The experiments were conducted using a glass flume at the Laboratory of Hydraulics of the Iran University of Science and Technology. The flume is 14 m long, 0.9 m wide and 0.6 m deep. The discharge was controlled by an electromagnetic flow meter installed at the entrance of the flume and was set for 31 L/s in the present study. The water level in the flume was adjusted by a tailgate located at the end of the flume and was set for a depth of 18.5 $\pm$ 0.3 cm in the present study. The distance between the flume entrance and vegetated flume zone was 6 m to ensure a fully developed flow in the region upstream of the vegetation patch.

The velocity measurements were conducted after the flow reached the steady state condition. The velocity profiles were measured using an acoustic Doppler velocimeter (ADV), placed at the centerline of each row of vegetation patch. There were 19–26 measuring points along each vertical line for velocity measurements, and the vertical distance between two adjacent measuring points was 4–10 mm. The sampling frequency and the measuring time of the ADV were 200 Hz and 120 s, respectively, resulting in 24,000 instantaneous velocities for point measurement.

To date, most previous studies have employed artificial simple rods of regular shapes to simulate natural vegetation patches. As real vegetation is flexible and irregular, this method might not represent the nature of vegetation's behavior [36]. On the other hand, the experiments showed that natural vegetation would lose its stiffness and develop a long-lasting curvature in the flow direction after couple of days. Therefore, a well-shaped synthetic plant was used to represent the natural vegetation, which was selected on the basis of a real world sample of patches in a gravel bed river.

Each model plant had three branches. Each branch had 12 leaves, and the diameter of the branch trunk was approximately 3 mm, as shown in Figure 1. The average height of the vegetation patch was 105 $\pm$ 5 mm, and the lateral and longitudinal spread widths of the leaves were approximately 9–19 and 11–22 mm, respectively. The model plants had a certain degree of flexibility and could swing in a flowing current similar to vegetation in a natural river. The vegetation patch was attached to a perforated board in a staggered arrangement. Four different layouts of vegetation patches were used to simulate both fully covered and non-fully covered channel beds. The dimensions of these vegetation patches, namely, length (L_v) $\times$ width (D_v), in this study were 120 $\times$ 90 cm, 90 $\times$ 60 cm, 60 $\times$ 45 cm and 40 $\times$ 30 cm, respectively. Table 1 summarizes the experimental runs. The material on the flume bed was a mixture of natural gravel similar to that in a natural gravel bed river (the Marbor River, Zagros Mountains region, Iran). The equivalent particle diameter, which 90% of the total particles were smaller than (d_{90}), was 18.8 mm, and d_{50} = 14 mm. The layout of the experimental device and employed materials are shown in Figure 1.

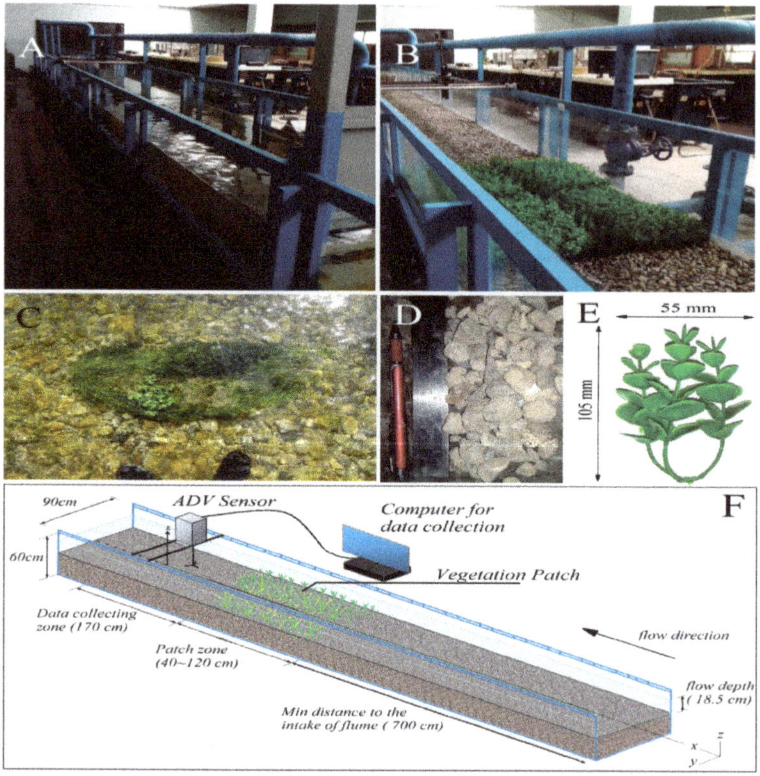

Figure 1. (**A**) The layout of the experimental device, (**B**) a sample of vegetation patches (120 × 90 cm, $D_v/D_c = 1$), (**C**) sample of vegetation patch in a natural gravel bed river (Marbor River, Iran), (**D**) a sample of bed materials, (**E**) a single synthetic plant used to simulate patch; (**F**) a sketch of the experimental setup (not to scale).

Table 1. Experimental parameters for the cases discussed in this paper.

Case	Q (Discharge, L/s)	n/m^2 (Number of Veg. per Square Meter)	L_v (Length of Patch, cm)	D_v (Width of Patch, cm)	H_v (Height of Patch, cm)
1	31 L/s	611.1	120	90	10
2	31 L/s	611.1	90	60	10
3	31 L/s	611.1	60	45	10
4	31 L/s	611.1	40	30	10
No. Veg.	31 L/s	-	-	-	-

4. Results and Discussion

The results are presented as follows: First, both the velocity profile and the TKE have been investigated to determine the flow layers that formed behind the vegetation patch. Then, spectral analysis and coherent Reynolds shear stress analysis have been carried out to determine the nature of turbulence and coherent Reynolds shear stresses in different flow layers. Consequently, the results of the quadrant analysis of bursting events are provided to compare and evaluate the coherent occurrences in the framework of the time domain. Then, by considering the particular characteristics of turbulence in the mixing layer, the temporal characteristics of turbulence in the mixing layer are investigated. Finally, by considering the results of previous sections, the transformation of coherent structures in the downstream region of a vegetation patch is provided.

4.1. Velocity Profile and TKE

For all experimental runs, three layers of flow can be observed based on the inflection points of the velocity profiles (Figure 2). The associated regions of these layers are known as the wake zone, the mixing layer and the Log–Law shear zone. These three regions resemble the results of trough patch velocimetry reported by other researchers [44–47]. For instance, in case 1 and for a distance from the vegetation edge $x/H_v = 8$, the wake zone formed at the water depth of $z/H_v = 0{\sim}0.7$ ($z/H = 0{\sim}0.56$), the mixing layer formed at the water depth of $z/H_v = 0.7{\sim}1.1$ ($z/H = 0.56{\sim}0.62$) and the Log–Law shear zone formed at the water depth of $z/H_v > 1.1$ ($z/H > 0.62$). The extent and thickness of the mixing layer reduced as the vegetation patch decayed. In this layer, a severe deflection is detectable in the velocity profile, which resembles an adverse pressure gradient. For case 1, this anomalous phenomenon covered a distance (from the vegetation edge) between $8 < x/H_v$ and $x/H_v \geq 17$. In case 2, this phenomenon covered a distance of $3 < x/H_v < 8$, and it was only visible around $x/H_v = 12$ in case 3. However, this phenomenon was absent in case 4. The wake zone was formed in the lower region beyond the patch. For case 1, the wake zone occurred at a distance from the vegetation edge of $x/H_v = 1$, with the upper limit of this layer at $z/H = 0.45$ or $z/H_v = 0.8$, which increased up to $z/H_v = 1$ for the smallest patch (case 4). In this layer, the velocity profiles tended to adopt a Log–Law shape beyond the distance of $x/H_v = 12$ for all cases. For a channel bed partially covered by a vegetation patch, the thickness of the wake layer varied along the flow direction, and increased slightly in such a way that it reached the lower boundary of the Log–Law shear region. Consequently, the mixing layer became narrower and even disappeared in the presence of smaller patches. However, for the case of a channel bed fully covered with a vegetation patch (case 1), the thickness of the mixing layer increased considerably at a distance (from the vegetation edge) of $x/H_v = 8$, and seemed to be effectively present at a distance of $x/H_v >> 17$.

While the vertical vortexes that normally generate behind a barrier have attracted a lot of attention from researchers, the development of horizontal vortexes (the von Karman Vortex Street) needs to be investigated too. Generally, the formation of von Karman vortexes is associated with the occurrence of peaks in the TKE values [23]. Thus, for each case, the TKE values have been calculated for the relative flow depths of $z/H_v = 0.5$ and $z/H_v = 1$, and also for the near-bed region behind the vegetation patch along the centerline of the flume (Figure 3). The variation in the streamwise velocity along the flow direction is also shown in this figure. In case 1, there was only one peak in the graph of $\mathrm{TKE}/(U_0)^2$ around $x/D_v = 0.9$ and $x/H_v = 7.5{\sim}8$, which is associated with severe deflection in the velocity profile. However, no flow escaped from the sides of the vegetation patch (since the channel bed was fully covered with vegetation). This peak in $\mathrm{TKE}/(U_0)^2$ clearly indicates the presence of a vertical vortex behind the vegetation patch and near the water's surface. While a sharp increase in TKE occured after a distance of $x/H_v = 4.5{\sim}5$, the length of the wake behind the patch can be considered as $L_{vv} = 5H_v$. Importantly, the anomalies in the mixing layer appeared in the same point as was reported [28] for a length of wake of $L_{vv} = (3.5{\sim}5)H_v$ with a wide submerged vegetation patch. However, Folkard (2005) reported a slightly lower range for a wide but highly submerged vegetation patch [29]. Overall, these results are comparable to those of Liu et al. (2018) for a wide and partially vegetation-covered bed with slight submergence [28].

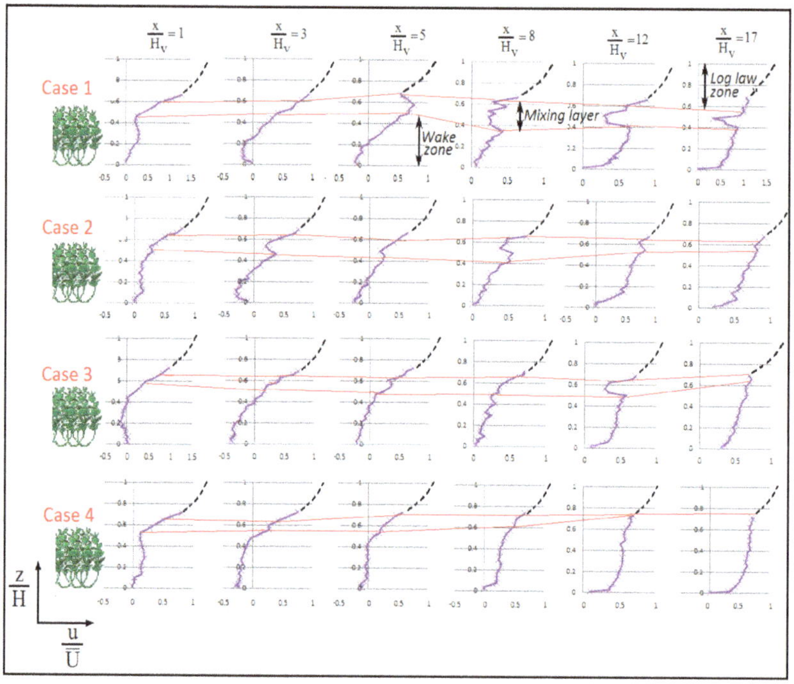

Figure 2. Streamwise velocity profiles along the centerline of the flume; x/H_v is the normalized distance from the downstream edge of the vegetation patch, and H is the depth of flow. The boundaries between zones affected by the wake and over the canopy flow are shown as red lines. The velocity of the Log–Law zone is shown as dashed and is not to scale.

For experiment cases 2, 3 and 4, the location of the minimum velocity was shifted towards the flow direction. However, since the patch width ratio of D_v/D_c was relatively high for case 2, L_{vv} was only shifted to the near-bed region. For cases 3 and 4, the shift of wake was also detectable at depths of $z/H_v = 0.5$ and $z/H_v = 1$. In these cases, the shift extent of the upper layers exceeded that of the near-bed zone. For all four cases, the minimum velocity was negative in the near-bed region, implying the occurrence of a vertical vortex. Considering the velocity values in the upper zone, it was observed that the center of the vortex (or rotation) moved upward if D_v/D_c decreased. In case 1, the center of vortex was located in the near-bed region and $z/H_v = 0.5$, while it was located between $z/H_v = 0.5$ and $z/H_v = 1$ in case 4 (smallest vegetation patch). This elevated center of rotation in case 4 pushed the larger part of flow down into the lower zones. Subsequently, the velocity pattern in the middle region became similar to that of the near-bed region (one can observe this by comparing the minimum velocity located between the near-bed region and $z/H_v = 0.5$). In addition, the larger radius of the rotation in the channel with a smaller patch width ratio of D_v/D_c caused the minimum streamwise velocity to be in the near-bed region rather than in the mixing layer. The reason for the upward movement of the rotation center in the case of a smaller patch in the channel may be the higher through-patch velocity. Regarding the larger values of L_{vv} due to the through-patch flow, this statement is consistent with the findings of other researchers [27,28,31]. In addition, for experiment cases 2, 3 and 4, the flow from the sides of the vegetation patches created the second peak in the TKE curve at the depth of $z/H_v = 1$. However, for cases 2 and 3, while the D_v/D_c values were still relatively high, the second peak in the TKE curve did not occur in the middle and near-bed region. Additionally, the location of the single peak in the TKE curve in the lower zones was shifted forward.

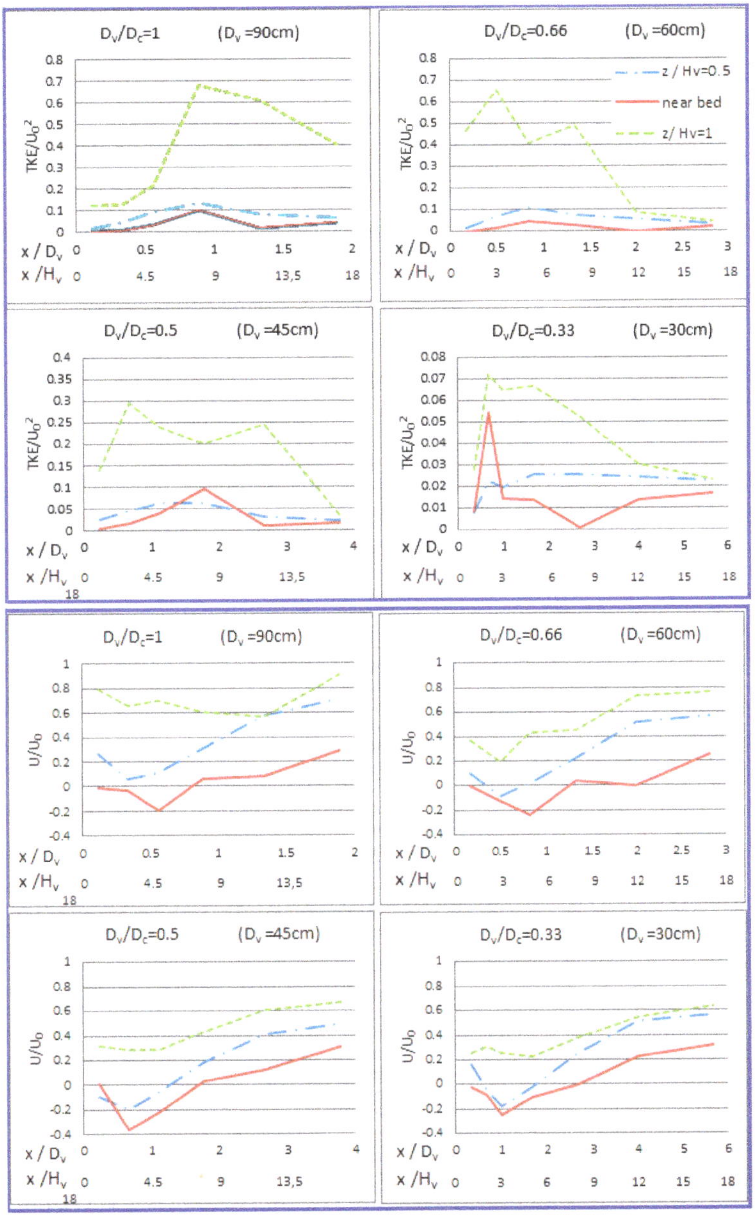

Figure 3. Normalized TKE and streamwise velocity behind 4 different patch layouts near channel bed ($z/H_v = 0$), at the middle height of patch ($z/H_v = 0.5$) and top of patch ($z/H_v = 1$).

Overall, the results of both the TKE and velocity analyses show that the three layers (wake, mixing and shear) were present in all cases of vegetation patches. However, the wake and shear layers were considerably affected by the flow passing from the sides of the vegetation patches, and as a consequence, different flow structures and associated length scales emerged, dependent on the upward movement of the center of the vertical rotation beyond the patch.

4.2. Spectral Analysis and Coherent Reynolds Shear Stress

The dominant frequency of the streamwise Reynolds shear stress ($f_{d_{u'w'}}$) was the peak of the power spectrum diagram calculated via $u'w'$ fluctuations (S_{uw}). Therefore, S_{uw} is calculated at different points for all experimental cases (Figure 4), and the results are used to calculate the coherent Reynolds shear stresses ($-\langle \widetilde{u}_c \widetilde{w}_c \rangle$). The results of the spectral analysis can be placed in the following three categories:

(1) The first category is the points where the turbulence was well matched with the von Karman isotropic–homogeneous turbulence, but no dominant frequency on S_{uw} was observable. At these points, the power spectra reached the slope of $-5/3$ in the inertial range. As there was no dominant frequency, $f_{d_{u'w'}}$ was too low and no coherent events occurred in the streamwise Reynolds shear stresses (the blank area in Figure 5 represents these points);

(2) The second category is the points where the turbulence was well matched with the von Karman isotropic–homogeneous turbulence. The dominant frequency on S_{uw} was observable in the form of an obvious peak around 0.07~0.3 Hz. At these points, the power spectra reached the slope of $-5/3$ in the inertial range. Where there was an obvious dominant frequency, $f_{d_{u'w'}}$ was used to calculate the streamwise coherent Reynolds shear stresses. Excluding points at the depth of $z/H_v \approx 1$, all other points for the coherent Reynolds shear stress belong to this group;

(3) The third category is the points where the turbulence was not compatible with the von Karman isotropic–homogeneous turbulence. The dominant frequency on S_{uw} was observable in the form of an obvious peak around 0.07~2 Hz. At these points, the power spectra reached the slope of -1 in the inertial range. According to the analysis provided by Tchen (1953), this slope is associated with the anisotropic turbulence characterized by a large vorticity and strong resonance [47]. These points were observed at a depth of $z/H_v \approx 1$ (mixing layer) for cases 1, 2 and 3, while they were absent in case 4. This finding confirms the upward movement of the rotation center in the presence of smaller patches in the channel (case 4), because the downward velocity beyond the patch was too weak to penetrate into the lower zones and the center of rotation developed in the upper zone. Such a weak vortex cannot produce the strong vertical vortexes required to alter the isotropic turbulence. A Matlab@ code was used to calculate the phase velocity and its deflection for the dominant frequency at each point. The coherent vertical momentum transfer triggered by coherent vertical vortexes continued to appear in both the near-canopy and near-bed regions in cases 1 to 3. However, near-bed coherent shear stress did not occur in case 4. It can be inferred that the flow regime in the near bed region of case 4 was dominated by the flow through the vegetation patch and the wall effect of the rough bed. The presence of the vegetation patch had no coherent effect on this region. In addition, for cases 1, 2 and 3, the majority of coherent occurrences were accompanied by a high spatial gradient of the Reynolds shear stress, which is associated with the boundary layer separation zone and large scale vortexes—a characteristic that was described by Lian (1990) [48]. However, in case 1, the occurrence of coherent Reynolds shear stresses was associated with very low Reynolds shear stresses with a low spatial gradient. Consequently, it can be referred that the coherent Reynolds shear stresses can be classified into two categories according to the origins of the coherence. The first and most prevalent category is the strong coherent Reynolds stress, which was prevalent in the mixing layers of cases 1, 2 and 3. The second category is the weak coherent Reynolds shear stress that occurred in the wake layer. Based on the results of the spectra analysis, the strong coherent shear stresses were associated with peaks of the anisotropic turbulent spectra. In contrast, the weak form of the coherent shear stress occurred in the presence of the peak isotropic turbulence.

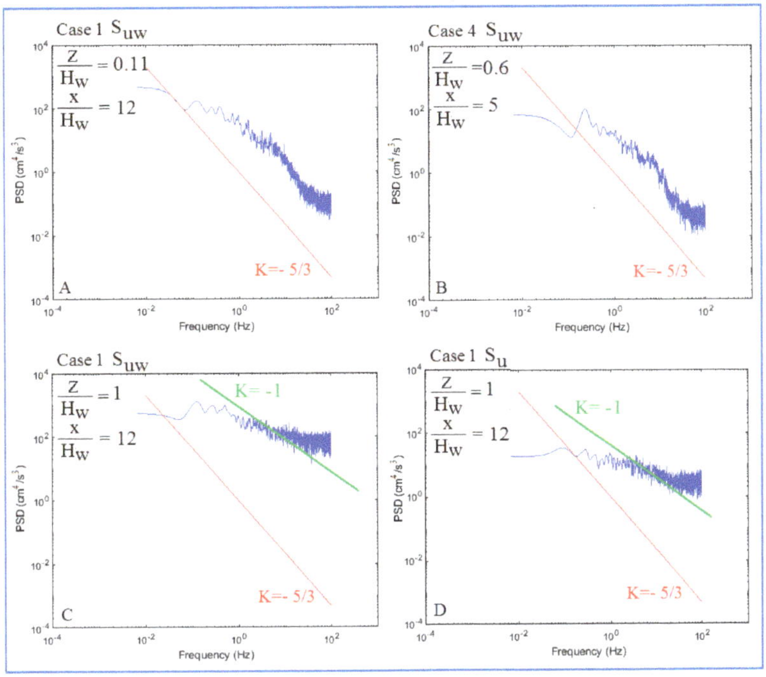

Figure 4. Samples of the power spectra behind the patch showing different classes of turbulence: (**A**) isotropic–homogenious turbulent flow without dominant frequency on S_{uw}, (**B**) isotropic–homogenious turbulent flow with dominant frequency on S_{uw}, observed below the height of the patch, (**C**) anisotropic turbulent flow with dominant frequency on S_{uw}, observed alongside the height of patch, (**D**) S_{uw} of the same anisotropic sample.

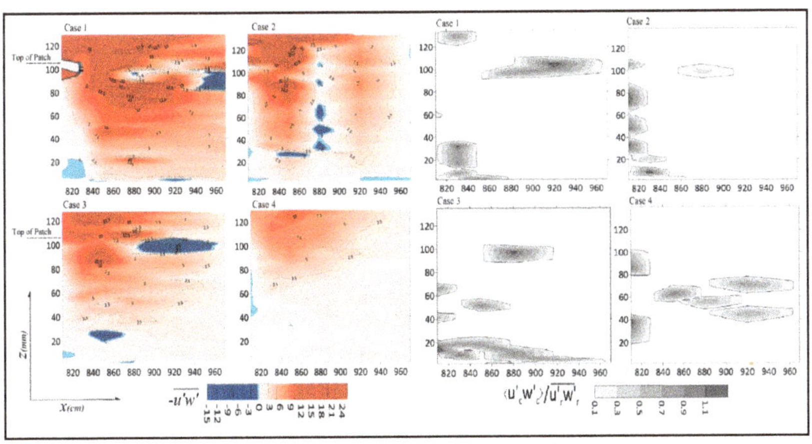

Figure 5. Values of $-\overline{u\prime w\prime}$ along the centerline of the channel beyond the vegetation patch (colored figures) beside values of τ_c/τ_r (graysclae). Note that the blank area in τ_c/τ_r is where no dominant frequency was detectable in S_{uw}. The end of the patch is located at x = 800.

On the other hand, for the patch width ratio of D_v/D_c = 0.5, a new type of coherent structure formed in the wake zone, and extended as the patch size reduced. With the patch width ratio of D_v/D_c = 0.33 (case 4), the greatest part of wake zone was covered by this structure. While, in this region, the turbulent spectra of instantaneous velocity resembled isotropic turbulence, the effect of the interaction between the mixing layer and the shear layer seems to be insignificant. On the other hand, the presence of strong flow from the patch sides could form strong von Karman vortexes in the horizontal plane. As reported by Siddique et al. (2008) and Liu et al. (2018), von Karman vortexes appeared in partially covered patches with $H_v/H > 0.55{\sim}0.7$ [28,30]. Consequently, for these regions, the frequency of vortex shedding was calculated as 0.07~0.11 Hz, which is equivalent to an average Strouhal number of about 0.25. These results are comparable with those reported by other researchers [27,28].

4.3. Quadrant Analysis of Bursting Events

The occurrence probabilities of the quadrant classes are shown in Figure 6. One can see from Figure 6 that in the presence of smaller patches, the ejection-dominated zone was shifted to a higher elevation at a shorter distance from the patch. This phenomenon was accompanied with lower Reynolds shear stress in the near-bed region. It was also associated with a marginally thicker wake layer in the presence of smaller patches, which was described in the previous section.

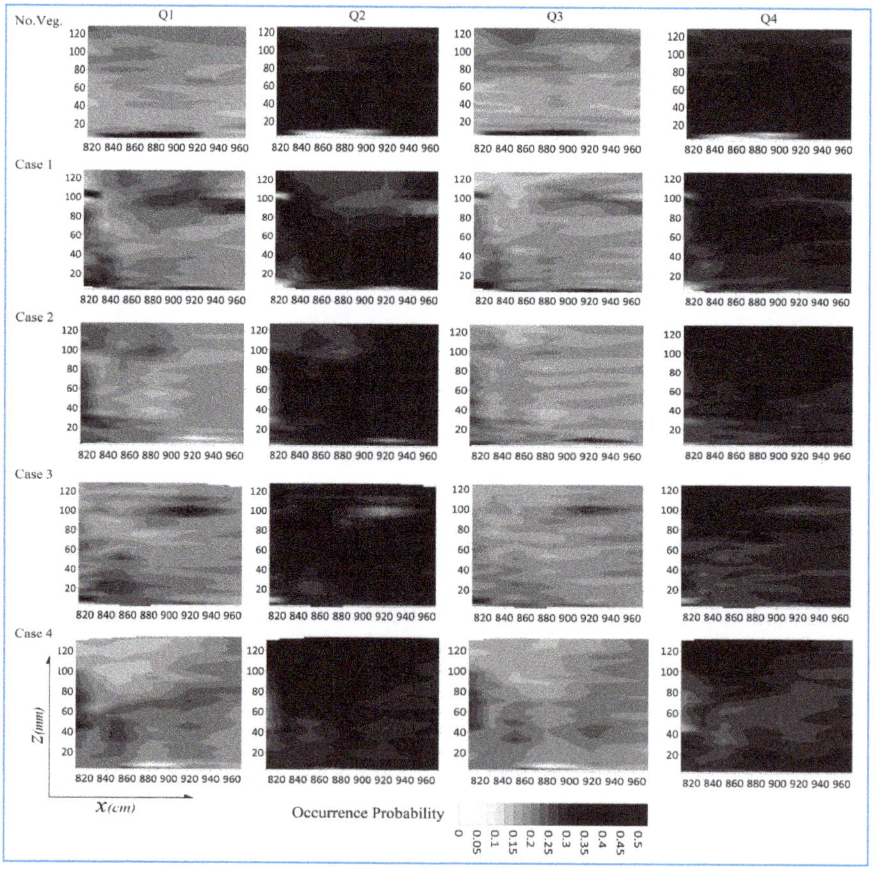

Figure 6. Probability of occurrence of quadrant classes behind the patch; the end of the patch is located at x = 800. The height of the patches is 105 ± 5 mm, and the values are shown up to level of the maximum domain of ADV capability (~140 mm).

In contrast, the sweep-dominated zone, observed above the top of the patch in case 1, faded with the decrease in the patch dimensions. As regards the Reynolds shear stresses of this region, the occurrences of the sweep were associated with the high Reynolds shear stress of the Log–Law layer in the downstream region of the patches. The same pattern was reported in the downstream region of dryland vegetation patches facing wind flow [42].

As strong coherent shear stresses were prevalent in the boundary between the mixing layer and the shear Log–Law layer, the quadrant analysis of the bursting events of this zone are outlined in Figure 7. The propinquity between the outward-dominated regions and regions with coherent Reynolds shear stresses is another notable result of the quadrant analysis. For all vegetation patches, the coherent shear stress in the mixing layer with a frequency of $f_{d_{u'w'}}$ appeared in exactly the same region in which the outward class of quadrant occurrences were dominant, where $f_{k=1} > 0.25$. This was a one-way relationship, and there were some points with $f_{k=1} > 0.25$ at which no coherent shear stress was observable. To assess this important outcome, for point x and quadrant class i, two conditional functions can be defined, as below:

$$\mathrm{A}(x,i) = \begin{cases} 1 \text{ if } f_{k=i} = \max\{f_{k=1..4}\} \\ 0 \text{ if } f_{k=i} \neq \max\{f_{k=1..4}\} \end{cases} \tag{11}$$

$$\mathrm{B}(x) = \begin{cases} 1 \text{ if } f_{d_{(u'w')}} \neq 0 \\ 0 \text{ if } f_{k=i} = 0 \end{cases} \tag{12}$$

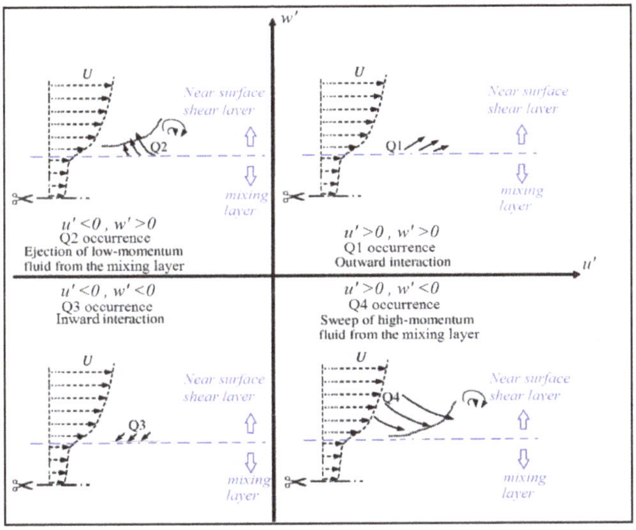

Figure 7. Conceptual illustration of the quadratic bursting events in the boundary between the mixing layer and the shear layer.

Subsequently, $\rho_{A,B}$ (the Pearson correlation coefficient between $\mathrm{A}(x,i)$ and (x)) can be used to determine which class of quadrant occurrences is well-matched with the coherent Reynolds shear stress. For a sample including 278 points for all four cases of patch setup, $\rho_{A,B}$ is 0.68 for Q1 (outward interaction). In comparison, this value is −0.45, 0.26 and −0.38 for Q2, Q3 and Q4, respectively. This result confirms that the outward interactions are the representative quadrant class of the coherent Reynolds shear stresses in the downstream region of vegetation patches, particularly for strong coherent occurrences. Figure 8 illustrates the values of f_k and $f_{du'w'}$ for the 278 points beyond the vegetation patches that were investigated in this research. According to the definition of the outward quadrant, the upward momentum flux within the mixing layer toward the shear layer must be considered (see Figure 7). This upward flux is also detectable in the velocity profiles (Figure 2), where anomalies in the mixing layer are exacerbated; the streamwise velocity was reduced in the mixing layer and increased in the bottom part of the shear layer. This phenomenon illustrates a particular behavior in the velocity time series, which will be discussed in the next section.

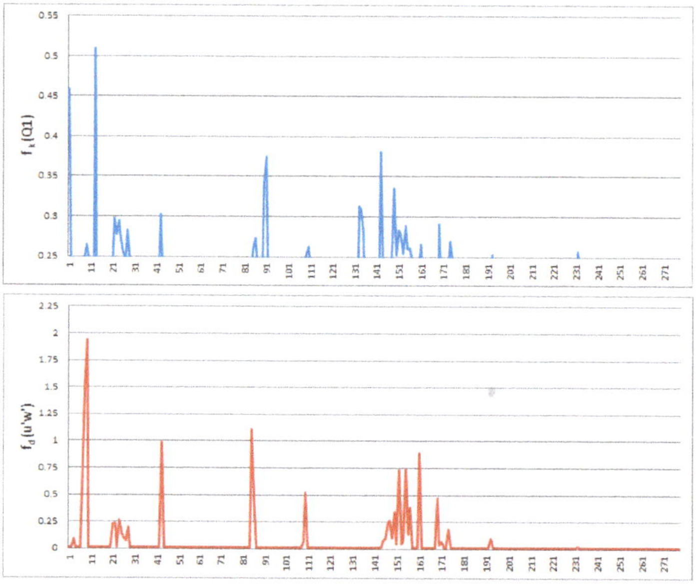

Figure 8. Values of $f_k(\mathrm{Q1})$ and $fd_{u'w'}$ for 278 points in the downstream region of the vegetation patches.

4.4. Temporal Characteristics of Turbulence in the Mixing Layer

As was described in the previous sections, the mixing layer beyond the vegetation patch is prone to anisotropic turbulence, characterized by outward interaction events and the existence of coherent Reynolds stresses. The deviation from an ideal isotropic turbulence originated from the turbulent intermittency. The turbulent intermittency was characterized by a preference for turbulence with large velocity gradients, which is reflected in the strongly non-Gaussian tails of the probability density functions of velocity differences. These tails are determined by extreme events [49]. Intermittency is also defined as an abnormality in turbulent flow initiated by the interaction between turbulent regions or the interaction between a turbulent region and a vicinal laminar behavior region. The second mechanism is the common source of intermittency in shear flows. Moreover, the occurrence of a highly intermittent region in the near-boundary regions was associated with the prevalence of coherent vortices [50]. The mechanism generates quasi-laminar intervals in the velocity time series of a turbulent flow, which postpones the non-viscous intermediate sub-range in the energy cascade. Thus, there was a slight slope in the energy spectra. Consequently, these intervals persist up to arbitrarily large Reynolds numbers through the energy cascade sequences, which produce extreme events in velocity time series [51]. The determination of a particular intermittency index is a challenging issue and requires the spatial analysis of the velocity time series measured at different locations at the same time. Clearly, there is a need for appropriate sensors, such as PIV and hot wires.

However, temporal analyses of velocity time series can also provide a proper description of the quantity and intensity of intermittent occurrences. As the intermittency is linked with the non-Gaussian tails of the probability density functions of velocity differences, the velocity time series for the mixing layer were subjected to a normality test. The traditional empirical rule of normality (or a simple normality test) states that, in a normally distributed data set, 68%, 95%, and 99.7% of the values lie within one, two, and three standard deviations (σ) of the mean (μ), respectively. Most of the points in the wake area could barely pass these restrictions; however, the normality test of Kolmogorov–Smirnov identified almost all points as non-normally distributed data sets. Regarding the methodology and application of the Kolmogorov–Smirnov normality test, please refer to Massey (1951) [52]. The results of the simple and Kolmogorov–Smirnov normality tests for case 1 are shown in Table 2. As the Kolmogorov–Smirnov normality test is very sensitive to non-Gaussian deviations, it even identifies semi-normal velocity time series for the wake layer as non-normal distributions. However, according to the concept of the Kolmogorov–Smirnov normality test, the intensity of non-normality can be observed on the basis of the deviation from the normal cumulative distribution. In this way, an error index can be defined as:

$$Error_{CDF} = 1 - \frac{CDF_E}{CDF_N} \tag{13}$$

where CDF_E refers to the cumulative distribution of the empirical velocity time series, and CDF_N is the cumulative distribution of a normal data set with the same μ and σ as the empirical data set. Figure 9 shows the variation in $Error_{CDF}$ for different depths at a distance of x/H_v = 8, where the mixing layer is well developed. One can see that the values of $Error_{CDF}$ are considerably higher at a depth of $0.8 < z/H_v < 1$. However, the most notable issue is that the max values reduce as the vegetation patch disappears. Comparing the results of a simple normality test to $Error_{CDF}$, it can be concluded that a threshold around $Error_{CDF}$ = 0.2 can be assumed for the lower limit of intermittency. This threshold is also confirmed by the results of the intermittent turbulence and coherent Reynolds shear stresses at a depth of $0.8 < z/H_v < 1$. Correspondingly, in case 4, where $Error_{CDF} < 0.2$ at all depths, no anisotropy was observed in the turbulent spectra. In contrast, the anisotropic turbulence was limited to the mixing layer in cases 1, 2 and 3. In the smaller z/H_v, the no intermittency is observed for all cases and the effect of bed roughness is the determining factor in the near bed region. Generally, in the near bed region of gravel bed channels, the effect of bed roughness exceeds the patch-produced vortexes [53–55].

Table 2. Results of the normality test of velocity time series of case 1 at a distance of x/H_v=8. Note that the unmatched values are highlighted.

z/H_v	% of Sample between μ − σ and μ + σ	% of Sample between μ − 2σ and μ + 2σ	% of Sample between μ − 3σ and μ + 3σ	Simple Normality Test	Kolmogorov–Smirnov Test
Reference Value	**0.68**	**95**	**0997**		
0.2	0.679	0.958	0.997	Gaussian	non-Gaussian
0.5	0.696	0.954	0.995	~Gaussian	non-Gaussian
0.8	0.754	0.948	0.981	non-Gaussian	non-Gaussian
1	0.710	0.938	0.998	non-Gaussian	non-Gaussian
1.2	0.753	0.928	0.993	non-Gaussian	non-Gaussian

According to the results of this research, the intermittency and anisotropy of the turbulence are associated with the non-Gaussian distribution of instantaneous velocity. This relationship was also identified by researchers, and is known as a fundamental characteristic of the anisotropic intermittent turbulence [49–51]. However, in the present research, this characteristic was detected in the mixing layer beyond vegetation patches with a relatively high blockage ratio ($D_v/D_c > 0.5$).

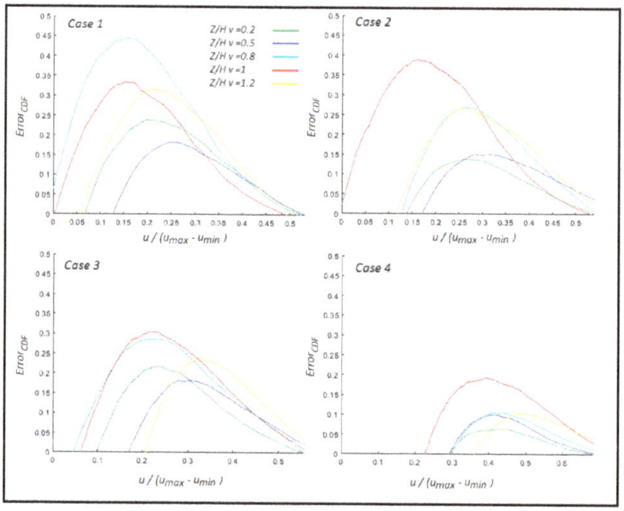

Figure 9. Values of $Error_{CDF}$ for different depths behind the decaying patch at a distance of x/H_v = 8.

In addition, the formation of intermittent turbulence in the downstream region of the patch has been investigated in this research. One can see from Table 1, by comparing to a normally distributed data set, that the non-Gaussian data sets are dependent on a greater concentration of data between $\mu - \sigma$ and $\mu + \sigma$. In addition, the values of the normalized standard deviation are higher in these velocity time series. Consequently, the variation in the probability density functions (PDF) of the instantaneous velocity was considered along the flow direction. The results have been compared with the turbulent spectra of the corresponding points. For the mixing layer of case 1, these results are shown in Figure 10. As can be seen in Figure 10, the intermittent anisotropic turbulence formed at a depth of $8 < x/H_v < 12$, which is the same range as the occurrence of the outward quadrant dominancy, leading to strong coherent Reynolds shear stresses and anomalies in streamwise velocity. This event can also be inferred from the flattened shape of PDF, which is associated with high standard deviations.

Furthermore, the intensity of the intermittency can also be evaluated in the energy domain. Researchers used an analysis method based on the filtering of TKE time series via a certain threshold [56,57]. These thresholds include $M_{TKE} - a\sigma_{TKE}$ and $M + a\sigma_{TKE}$, where M_{TKE} is the mean, "a" is a constant multiplayer (commonly 1 and 3) and σ_{TKE} is the standard deviation of the TKE time series in 2D space. The values that exceed this range are considered as strong occurrences characterizing intermittent events. Although the ratio of filtered samples describes the quantity of intermittent events, it cannot provide enough information about the intensity of the intermittency. It is reported that the TKE of extreme events is a suitable indicator of the intermittency [58]. Since extreme events are associated with coherent occurrences in the mixing layer, the Turbulent Kinetic Energy of these extreme events (TKE_i) can be used for this propose. This approach is used in this paper for calculating TKE and TKE_i in a 3D space. Overall, Figure 11 summarizes the results regarding the formation of intermittent turbulence in the mixing layer behind the vegetation patch by means of demonstrating the filtered time series of the instantaneous velocity and the variation in the normalized standard deviation of TKE and normalized TKE_i.

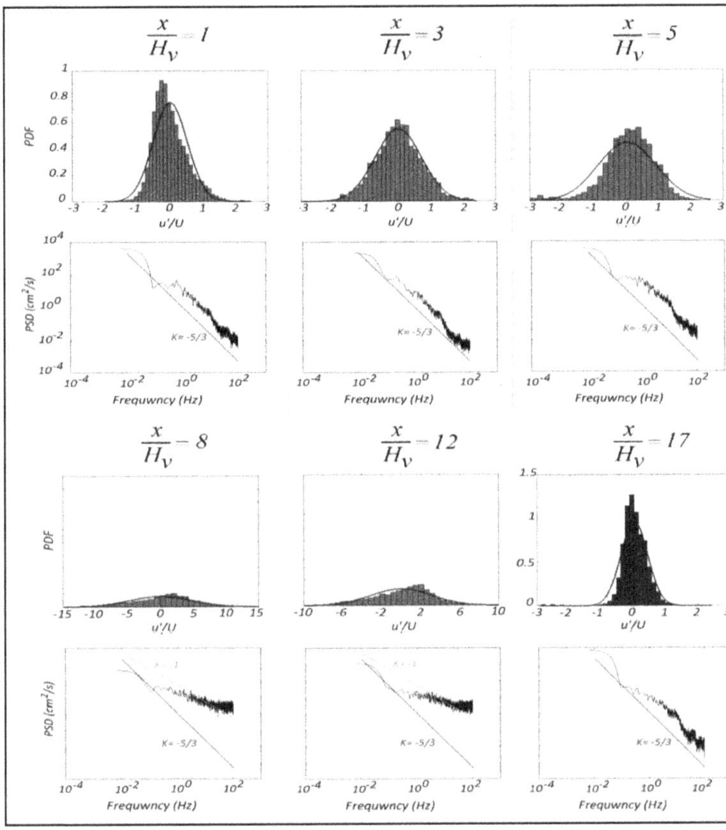

Figure 10. Probability density function and the turbulent spectra for the mixing layer of Case 1.

4.5. Transformation of Coherent Structures beyond a Vegetation Patch

Considering the results of the previous sections, the transformation of coherent structures beyond a vegetation patch can be summarized as below:

(1) In the downstream region of a fully channel-spanning vegetation patch, the coherent structures are observable just behind the patch. These structures originate from the stem-scale vortexes that are formed by the leaking flow passing through the vegetation patch [24,28]. As the patch width ratio of D_v/D_c reduces to 0.66, the leaking flow increases. Consequently, the stem-scale coherent structures are observed in a larger area (Figure 5). With the patch width ratio of D_v/D_c = 0.5, these structures spread to the near-bed region and create a large area of coherent shear stress, in which a Q1-dominant core is surrounded with a sweep ejection-dominant region. However, with a patch width ratio of D_v/D_c = 0.33, the area of stem-scale coherent structures reduces suddenly, and is limited to $x/D_v < 3$ in the top and middle zones of the wake layer behind the patch;

(2) In the mixing layer of a fully channel-spanning vegetation patch, coherent shear stresses are associated with largely intermittent fluctuations in the instantaneous velocity forming anisotropic turbulence at $8D_v < x < 12D_v$. In addition, this structure is accompanied by an outward interaction. The intensity and extent of this structure are reduced in the presence of smaller patches, and disappear with a patch width ratio of D_v/D_c = 0.33. However, Q1 dominancy is also detectable in the mixing regions of small patches.

(3) With a patch width ratio of $D_v/D_c = 0.5$, a new type of coherent shear stresses emerges in the wake zone. These structures grow as the dimensions of the vegetation patch reduce, and they come to cover most area of the wake zone with a patch width ratio of $D_v/D_c = 0.33$.

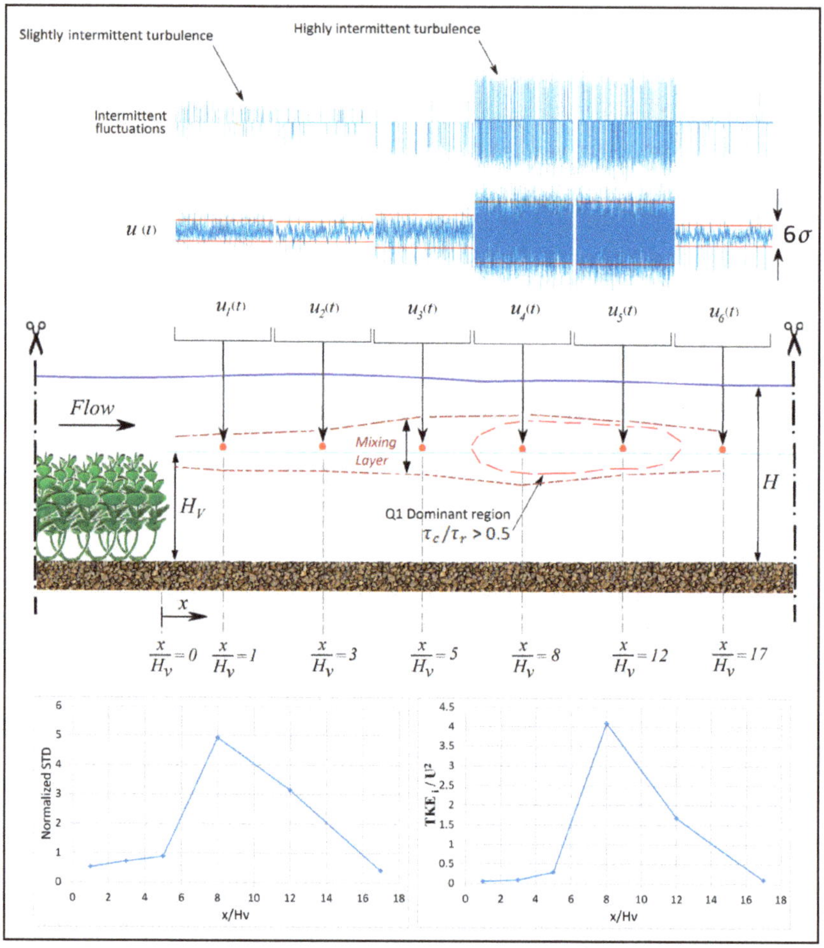

Figure 11. Formation of the intermittent fluctuations in the mixing layer beyond the patch (case 1). The filtered velocity time series is shown as intermittent fluctuations above the original velocity time series. Variations in the normalized standard deviation of TKE and normalized intermittent Turbulent Kinetic Energy are shown at the bottom.

In summary, the flow in the downstream region of a decaying vegetation patch alters as the size of vegetation patch decreases. This change includes not only the intensity of occurrences, but also the nature of structure. This means that the dominant coherent structures are transferred from the intermittent anisotropic turbulent fluctuations in the mixing layer of a fully covered patch into the isotropic turbulence associated with the von Karman vortexes on the horizontal plane of the wake layer in the presence of the smallest patch.

5. Conclusions

In the present study, the changes in turbulent structures in the regions downstream of vegetation patches have been investigated. The model vegetation was selected carefully to simulate aquatic vegetation patches in natural rivers. Velocity profile, TKE, turbulent power spectra and quadrant analysis have been used via various approaches to investigate the features and intensity of the turbulent structures. Three different flow layers were detected in the downstream regions of vegetation patches, including the wake layer, the mixing layer and the shear layer. Overall, the results of the TKE and velocity analysis show that those three flow layers (wake, mixing and shear) occurred in all cases of patch setup. However, the wake and shear layers were considerably affected by the flow passing around the sides of the patches. Consequently, different flow structures and associated length scales can be formed, which are dependent on the upward movement of the center of the vertical rotation beyond the patch. The results of spectral analysis indicate that the strong coherent shear stresses are associated with the peaks of the anisotropic turbulent spectra. In contrast, the weak form of coherent shear stresses occurs in the presence of peak isotropic turbulence. The presence of von Karman vortexes near partially covered patches at a depth of $H_v/H > 0.55\sim0.7$ shows that the frequency of vortex shedding was around 0.07–0.11 Hz, which is equivalent to an average Strouhal number of about 0.25. The main characteristics of turbulent structures beyond a wide patch are associated with the highly intermittent anisotropic turbulent events in the mixing layer, and appear at a depth of $z/H_v = 0.7\sim1.1$ and distance of $x/H_v = 8\sim12$. The outward interactions are the representative quadrant class of coherent Reynolds shear stresses in the regions downstream of vegetation patches—particularly the strong coherent occurrences. The intensity and extent of these structures decrease as the size of the patch reduces. In addition, the intermittency and anisotropy of the turbulence are associated with the non-Gaussian distribution of the instantaneous velocity detected in the mixing layer beyond patches with a relatively high patch width ratio of $D_v/D_c > 0.5$. Finally, when the size of the patch reduces, von Karman vortexes appear in the wake layer, and form the dominant flow structures in the downstream region of the vegetation patch. It should be noted that the results of this research are limited by our use of a certain type of vegetation, and so it is strongly suggested that other types of the vegetation patches often found in natural rivers be considered in future research.

Author Contributions: M.K.; laboratory works, methodology, software and writing—original draft preparation, H.A.; laboratory supervisory, methodology and validation, J.S.; methodology and writing reviewing. All authors have read and agreed to the published version of the manuscript.

Funding: This research received no external funding.

Institutional Review Board Statement: This study did not involve humans or animals.

Informed Consent Statement: Not applicable.

Data Availability Statement: The data are available upon request.

Acknowledgments: Special thanks to G. Goodrati Amiri for his contribution in providing the laboratory devices and Esmaeel Dodangeh for his contribution in laboratory setup.

Conflicts of Interest: The authors declare no conflict of interest.

References

1. Carpenter, S.R.; Lodge, D.M. Effects of submersed macrophytes on ecosystem processes. *Aquat. Bot.* **1986**, *26*, 341–370. [CrossRef]
2. Martin, C.W.; Valentine, J.F. Eurasian milfoil invasion in estuaries: Physical disturbance can reduce the proliferation of an aquatic nuisance species. *Mar. Ecol. Prog. Ser.* **2012**, *449*, 109–119. [CrossRef]
3. Short, F.T.; Kosten, S.; Morgan, P.A.; Malone, S.; Moore, G.E. Impacts of climate change on submerged and emergent wetland plants. *Aquat. Bot.* **2016**, *135*, 3–17. [CrossRef]
4. Zhang, Y.; Jeppesen, E.; Liu, X.; Qin, B.; Shi, K.; Zhou, Y.; Thomaz, S.M.; Deng, J. Global loss of aquatic vegetation in lakes. *Earth-Sci. Rev.* **2017**, *173*, 259–265. [CrossRef]
5. Gillard, M.; Thiébaut, G.; Deleu, C.; Leroy, B. Present and future distribution of three aquatic plants taxa across the world: Decrease in native and increase in invasive ranges. *Biol. Invasions* **2017**, *19*, 2159–2170. [CrossRef]

6. O'Hare, M.T.; Aguiar, F.C.; Asaeda, T.; Bakker, E.; Chambers, P.A.; Clayton, J.S.; Elger, A.; Ferreira, T.M.; Gross, E.M.; Gunn, I.D.M.; et al. Plants in aquatic ecosystems: Current trends and future directions. *Hydrobiologia* **2018**, *812*, 1–11. [CrossRef]
7. Okamoto, T.-A.; Nezu, I. Spatial evolution of coherent motions in finite-length vegetation patch flow. *Environ. Fluid Mech.* **2013**, *13*, 417–434. [CrossRef]
8. Afzalimehr, H.; Riazi, P.; Jahadi, M.; Singh, V.P. Effect of vegetation patches on flow structures and the estimation of friction factor. *ISH J. Hydraul. Eng.* **2019**, *1–11*. [CrossRef]
9. Kazem, M.; Afzalimehr, H.; Sui, J. Formation of Coherent Flow Structures beyond Vegetation Patches in Channel. *Water* **2021**, *13*, 2812. [CrossRef]
10. Killgore, K.J.; Dibble, E.D.; Hoover, J.J. *Relationships between Fish and Aquatic Plants: A Plan of Study*; Miscellaneous Paper; U.S. Army Corps of Engineers: Vicksburg, MI, USA, 1993; pp. A-91–A-93.
11. Lauridsen, T.; Buenk, I. Diel changes in the horizontal distribution of zooplankton in the littoral zone of two shallow eutrophic lakes. *Archiv. Hydrobiol.* **1996**, *137*, 161–176. [CrossRef]
12. Looby, A.; Reynolds, L.K.; Adams, C.R.; Martin, C.W. Submerged Aquatic Vegetation Patch Size Affects Fish Communities in a Turbid-Algal Lake. *Front. Conserv. Sci.* **2021**, *2*, 12. [CrossRef]
13. Neary, V.S.; Constantinescu, S.G.; Bennett, S.J.; Diplas, P. Effects of Vegetation on Turbulence, Sediment Transport, and Stream Morphology. *J. Hydraul. Eng.* **2012**, *138*, 765–776. [CrossRef]
14. Follett, E.; Nepf, H.M. Sediment patterns near a model patch of reedy emergent vegetation. *Geomorphology* **2012**, *179*, 141–151. [CrossRef]
15. Huai, W.-X.; Li, S.; Katul, G.G.; Liu, M.-Y.; Yang, Z.-H. Flow dynamics and sediment transport in vegetated rivers: A review. *J. Hydrodyn.* **2021**, *33*, 400–420. [CrossRef]
16. Zhang, J.; Wang, W.-J.; Shi, H.; Wang, W.; Li, Z.; Tang, X.; Xia, Z. An analytical two-layer model for velocity distribution in open-channel flows with submerged flexible canopies considering multiply fluids mechanics. *J. Hydrol.* **2021**, *603*, 127102. [CrossRef]
17. Marjoribanks, T.I.; Paul, M. Modelling flow-induced reconfiguration of variable rigidity aquatic vegetation. *J. Hydraul. Res.* **2021**, *1–16*. [CrossRef]
18. D'Ippolito, A.; Calomino, F.; Alfonsi, G.; Lauria, A. Flow Resistance in Open Channel Due to Vegetation at Reach Scale: A Review. *Water* **2021**, *13*, 116. [CrossRef]
19. Li, S.; Shi, H.; Xiong, Z.; Huai, W.; Cheng, N. New formulation for the effective relative roughness height of open channel flows with submerged vegetation. *Adv. Water Resour.* **2015**, *86*, 46–57. [CrossRef]
20. Zhao, F.; Huai, W. Hydrodynamics of discontinuous rigid submerged vegetation patches in open-channel flow. *HydroResearch* **2016**, *12*, 148–160. [CrossRef]
21. Hopkinson, L.C. Flow through vegetation patches with multiple layers: A quadrant analysis. In *Crossing Boundaries, Proceedings of the World Environmental and Water Resources Congress, Albuquerque, NM, USA, 20–24 May 2012*; American Society of Civil Engineers: Reston, VI, USA, 2012; pp. 1406–1415.
22. Devi, T.B.; Kumar, B. Channel Hydrodynamics of Submerged, Flexible Vegetation with Seepage. *J. Hydraul. Eng.* **2016**, *142*, 04016053. [CrossRef]
23. Chen, Z.; Ortiz, A.; Zong, L.; Nepf, H. The wake structure behind a porous obstruction and its implications for deposition near a finite patch of emergent vegetation. *Water Resour. Res.* **2012**, *48*. [CrossRef]
24. Liu, C.; Nepf, H. Sediment deposition within and around a finite patch of model vegetation over a range of channel velocity. *Water Resour. Res.* **2016**, *52*, 600–612. [CrossRef]
25. Ortiz, A.C.; Ashton, A.; Nepf, H. Mean and turbulent velocity fields near rigid and flexible plants and the implications for deposition. *J. Geophys. Res. Earth Surf.* **2013**, *118*, 2585–2599. [CrossRef]
26. Zong, L.; Nepf, H. Flow and deposition in and around a finite patch of vegetation. *Geomorphology* **2010**, *116*, 363–372. [CrossRef]
27. Zong, L.; Nepf, H. Vortex development behind a finite porous obstruction in a channel. *J. Fluid Mech.* **2012**, *691*, 368–391. [CrossRef]
28. Liu, C.; Hu, Z.; Lei, J.; Nepf, H. Vortex structure and sediment deposition in the wake behind a finite patch of model sub-merged vegetation. *J. Hydraul. Eng.* **2018**, *144*, 04017065. [CrossRef]
29. Folkard, A.M. Hydrodynamics of model *Posidonia oceanica* patches in shallow water. *Limnol. Oceanogr.* **2005**, *50*, 1592–1600. [CrossRef]
30. Sadeque, M.A.; Rajaratnam, N.; Loewen, M.R. Effects of Bed Roughness on Flow around Bed-Mounted Cylinders in Open Channels. *J. Eng. Mech.* **2009**, *135*, 100–110. [CrossRef]
31. Perera, M.D.A.E.S. Shelter behind two-dimensional solid and porous fences. *J. Wind Eng. Ind. Aerodyn.* **1981**, *8*, 93–104. [CrossRef]
32. Sui, J.; Faruque, M.A.; Balachandar, R. Local Scour Caused by Submerged Square Jets under Model Ice Cover. *J. Hydraul. Eng.* **2009**, *135*, 316–319. [CrossRef]
33. Jafari, R.; Sui, J. Velocity Field and Turbulence Structure around Spur Dikes with Different Angles of Orientation under Ice Covered Flow Conditions. *Water* **2021**, *13*, 1844. [CrossRef]
34. Kabiri, F.; Afzalimehr, H.; Sui, J. Flow structure over a wavy bed with vegetation cover. *Int. J. Sediment Res.* **2017**, *32*, 186–194. [CrossRef]
35. Hussain, A.F. Role of coherent structures in turbulent shear flows. *Proc. Indian Acad. Sci. Sect. C Eng. Sci.* **1981**, *4*, 129–175.

36. Shahmohammadi, R.; Afzalimehr, H.; Sui, J. Impacts of turbulent flow over a channel bed with a vegetation patch on the incipient motion of sediment. *Can. J. Civ. Eng.* **2018**, *45*, 803–816. [CrossRef]
37. Afzalimehr, H.; Moghbel, R.; Gallichand, J.; Sui, J. Investigation of turbulence characteristics in channel with dense vegetation. *Int. J. Sediment Res.* **2011**, *26*, 255–282. [CrossRef]
38. Przyborowski, Ł.; Łoboda, A.M.; Bialik, R.J. Effect of two distinct patches of *Myriophyllum* species on downstream turbulence in a natural river. *Acta Geophys.* **2019**, *67*, 987–997. [CrossRef]
39. Huai, W.-X.; Zhang, J.; Katul, G.G.; Cheng, Y.-G.; Tang, X.; Wang, W.-J. The structure of turbulent flow through submerged flexible vegetation. *J. Hydrodyn.* **2019**, *31*, 274–292. [CrossRef]
40. Termini, D. Experimental Analysis of Horizontal Turbulence of Flow over Flat and Deformed Beds. *Arch. Hydro-Eng. Environ. Mech.* **2015**, *62*, 77–99. [CrossRef]
41. Termini, D.; Sammartano, V. Experimental observation of horizontal coherent turbulent structures in a straight flume. In Proceedings of the River, Coastal and Estuarine Morphodynamics—RCEM, Santa Fe City, Argentina, 21–29 September 2009.
42. Mayaud, J.R.; Wiggs, G.F.; Bailey, R.M. Dynamics of skimming flow in the wake of a vegetation patch. *Aeolian Res.* **2016**, *22*, 141–151. [CrossRef]
43. Wang, H.; Peng, G.; Chen, M.; Fan, J. Analysis of the Interconnections between Classic Vortex Models of Coherent Structures Based on DNS Data. *Water* **2019**, *11*, 2005. [CrossRef]
44. El-Hakim, O.; Salama, M.M. Velocity Distribution Inside and Above Branched Flexible Roughness. *J. Irrig. Drain. Eng.* **1992**, *118*, 914–927. [CrossRef]
45. Nezu, I.; Sanjou, M. Turburence structure and coherent motion in vegetated canopy open-channel flows. *HydroResearch* **2008**, *2*, 62–90. [CrossRef]
46. Shivpure, V.; Devi, T.B.; Kumar, B. Turbulent characteristics of densely flexible submerged vegetated channel. *ISH J. Hydraul. Eng.* **2016**, *22*, 220–226. [CrossRef]
47. Tchen, C. On the spectrum of energy in turbulent shear flow. *J. Res. Natl. Inst. Stand. Technol.* **1953**, *50*, 51. [CrossRef]
48. Lian, Q.X. A visual study of the coherent structure of the turbulent boundary layer in flow with adverse pressure gradient. *J. Fluid Mech.* **1990**, *215*, 101–124. [CrossRef]
49. Staicu, A.D. Intermittency in Turbulence. Ph.D. Thesis, Technische Universiteit Eindhoven, Eindhoven, The Netherlands, 2002. [CrossRef]
50. Coscarella, F.; Penna, N.; Servidio, S.; Gaudio, R. Turbulence anisotropy and intermittency in open-channel flows on rough beds. *Phys. Fluids* **2020**, *32*, 115127. [CrossRef]
51. Avila, M.; Hof, B. Nature of laminar-turbulence intermittency in shear flows. *Phys. Rev. E* **2013**, *87*, 063012. [CrossRef]
52. Massey, F.J., Jr. The Kolmogorov-Smirnov test for goodness of fit. *J. Am. Stat. Assoc.* **1951**, *46*, 68–78. [CrossRef]
53. Afzalimehr, H.; Sui, J.; Moghbel, R. Hydraulic parameters in channels with wall vegetation and gravel bed–field observations and experimental studies. *Int. J. Sediment Res.* **2010**, *25*, 81–90. [CrossRef]
54. Afzalimehr, H.; Barahimi, M.; Sui, J. Non-uniform flow over cobble bed with submerged vegetation strip. *Proc. ICE–Water Manag.* **2019**, *172*, 86–101. [CrossRef]
55. Fazlollahi, A.; Afzalimehr, H.; Sui, J. Impacts of pool and vegetated banks on turbulent flow characteristics. *Can. J. Civil Eng.* **2015**, *42*, 979–986. [CrossRef]
56. Cox, D.T.; Kobayashi, N. Identification of intense, intermittent coherent motions under shoaling and breaking waves. *J. Geophys. Res. Space Phys.* **2000**, *105*, 14223–14236. [CrossRef]
57. Cox, D.T.; Anderson, S.L. Statistics of Intermittent Surf Zone Turbulence and Observations of Large Eddies using PIV. *Coast. Eng. J.* **2001**, *43*, 121–131. [CrossRef]
58. McCaffrey, K.; Fox-Kemper, B.; Hamlington, P.E.; Thomson, J. Characterization of turbulence anisotropy, coherence, and intermittency at a prospective tidal energy site: Observational data analysis. *Renew. Energy* **2015**, *76*, 441–453. [CrossRef]

water

MDPI

Article

Formation of Coherent Flow Structures beyond Vegetation Patches in Channel

Masoud Kazem [1], Hossein Afzalimehr [1,*] and Jueyi Sui [2]

[1] Faculty of Civil Engineering, Iran University of Science and Technology, Tehran 16846-13114, Iran; masoud_kazem@cmps2.iust.ac.ir

[2] School of Engineering, University of Northern British Columbia, Prince George, BC V2N 4Z9, Canada; jueyi.sui@unbc.ca

* Correspondence: hafzali@iust.ac.ir; Tel.: +98-913-2175524

Abstract: By using model vegetation (e.g., synthetic bars), vortex structures in a channel with vegetation patches have been studied. It has been reported that vortex structures, including both the vertical and horizontal vortexes, may be produced in the wake in the channel bed with a finite-width vegetation patch. In the present experimental study, both velocity and TKE have been measured (via Acoustic Doppler Velocimeter—ADV) to study the formation of vortexes behind four vegetation patches in the channel bed. These vegetation patches have different dimensions, from the channel-bed fully covered patch to small-sized patches. Model vegetation used in this research is closely similar to vegetation in natural rivers with a gravel bed. The results show that, for a channel with a small patch ($L_V/D_c = 0.44$ and $D_V/D_c = 0.33$; where L_V and D_V are the length and width of patch and D_c is the channel width, respectively), both the flow passing through the patch and side flow around the patch have a considerable effect on the formation of flow structures beyond the patch. The results of further analysis via 3D classes of the bursting events show that the von Karman vortex street splits into two parts beyond the vegetation patch as the strong part near the surface and the weak part near the bed; while the middle part of the flow is completely occupied by the vertical vortex formed at a distance of 0.8–1 H_v beyond the vegetation patch, and thus, the horizontal vortexes cannot be detected in this region. The octant analysis is conducted for the coherent shear stress analysis that confirms the results of this experimental study.

Keywords: coherent flow structures; vegetation patch; von Karman vortex; octant analysis

Citation: Kazem, M.; Afzalimehr, H.; Sui, J. Formation of Coherent Flow Structures beyond Vegetation Patches in Channel. *Water* **2021**, *13*, 2812. https://doi.org/10.3390/w13202812

Academic Editor: Achim A. Beylich

Received: 2 September 2021
Accepted: 5 October 2021
Published: 10 October 2021

1. Introduction

Up to date, a lot of results regarding the interaction between flow and vegetation in channels have been reported. The effects of vegetation patches on river morphology, sediment transport and hydro-environment sequences have been explored by many researchers, i.e., [1–5]. The majority of the studies regarding this topic have been conducted in laboratory flumes by using submerged model vegetation patches. Research activities have been focused on the impacts of the canopy of relatively long vegetation patches on the fully-developed flow and turbulent characteristics. Some research works have been carried out to study the flow structures inside the sparse vegetation patches. In many natural rivers, however, the aquatic vegetation appears in patches with finite length. In such flows over a vegetated bed, the shear layer is unable to form at the upstream edge of the vegetation patch, and the coherent motions develop downstream [6]. The deposition process of fine sediment particles appeared due to a lower velocity in the wake behind the vegetation patch. Thus, it created a better condition for the growth of the vegetation patch further downstream [2,7]. Furthermore, turbulent structures developed in the region downstream of vegetation patches or rocks can provide a particular condition for the nutrition and reproduction of fishes and aquatic organisms [8,9], namely the "hotspot" for aquatic systems.

With the presence of emergent vegetation in channel beds, researchers have tried to study the flow structures in a variety of aspects. The main characteristics of the flow structures with the presence of emergent vegetation are the obvious fluctuation in the lateral components of velocity in the region downstream of the vegetation patch, which resulted from the instabilities of flow at the side edges of the vegetation patches, the phenomenon that is known as the von Kármán Vortex Street. However, the von Karman vortex street is only detectable for relatively dense vegetation patches [10]. In reality, it is possible for the flow to pass through the vegetation patch (termed as "through-patch flow" in this study), and this "through-patch flow" may cause a spatial delay to form the vortexes. The results indicated that for very sparse vegetation patches in a channel bed, the absence of those vortexes has been reported. For a solid barrier, however, there is no delay in the formation of vortexes. As a consequence, the distance between the trailing edge of the barrier and the vortexes (L_{kv}) is equal to zero. For a relatively dense vegetation patch with a diameter of D, the velocity of the through-patch flow is negligible, and the distance between the trailing edge of the barrier and the vortexes: $L_{kv} \approx 2.5$ D [3,10]. For slightly submerged vegetation patches in a channel bed, the von Karman vortex street is detectable, but it disappears in highly submerged patches (h/H < 0.55; where h is the height of patch and H is the flow depth), particularly when the patch has a low ratio of height to width [11]. Some researchers pointed out that, in the case that the von Karman vortex is present in the fully submerged vegetation patches, the vortex appeared around the trailing edges, namely $L_{kv} = 0$ [11]. The dominant structure of a wide submerged path appears in the vertical plane, and a vertical recirculation zone is created during the formation of vortexes [3,12]. For a solid submerged barrier, the vortices occur at the trailing edge of the barrier and the length of the wake behind the barrier is zero ($L_{vv} = 0$) [13]. However, for a sparsely distributed vegetation patch in a channel bed, the through-patch flow may delay the formation process of the recirculating region ($L_{vv} \neq 0$). These statements indicate that both L_{vv} and L_{kv} are affected by the dimensions of the vegetation patch, specifically the height (H_v) and diameter (D_v). In addition, the through-patch flow, which enters into the wake region, can increase the values considerably [13–15]. Additionally, the increase in the turbulent kinetic energy $TKE = 0.5\left(\overline{u'^2} + \overline{v'^2} + \overline{w'^2}\right)$, which is the kinetic energy of the fluctuating part of the velocity, is linked to the von Karman Vortex Street [14]. The velocity fluctuations are defined by:

$$\begin{aligned} u' &= u_t - \overline{u} \\ v' &= v_t - \overline{v} \\ w' &= w_t - \overline{w} \end{aligned} \tag{1}$$

where, $\overline{u}$, $\overline{v}$ and $\overline{w}$ are defined as the average of velocities in x, y and z directions, respectively, and can be determined in the time domain as the following:

$$\begin{aligned} \overline{u} &= \lim_{T\to\infty} \tfrac{1}{T} \int_0^T u_t dt \\ \overline{v} &= \lim_{T\to\infty} \tfrac{1}{T} \int_0^T v_t dt \\ \overline{w} &= \lim_{T\to\infty} \tfrac{1}{T} \int_0^T w_t dt \end{aligned} \tag{2}$$

where T is the duration of velocimetry.

Liu et al. (2018) [3] claimed that the presence of the von Karman vortex street was visible in the power spectrum, particularly behind vegetation patches with a high submergence ratio [3]. For example, for a vegetation patch with a diameter of 10 cm, they found that the peak frequencies ranged from 0.08 to 0.12 Hz in the near-bed region for the submergence ratio between 0.5 and 0.79. Similar results have also been reported by other researchers [10]. Overall, the majority of the previous researches have been focused on the parameters used to detect and measure vortexes behind the vegetation patch based on laboratory experiments using model vegetation (e.g., thin rods). The objectives of this study are (1) to assess the formation process of both vertical and horizontal vortexes behind vegetation patches with finite dimensions which are similar to those in natural rivers and

(2) to characterize the coherent flow structure beyond a vegetation patch via analyzing the bursting events and coherent shear stresses.

2. Coherent Shear Stress

The determination of coherent structures in channels with the presence of barriers has been attracting attention from many researchers in recent years [16–18]. According to the triple decomposition approach, the instantaneous velocity can be written as [19]:

$$u(x,t) = \overline{u}(x) + \widetilde{u}_c(x,t) + u_r(x,t) \tag{3}$$

where, u_r is the deflection of the velocity, which is known as u' in the classic approach, and $\overline{u}(x)$ is the classic average of the velocity during the time period of T. In addition to the classic averaging method, the periodic phase average can be defined as:

$$\langle u(x,t)\rangle = \lim_{N\to\infty} \frac{1}{T_P} \sum_{1}^{N} u_i(x, t + iT_P) \tag{4}$$

where, T_p is the period of occurrence of a coherent structure and is equal to $1/f_d$ where f_d is the dominant frequency of the occurrence. For a time series with length T, N is the number of cycles with period of T_p that can be calculated as $N = \frac{T}{Tp}$. Following these definitions, the coherent velocity deflection is:

$$\widetilde{u}_c(x,t) = \langle u(x,t)\rangle - \overline{u}(x) \tag{5}$$

Considering same approach for other components of velocity, the coherent Reynolds shear stress can be written as:

$$\tau_c = -\langle \widetilde{u}_c \widetilde{v}_c \rangle \tag{6}$$

The total coherent and non-coherent Reynolds shear stress is defined as follows:

$$\tau_r = -\overline{\overline{u}_r \overline{v}_r} \tag{7}$$

3. Decomposition of Bursting Events

The decomposition of bursting events is also widely applied to determine the dominant turbulent events with the presence of both emerged and submerged vegetation patches in a channel bed [20–24]. In most of the research, quadrant analysis is generally used to detect the ejection and sweep events in the vertical plane. However, for detecting the dominant events in the horizontal plane, the three-dimensional octant analysis is recommended because the two-dimensional analysis of bursting events is unable to define the entrainment process when there is a fully three-dimensional flow in nature [25,26].

The sign of the velocity fluctuations in three dimensions is used to classify the bursting events in the octant analysis. Overall, the eight class of bursting events in three-dimensional octant analysis is defined in Table 1. Keshavarzi et al. (2014) used a different naming system, including internal and external events [26]. However, the naming system in this paper is based on the classic quadrant events in which the attitude of the lateral component of velocity is stated as Right wing for events including $\overline{v} > 0$ and Left wing for events including $\overline{v} < 0$. To study sediment transport, the octant analysis has a distinct advantage over the classic quadrant. Since the sweep events are responsible for erosion, the quadrant analysis can only determine the occurrence of the event, while in octant analysis, there is a difference between the right-wing sweep and left-wing sweep. The vortices in a horizontal plane are in different directions: clockwise for the left-wing sweep and counterclockwise for the right-wing sweep, so the sediment removal direction is different for these sweep events. The octant events and right-wing and left-wing classes are shown in Figure 1.

Table 1. Three-dimensional analysis of bursting events in the octant analysis.

Classes of Bursting Events	Class Name	Sign of Fluctuating Velocities		
		u'	v'	w'
Right-wing outward interaction	PPP	+	+	+
Right-wing sweep	PPN	+	+	−
Left-wing outward interaction	PNP	+	−	+
Left-wing sweep	PNN	+	−	−
Right-wing ejection	NPP	−	+	+
Right-wing inward interaction	NPN	−	+	−
Left-wing ejection	NNP	−	−	+
Left-wing inward interaction	NNN	−	-	−

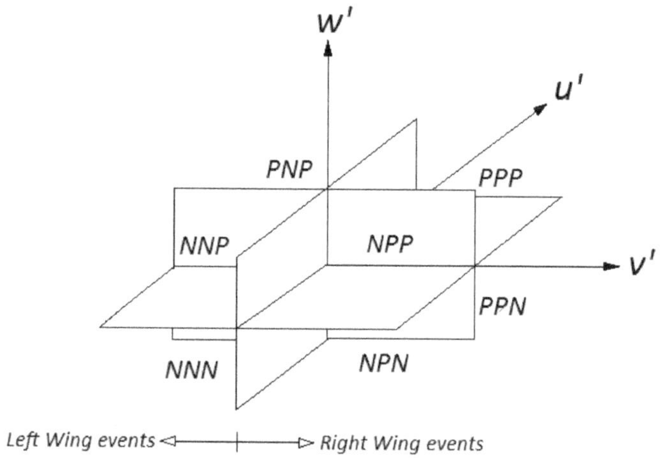

Figure 1. Bursting events in the octant analysis and right-wing and left-wing classification of the events (PNN event which is located in the left wing is not seen in this angle of view).

In this study, two categories of outputs are considered for the octant analysis: (1) the occurrence probability of the bursting events and (2) the transition probability of the bursting events. The calculation of the occurrence probability of events is similar to the approach of the quadrant analysis and is equal to the normalized occurrence frequency (f_k) for a particular class of events related to different classes of events [27]:

$$f_k = \frac{n_k}{N} \tag{8}$$

$$N = \sum_{1}^{8} n_k \; k = 1, 2, 3, \ldots 8$$

For the quadrant analysis, the majority of previous researches used a "hole" region in which the event must be filtered and not be considered. For the octant analysis, this threshold can be written as:

$$\left|u'(t)v'(t)w'(t)\right| > C_H\left|\overline{u'v'w'}\right| \tag{9}$$

As a threshold parameter C_H, it omits the events with low intensity below a certain limit, which is scaled by the average of velocity fluctuations. High values of C_H indicate a selection of the strong events but the total number of instantaneous events $(u'(t)\; v'(t)\; w\prime(t))$ decreases so much that the contribution region from each quadrant became meaningless. For the quadrant analysis, the threshold level of $C_H = 1$ is suggested to be used to reach a good compromise between the clear identification of the events and the preservation of a number of instantaneous events of each class [27–29]. Keshavarzi et al. (2014) considered the threshold level of $C_H = 0$ for assessing the coherent structure around a circular bridge pier via the octant analysis [26].

While the occurrence probability provided a vision for detecting the dominant events of a three-dimensional velocity time series, it does not provide any information about the stability of the events. The transition probability ($P_{M \to N}$) provides the occurrence probability of an event belonging to class N in the time step of ($t + \Delta t$), but the occurred event belongs to class M if only in the previous step of (t). Therefore, for a particular class of events, for example, class N, the class is fully stable if $P_{N \to N} = 1$ and fully unstable if $P_{N \to N} = 0$ [26]. The intermediate values must be compared to other classes to see to what extent a particular class of events is stable. The 8×8 matrix of the transition probability for an octant analysis is shown in Figure 2:

$$\begin{bmatrix} P_{PPP \to PPP} & P_{PPP \to PPN} & P_{PPP \to PNP} & P_{PPP \to PNN} & P_{PPP \to NPP} & P_{PPP \to NPN} & P_{PPP \to NNP} & P_{PPP \to NNN} \\ P_{PPN \to PPP} & P_{PPN \to PPN} & P_{PPN \to PNP} & P_{PPN \to PNN} & P_{PPN \to NPP} & P_{PPN \to NPN} & P_{PPN \to NNP} & P_{PPN \to NNN} \\ P_{PNP \to PPP} & P_{PNP \to PPN} & P_{PNP \to PNP} & P_{PNP \to PNN} & P_{PNP \to NPP} & P_{PNP \to NPN} & P_{PNP \to NNP} & P_{PNP \to NNN} \\ P_{PNN \to PPP} & P_{PNN \to PPN} & P_{PNN \to PNP} & P_{PNN \to PNN} & P_{PNN \to NPP} & P_{PNN \to NPN} & P_{PNN \to NNP} & P_{PNN \to NNN} \\ P_{NPP \to PPP} & P_{NPP \to PPN} & P_{NPP \to PNP} & P_{NPP \to PNN} & P_{NPP \to NPP} & P_{NPP \to NPN} & P_{NPP \to NNP} & P_{NPP \to NNN} \\ P_{NPN \to PPP} & P_{NPN \to PPN} & P_{NPN \to PNP} & P_{NPN \to PNN} & P_{NPN \to NPP} & P_{NPN \to NPN} & P_{NPN \to NNP} & P_{NPN \to NNN} \\ P_{NNP \to PPP} & P_{NNP \to PPN} & P_{NNP \to PNP} & P_{NNP \to PNN} & P_{NNP \to NPP} & P_{NNP \to NPN} & P_{NNP \to NNP} & P_{NNP \to NNN} \\ P_{NNN \to PPP} & P_{NNN \to PPN} & P_{NNN \to PNP} & P_{NNN \to PNN} & P_{NNN \to NPP} & P_{NNN \to NPN} & P_{NNN \to NNP} & P_{NNN \to NNN} \end{bmatrix}$$

Figure 2. The transition probability matrix for the octant analysis; the values on the main diagonal show the probability of self-repeating of an event in the same octant class.

4. Experimental Setup

Experiments have been conducted using a glass flume in the hydraulic laboratory at the Iran University of Science and Technology. The flume is 14 m long, 0.9 m wide and 0.6 m deep. The discharge is controlled by an electromagnetic flow meter installed at the entrance of the flume and is set for 31 L/s in the present study. The water level in the flume is adjusted by a tailgate located at the end of the flume and is set for a depth of 18.5 ± 0.3 cm in the present study. The distance between the flume entrance and vegetated zone is 6 m to ensure a fully developed flow in the region upstream of the vegetation patch.

The velocity measurements have been conducted after the flow reaches the steady state condition. The velocity profiles have been measured using an Acoustic Doppler velocimeter (ADV), placed at the centerline of each row of vegetation patch. There are 19–26 measuring points along each vertical line for velocity measurements, and the vertical distance between two adjacent measuring points is 4–10 mm. The sampling frequency and the measuring time of the ADV are 200 Hz and 120 s, respectively, resulting in 24,000 instantaneous velocities for point measurement.

Up to date, the majority of the previous studies have employed artificial thin rods of regular shapes to simulate vegetation patches in natural rivers. While real vegetation is flexible and irregular, the model vegetation by using the artificial thin rods of regular shapes may not represent the nature of vegetation behavior [22]. Additionally, experiments showed that the natural vegetation would lose its stiffness and get a long-lasting curvature towards the flow direction during an experimental run. Thus, in this study, a well-shaped synthetic plant is used to model the natural vegetation. This selection is based on a real-world sample of vegetation patches in a gravel-bed river.

Each model plant has three branches. Each branch has 12 leaves, and the diameter of the branch trunk is approximately 3 mm, as showen in Figure 3. The average height of the vegetation patch is 105 ± 5 mm, and the lateral and longitudinal spread widths of the leaves are approximately 9–19 and 11–22 mm, respectively. The model plants have a certain degree of flexibility and can swing in a flowing current similar to the vegetation in a natural river. The vegetation patch is attached on a perforated board with a staggered arrangement. Four different layouts of vegetation patches are used to simulate both fully covered and non-fully covered channel beds. The dimensions of these vegetation patches, namely, length (L_V) × width (D_V) in this study are 120 × 90 cm, 90 × 60 cm, 60 × 45 cm and 40 × 30 cm, respectively. Note that the length of the vegetation patch (L_V) is along the longitudinal axis of the flume, and the width (D_V) is in the transverse direction (perpendicular to the flume). Parameters for all experimental runs are summarized in Table 2.

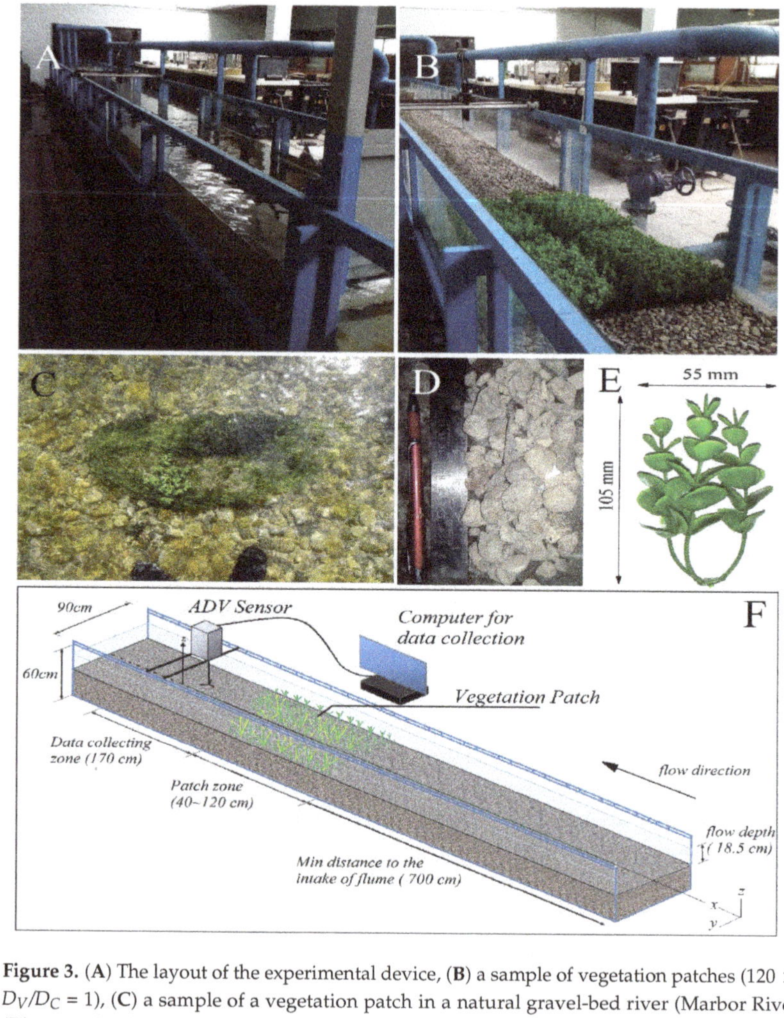

Figure 3. (**A**) The layout of the experimental device, (**B**) a sample of vegetation patches (120 × 90 cm, $D_V/D_C = 1$), (**C**) a sample of a vegetation patch in a natural gravel-bed river (Marbor River, Iran), (**D**) a sample of bed materials, (**E**) a single synthetic plant used to simulate patch, and (**F**) a sketch of the experimental set-up (Not to scale).

Table 2. Experimental parameters for the four cases discussed in this study.

Case	Q (Discharge, L/s)	n/m^2 (Number of Veg. Per Square Meter)	L_V (Length of Patch, cm)	D_V (Width of Patch, cm)	H_V (Height of Patch, cm)
1	31	611.1	120	90	10
2	31	611.1	90	60	10
3	31	611.1	60	45	10
4	31	611.1	40	30	10

The bed material on the flume bed was a mixture of natural gravel similar to that in a natural gravel-bed river (the Marbor River, Zagros Mountains region, Iran). The equivalent particle diameter that 90% of the total particles are smaller than this size (d_{90}) is 18.8 mm, and $\sqrt{\frac{D_{84}}{D_{16}}} < 1.4$.

5. Results and Discussions

The results of present study are summarized as follows: (1) the development of the vertical and horizontal vortexes behind the vegetation patch, (2) the occurrences of the dominant octant behind the vegetation patch and (3) the coherent shear stresses behind the vegetation patch. In the first part, the results regarding four cases of different vegetation layouts are provided and discussed to evaluate the impacts of the presence of a vegetation patch on the horizontal and vertical vortexes. Then, the spatial and temporal distributions of the dominant vortexes have been evaluated based on the data collected from case 4 with (dimension of a vegetation patch: 40 × 30 cm). The results and data are presented as follows: in the graphs for describing both the TKE and velocity, the vertical distances (z) are normalized by H_V, and the longitudinal distances (x) are normalized by D_V and H_V; both TKE and velocity values are also normalized by U_0^2 and U_O, respectively, where U_O is the average velocity through the flume cross-section. All velocity time series were filtered with SNR > 15 dB and Correlation >70% according to the manual of the manufacturer and also according to the previous researches.

5.1. Development of Vertical and Horizontal Vortexes behind the Patch

While a vertical vortex is normally generated behind a barrier, the development of horizontal vortexes (von Karman Vortex Street) also needs to be investigated. As mentioned before, the formation of the von Karman vortexes is associated with the occurrence of peaks in the TKE values [14]. Thus, for each case, TKE values are calculated for the relative flow depths of $z/H_v = 0.5$, $z/H_v = 1$ and also for the near-bed region behind the vegetation patch along the centerline of the flume (Figure 4). For case 1 (fully covered by vegetation across the channel bed), there exists only one peak in the graph of $\mathrm{TKE}/\mathrm{U}_0^2$ around $x/D_V = 0.9$ and $x/H_V = 7.5$. However, there is no flow escaping from the sides of the patch (since channel bed is fully covered with vegetation); this peak of $\mathrm{TKE}/\mathrm{U}_0^2$ definitely indicates the presence of a vertical vortex behind the patch and near the water surface. In this case, the minimum velocity occurred around $x/D_V = 0.3$~0.6 and $x/H_V = 3.5$~5 and near the channel bed (solid line in Figure 4). According to Liu et al. (2018) [3], the length of the wake for a wide submerged vegetation patch, L_{VV} = (3.5~5) H_V. Therefore, the calculated wake length has a higher value than that claimed by Folkard (2005) for a wide but highly submerged vegetation patches. However, our results are comparable to the values stated by Liu et al. (2018) for a non-fully covered bed with slightly submerged patches [3,12].

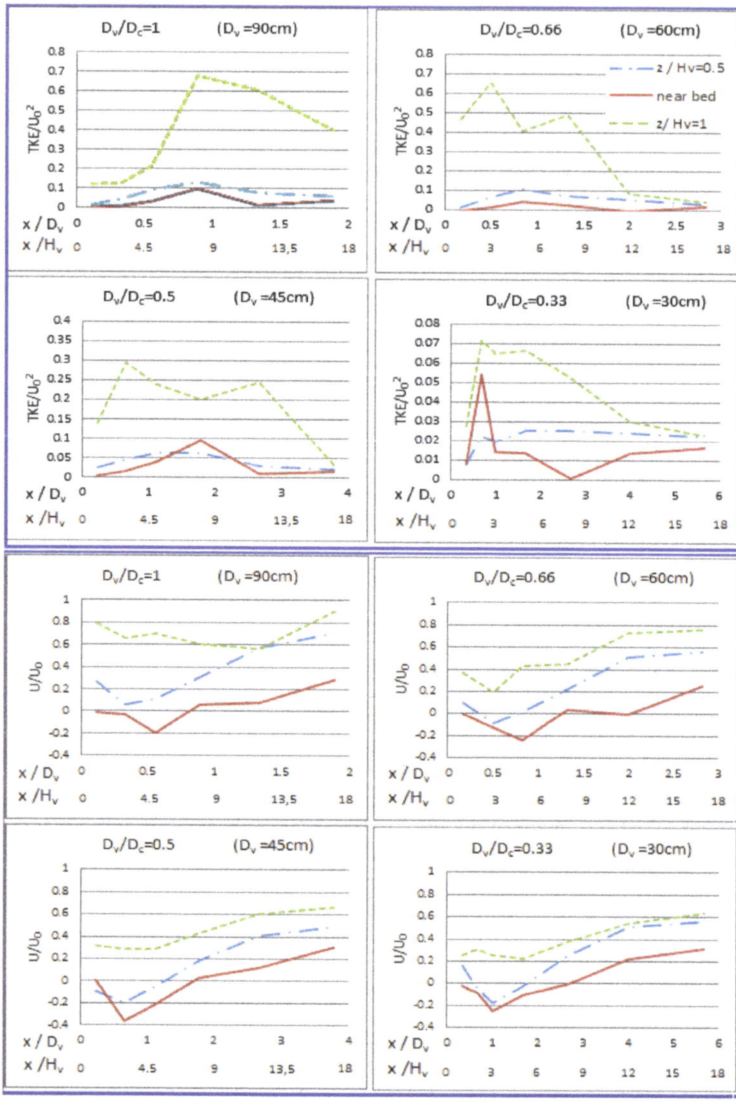

Figure 4. Normalized TKE and streamwise velocity behind 4 different patch layouts near channel bed (z/H_v = 0), at the middle height of patch (z/H_v = 0.5) and top of patch (z/H_v = 1).

For experiment cases 2, 3 and 4, there is a shift in the minimum of velocity towards the flow direction. However, since D_V/D_C is relatively high for case 2 (90 × 60 cm), the wake behind the vegetation patch (L_{vv}) is only shifted in the near-bed region. For cases 3 and 4, the shift of the wake is also detectable at the depths of z/H_V = 0.5 and z/H_v = 1. In these cases, the shift of the upper zones exceeds the shift of the near-bed zone. For all four cases, the minimum velocity is negative in the near-bed region, implying a vertical vortex. Considering the values of the corresponding velocities in the upper zone, it is observed that the center of the vortex (or rotation) moves upward if D_V/D_C decreases. In case 1, the center of the vortex is definitely between the near-bed region and z/H_V = 0.5, while it is located between z/H_V = 0.5 and z/H_V = 1 in case 4 (smallest vegetation patch).

This higher center of rotation in case 4 (smallest vegetation patch) pushes down the larger part of the flow into the lower zones. Subsequently, the velocity pattern in the middle region becomes closely similar to that of the near-bed region (one can observe this by comparing the minimum velocity in the near-bed region and $z/H_V = 0.5$). In addition, the larger radius of the rotation in smaller D_V/D_C indicates the reason for the minimum streamwise velocity in the near-bed zone rather than the upper zone. The reason for the upward movement of the rotation center in the channel with a smaller patch might be a result of a higher through-patch velocity. This statement is consistent with findings of Perera (1981), Liu et al. (2018) and Zong and Nepf (2012) regarding the larger values of L_{vv} due to the through-patch flow [3,10,13].

In addition, the flow from the sides of vegetation patches creates the second peak in the TKE curve at the depth of $z/H_V = 1$ in experiment cases 2, 3 and 4. However, in cases 2 and 3, while the D_V/D_C values are still relatively high, the second peak in the TKE curve does not occur at the middle and near-bed region. Additionally, the single peaks in the TKE curve in the lower zones are shifted forward.

Only in case 4 ($D_V/D_C = 0.33$), a second peak in the TKE curve is detectable at the depth of $z/H_V = 0.5$. In this case, there is a sharp decline in TKE in the near-bed region, and the first peak in the upper zone occurred at the same point that the single peak occurs in the near-bed zone. The absence of the second peak in the TKE in the lower zone may result from the upward movement of the center of the vertical vortex and the increase in the radius of the rotation, which leads to the blockage of flow in this region. In both the near-bed and the $z/H_V = 0.5$ zones, the extremum values of TKE occurred where the flow velocity approaches zero. The results show that the strong lateral components of velocity exist in the zone of $z/H_V = 0.5$, even when the streamwise component of the velocity is zero. However, after the second extremum, the TKE decreases in zones of $z/H_V = 0.5$ and $z/H_V = 1$ but increases in the near-bed zone.

In cases 2, 3 and 4, it seems that despite large values of turbulent energy in the upper zones, these values decrease rapidly after the second peak ($x/D_V = 1.5$–2.7 depending on the patch layout case). However, in patch layout case 4, the turbulent energy increases in the near-bed zone.

Overall, it is found that the vertical vortex behind the slightly submerged and non-fully covered vegetation patches scatters the von Karman vortexes downstream. However, for a relatively smaller vegetation patch (i.e., the patch-layout case 4), the lateral fluctuations may continue downstream and form a weak but relatively stable structure.

5.2. Dominant Octant Occurrences behind the Vegetation Patch

The probability and stability of the octant occurrences for the patch-layout case 4 are presented in Figures 5 and 6, respectively. The results of the probability of occurrence show that there are three distinct zones for almost all classes.

The first zone is shaped just behind the vegetation patch ($x < 820$ or $x/D_V < 0.66$) and located at a depth between the half-height and the vegetation tip. The dominant classes of this zone are PPP, PNP, NPN and NNN, which are outward and inward interactions. However, the classes of PPP and NNN are not stable comparing to the dominant classes of PNP and NPN in the same zone. Considering the small area of the zone, each PNP and NPN indicates a couple of small contra-rotating vortexes in the stem-scale shaped by the through-patch flow. The presence of the stem-scale turbulent region is also confirmed by and Liu et al. (2018) [3].

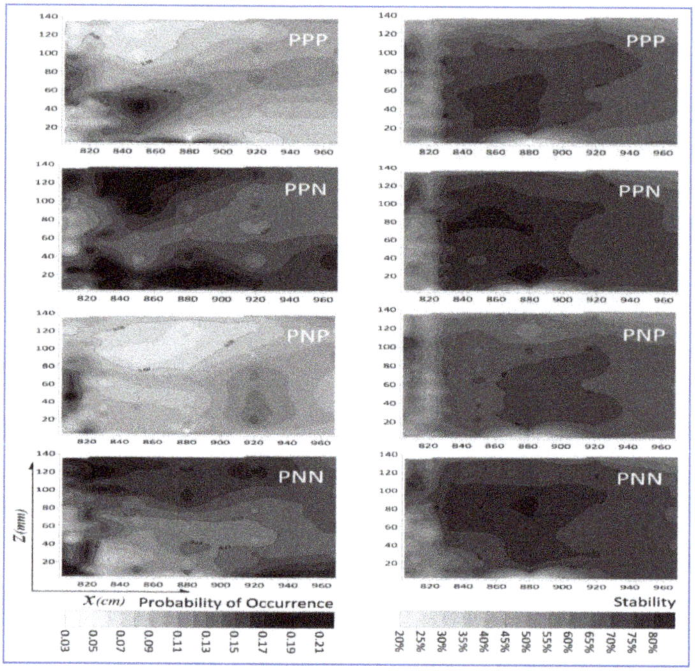

Figure 5. Probability of the occurrence and stability of octant classes PPP, PPN, PNP and PPP behind the vegetation patch. The end of the patch is located in x = 800. The height of patches is 105 ± 5 mm. The values are shown up to the level of the maximum domain of ADV capability (~140 mm).

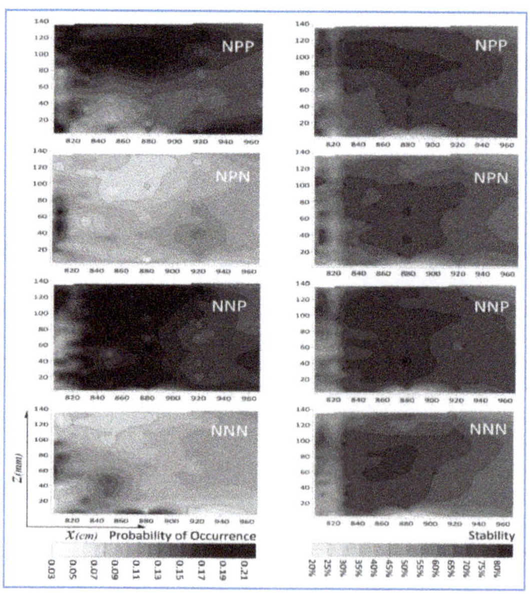

Figure 6. Probability of the occurrence and stability of octant classes NPP, NPN, NNP and NPP behind the vegetation patch.

The second zone is shaped from the top edge of the vegetation patch and covers all upper parts of the flow ($z > 80$ mm or $z/H_V > 0.75$). The dominant classes of this zone are PPN, PNN, NPP and NNP, which are outward and inward interactions. However, the classes of PPN, PNN and NPP are more dominant and also more stable in comparison to the class of NNP. In this zone, the projections of PPN and NNP on the vertical plane are contrarotating vertical vortexes that create sweep and ejection events, respectively, and erupting momentum in the right and left sides of the plane as the secondary flow in addition to the vertical momentum transition [30]. The occurrence of the sweep is responsible for the horizontal transfer of momentum towards the right side of the channel, while the occurrence of the ejection does it towards the left side. Obviously, in this procedure, the signs of both v' and u' vary simultaneously and create a fluctuation inward-outward series of occurrences in the horizontal plane. This fluctuation is formed from the top edge of the vegetation patch and creates the von Karman Vortex Street in the upper layer of the flow. The couple of PNN and NPP also act similarly but in the form of a fluctuating sweep-ejection series. However, because of the relatively less stability of the class PNN, the class NPP is not balanced in this zone. Therefore, it can be concluded that the dominant class of this zone in the vertical plane is in the form of ejection events created by NPP with a small margin. This is confirmed by Okamoto and Nezu (2013), who did a 2D quadrant analysis and showed this small margin of ejection dominancy in a similar zone for a submerged patch [6].

The third zone is shaped beyond the vegetation patch ($z/H_V < 0.75$). The spatial distribution of dominant classes in this zone is more complicated. Even though some spots look like the PPP and NNP events, no obvious dominant class of occurrence is detectable in the middle and upper part of this zone. However, in the lower region of this zone (the near-bed zone), both NNP and PPN are dominant and relatively stable for $1 < x/D_V < 4$ ($\sim 830 < x < \sim 920$). There is also a region with a high probability for PPP, but it is very unstable. Both NPP and PNN are also dominant and relatively stable for $x/D_V > 4$. Similar to the second zone, the von Karman Vortex Street is detectable in this zone. In the zone $x/D_V = 2.7$ ($x = 880$), where PPN and NNP have the highest rate of occurrence, the TKE drops to nil. However, the TKE increases slightly for $x/D_V > 4$ (see Figure 4), which is related to the change of the inward-outward series to the sweep-ejection series in the horizontal plane. For $x/D_V > 4$ and in the vertical plane, both sweep and ejection events are almost in the same prevalence, similar to $1 < x/D_V < 4$, where the outward and inward occurrences are comparable with results of a 2D quadrant analysis by Okamoto and Nezu (2013) [6]. For $x/D_V < 1$, due to the stem-scale effects of the through-patch flow, a small region of PNN-NPP is detectable in the near-bed zone.

5.3. Coherent Shear Stresses behind Thevegetation Patch

The bursting events and TKE analysis show that strong vertical vortexes lead to the diminution of the von Karman Vortex Street in the middle depth of flow. However, a part of horizontal vortexes is pushed down to the near-bed zone, and a considerable part moves up to near the surface zone. As the presence of coherent structures is associated with the presence of dominant frequency in the velocity time series, power spectra of the u component are calculated for different regions of flow behind the vegetation patch (Figure 7). The absence of peaks in the spectra in the middle region of the flow confirms that the von Karman vortex street is pushed towards the near-bed and near-surface zones. The frequency of vortex shedding in the stem-scale zone was considerably higher than that in the non-stem scale regions, as expected. The frequency of the vortex shedding in both the near-bed zone and near-surface zone is 0.07–0.11 Hz. Then, both $-\langle\widetilde{u}_c\widetilde{v}_c\rangle$ and $-\overline{\widetilde{u}_r\widetilde{v}_r}$ are calculated for regions where the von Karman vortex street is detected (Figure 8). A Matlab@ code was used to calculate the phase velocity and its deflection for the dominant frequency at each point. The coherent horizontal momentum transfer, which is triggered by the coherent horizontal vortexes, continues to happen in both the near-surface zone and near-bed zone. However, the magnitude of the phenomena is considerably low in the near-bed zone. In addition, the presence of the von Karman vortex in the near-bed zone is associated with the dominancy of NNP and PPN events that reveals the relatively strong horizontal momentum transfer events in this region. In contrast, the traditional 2D quadrant analysis is not capable of detecting this behavior of the flow in such cases. The previous research, such as Siddique et al. (2008) and Liu et al. (2018), indicate the presence of von Karman vortexes occurred in Hv/H > 0.55~0.7 [3,11]. However, they used TKE variation and visual approaches to detect the formation of von Karman vortexes in the horizontal plane. In contrast, the coherent shear stress analysis reveals that although the formation of horizontal vortexes in the near-bed zone is postponed by a strong vertical vortex behind the patch, the formation of the near-bed horizontal vortexes is not disturbed by the vertical vortex in a small patch. All results indicate that the von Karman vortexes must be also considered as a vertically distributed phenomenon that may have different behaviors in altered depths.

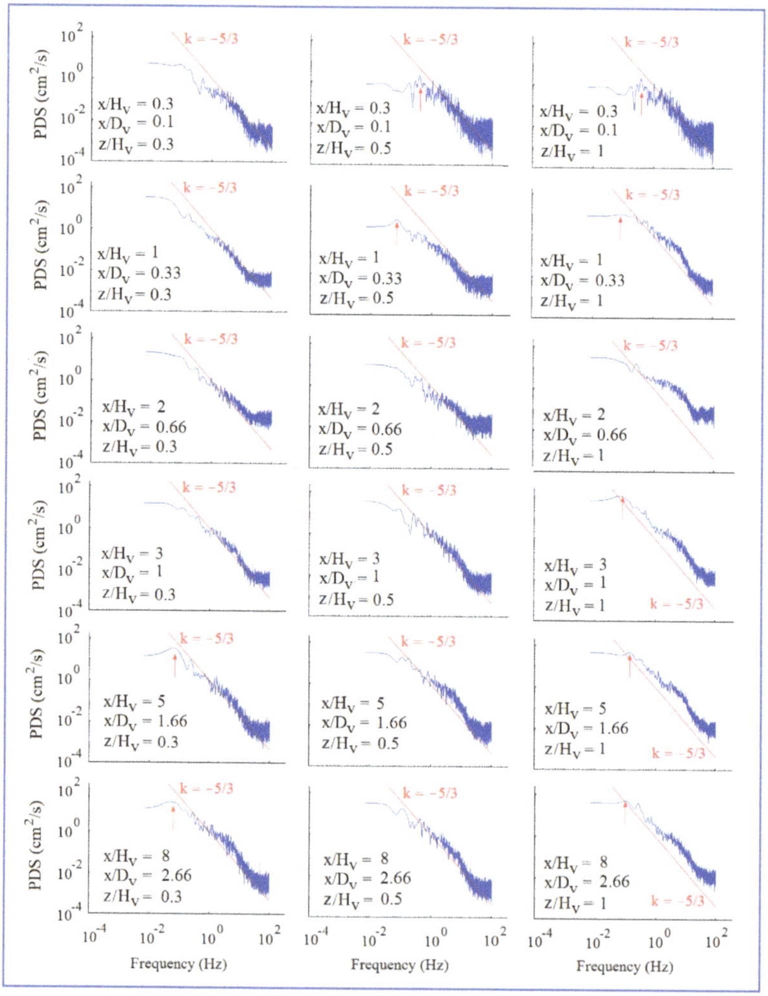

Figure 7. Power spectra behind vegetation patch for case 4; the arrow indicates the peak frequency associated with vortex shedding behind the patch, which varied from 0.07 to 0.11 Hz for main vortexes and is around 0.4 Hz for stem-scale vortexes.

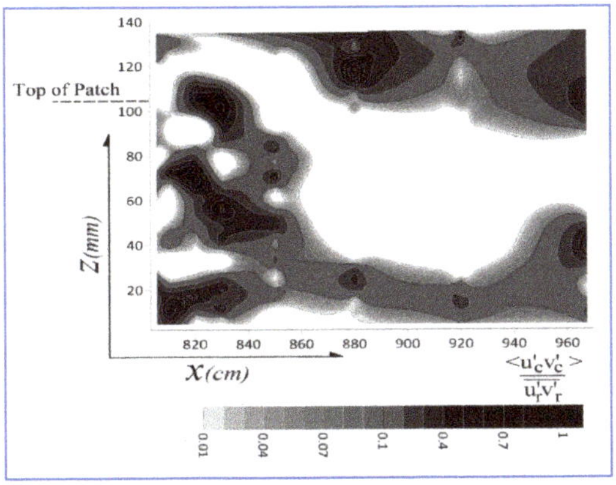

Figure 8. Values of τ_c/τ_r in the near-bed and near-surface zones; note that the white-colored area is where that no dominant frequency is detectable for velocity time series.

The frequency of the vortex shedding in the downstream of a barrier is generally calculated via Strouhal number. This dimensionless parameter is defined as St = f L/U, where f is the frequency of the vortex shedding, L is the characteristic length of the barrier and U is the upstream velocity. While the Strouhal number is calculable via Reynolds number and the shape of the barrier, the reported values for the vegetation patches are around 0.2 [5,31]. Therefore, the frequency of the vortex shedding is available from the pre-calculated Strouhal number for a particular characteristic length and upstream velocity. The dimensionless Strouhal number is calculated for the frequency of the bursting events and v (Figure 9). Obviously, the dominant frequency of v is comparable with some classes of the bursting events, particularly NNP and PPN for $x/D_V > 2.5$. However, the Strouhal number for all illustrated are similar beyond $x/D_V > 4$ (St $\approx$ 0.25). While the dominant frequency was 0.07–0.3 Hz, the period of the von Karman vortexes is evaluated as 3.5–15 s. In contrast, Liu et al. (2018) considered a period of 8–13 s [3]. Therefore, a 30 s video was taken to evaluate the length of the wake behind the vegetation patch (Figure 10). The wake length is estimated at around (0.8–1) H_V, which is the same as that of previous researches.

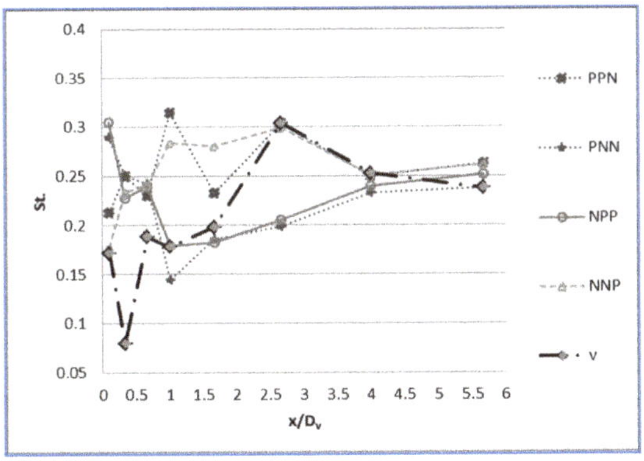

Figure 9. Strouhal Number calculated via the frequency of the bursting events and v.

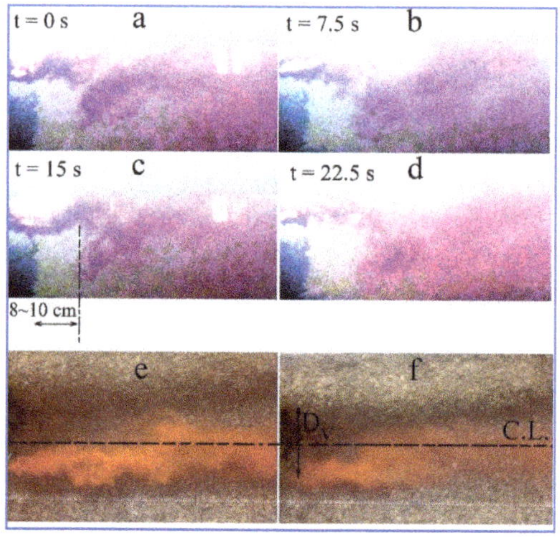

Figure 10. (**a–d**) Formation of the vertical vortexes and the wake zone behind the vegetation patch for case 4; (**e**) formation of the horizontal vortexes behind the patch with dike injection at z/Hv = 1 and (**f**) near the bed. Note that the injection point is located at the right edge of the patch for the horizontal vortexes.

6. Conclusions

In this paper, the model vegetation patches are used to model vegetation patches in natural rivers. The characteristics of the formation of vortex structures in the wake of submerged vegetation patches have been studied. The results of the TKE and streamwise velocity indicate that both horizontal and vertical vortexes formed behind the vegetation patch for a channel bed that was partially covered with vegetation. However, strong vertical vortexes lead to the diminution of the von Karman Vortex Street. With the presence of the smallest vegetation patch in the channel bed (case 4, $L_V/D_c = 0.44$ and $D_V/D_c = 0.33$), the through-patch flow near the channel bed leads to the formation of vortex toward the downstream; however, the von Karman vortex street is presented near the channel bed. Although previous researches indicated the formation of von Karman vortexes beyond some patches with particular geometric scales, the vertical distribution of horizontal vortexes has not been investigated as a particular topic. The results of a 3D analysis of the bursting events show that the von Karman vortex street is fragmented in two parts in the near-bed zone (the weak part) and near-surface zone (the strong part), where the horizontal vortexes are associated with a particular bursting event in each zone. The formation of the near-surface von Karman vortexes is affected considerably by the vertical vortexes behind the patch; however, the near-bed horizontal vortexes almost remain intact.

The results of this study are confirmed by the behavior of the power spectra as well as the coherent shear stresses. The Strouhal number of about 0.25 is also calculated based on the frequencies of the associated bursting events and fluctuations of the lateral velocity component.

Author Contributions: M.K.: Laboratory works, methodology, software and writing—original draft preparation; H.A.: Laboratory supervisory, methodology and validation; J.S.: methodology and writing reviewing. All authors have read and agreed to the published version of the manuscript.

Funding: This research received no external funding.

Institutional Review Board Statement: Not applicable.

Informed Consent Statement: Not applicable.

Data Availability Statement: The data are available in the case that it is required.

Acknowledgments: Special thanks to G. Ghodrati Amiri for providing the laboratory devices and Esmaeel Dodangeh for his contribution in laboratory setup.

Conflicts of Interest: The authors declare no conflict of interest.

References

1. Nikora, V.; Larned, S.; Nikora, N.; Debnath, K.; Cooper, G.; Reid, M. Hydraulic resistance due to aquatic vegetation in small streams: Field study. *J. Hydraul. Eng.* **2008**, *134*, 1326–1332. [CrossRef]
2. Gurnell, A.M.; Bertoldi, W.; Corenblit, D. Changing river channels: The roles of hydrological processes, plants and pioneer fluvial landforms in humid temperate, mixed load, gravel bed rivers. *Earth-Sci. Rev.* **2012**, *111*, 129–141. [CrossRef]
3. Liu, C.; Hu, Z.; Lei, J.; Nepf, H. Vortex structure and sediment deposition in the wake behind a finite patch of model submerged vegetation. *J. Hydraul. Eng.* **2018**, *144*, 04017065. [CrossRef]
4. Caroppi, G.; Västilä, K.; Järvelä, J.; Rowiński, P.M.; Giugni, M. Turbulence at water-vegetation interface in open channel flow: Experiments with natural-like plants. *Adv. Water Resour.* **2019**, *127*, 180–191. [CrossRef]
5. Afzalimehr, H.; Riazi, P.; Jahadi, M.; Singh, V.P. Effect of vegetation patches on flow structures and the estimation of friction factor. *ISH J. Hydraul. Eng.* **2019**, 1–11. [CrossRef]
6. Okamoto, T.A.; Nezu, I. Spatial evolution of coherent motions in finite-length vegetation patch flow. *Environ. Fluid Mech.* **2013**, *13*, 417–434. [CrossRef]
7. Vandenbruwaene, W.; Temmerman, S.; Bouma, T.J.; Klaassen, P.C.; De Vries, M.B.; Callaghan, D.P.; Van Steeg, P.; Dekker, F.; Van Duren, L.A.; Martini, E.; et al. Flow interaction with dynamic vegetation patches: Implications for biogeomorphic evolution of a tidal landscape. *J. Geophys. Res. Earth Surf.* **2011**, *116*. [CrossRef]
8. Bretón, F.; Baki, A.B.M.; Link, O.; Zhu, D.Z.; Rajaratnam, N. Flow in nature-like fishway and its relation to fish behavior. *Can. J. Civ. Eng.* **2013**, *40*, 567–573. [CrossRef]
9. Golpira, A.; Baki, A.B.; Zhu, D.Z. Turbulent Events around an Intermediately Submerged Boulder under Wake Interference Flow Regime. *J. Hydraul. Eng.* **2021**, *147*, 06021005. [CrossRef]
10. Zong, L.; Nepf, H. Vortex development behind a finite porous obstruction in a channel. *J. Fluid Mech.* **2012**, *691*, 368–391. [CrossRef]
11. Sadeque, M.A.; Rajaratnam, N.; Loewen, M.R. Flow around cylinders in open channels. *J. Eng. Mech.* **2008**, *134*, 60–71. [CrossRef]
12. Folkard, A.M. Hydrodynamics of model Posidonia oceanica patches in shallow water. *Limnol. Oceanogr.* **2005**, *50*, 1592–1600. [CrossRef]
13. Perera, M.D.A.E.S. Shelter behind two-dimensional solid and porous fences. *J. Wind Eng. Ind. Aerodyn.* **1981**, *8*, 93–104. [CrossRef]
14. Chen, Z.; Ortiz, A.; Zong, L.; Nepf, H. The wake structure behind a porous obstruction and its implications for deposition near a finite patch of emergent vegetation. *Water Resour. Res.* **2012**, *48*, 9. [CrossRef]
15. Chen, Z.; Jiang, C.; Nepf, H. Flow adjustment at the leading edge of a submerged aquatic canopy. *Water Resour. Res.* **2013**, *49*, 5537–5551. [CrossRef]
16. Kabiri, F.; Afzalimehr, H.; Sui, J. Flow structure over a wavy bed with vegetation cover. *Int. J. Sediment Res.* **2017**, *32*, 186–194. [CrossRef]
17. Sui, J.; Faruque, M.A.; Balachandar, R. Local scour caused by submerged square jets under model ice cover. *J. Hydraul. Eng.* **2009**, *135*, 316–319. [CrossRef]
18. Jafari, R.; Sui, J. Velocity Field and Turbulence Structure around Spur Dikes with Different Angles of Orientation under Ice Covered Flow Conditions. *Water* **2021**, *13*, 1844. [CrossRef]
19. Hussain, A.F. Role of coherent structures in turbulent shear flows. *Proc. Indian Acad. Sci. Sect. C Eng. Sci.* **1981**, *4*, 129–175.
20. Afzalimehr, H.; Moghbel, R.; Gallichand, J.; Sui, J. Investigation of turbulence characteristics in channel with dense vegetation. *Int. J. Sediment Res.* **2011**, *26*, 269–282. [CrossRef]
21. Devi, T.B.; Kumar, B. Channel hydrodynamics of submerged, flexible vegetation with seepage. *J. Hydraul. Eng.* **2016**, *142*, 04016053. [CrossRef]
22. Huai, W.X.; Zhang, J.; Katul, G.G.; Cheng, Y.G.; Tang, X.; Wang, W.J. The structure of turbulent flow through submerged flexible vegetation. *J. Hydrodyn.* **2019**, *31*. [CrossRef]
23. Zamani, M.; Afzalimehr, H.; Jahadi, M.; Singh, V.P. Flow structure over bed form with flexible vegetation patches. *ISH J. Hydraul. Eng.* **2021**, 1–7. [CrossRef]
24. Przyborowski, Ł.; Łoboda, A.M. Identification of coherent structures downstream of patches of aquatic vegetation in a natural environment. *J. Hydrol.* **2021**, *596*, 126123. [CrossRef]
25. Keshavarzi, A.R.; Gheisi, A.R. Stochastic nature of three dimensional bursting events and sediment entrainment in vortex chamber. *Stoch. Environ. Res. Risk Assess.* **2006**, *21*, 75–87. [CrossRef]
26. Keshavarzi, A.; Melville, B.; Ball, J. Three-dimensional analysis of coherent turbulent flow structure around a single circular bridge pier. *Environ. Fluid Mech.* **2014**, *14*, 821–847. [CrossRef]
27. Termini, D. Experimental Analysis of Horizontal Turbulence of Flow over Flat and Deformed Beds. *Arch. Hydro-Eng. Environ. Mech.* **2015**, *62*, 77–99. [CrossRef]

28. Termini, D.; Sammartano, V. Analysis of the role of turbulent structure in bed-forms formation in a rectilinear flume. In Proceedings of the 32nd Congress of IAHR, "Harmonizing the Demands of Art and Nature in Hydraulics", Venice, Italy, 1–6 July 2007.
29. Termini, D.; Sammartano, V. Experimental observation of horizontal coherent turbulent structures in a straight flume. In Proceedings of the 2019 River Coast. Estuarine Morphodynamics—RCEM 2019, Santa Fe City, Argentina, 21–29 September 2009.
30. Wang, H.; Peng, G.; Chen, M.; Fan, J. Analysis of the interconnections between classic vortex models of coherent structures based on DNS data. *Water* **2019**, *11*, 2005. [CrossRef]
31. Graf, W.H.; Altinakar, M.S. *Fluvial Hydraulics: Flow and Transport Processes in Channels of Simple Geometry*; No. 551.483 G7; Wiley: Hoboken, NJ, USA, 1998.

water

MDPI

Article

Investigation of the Effect of Vegetation on Flow Structures and Turbulence Anisotropy around Semi-Elliptical Abutment

Seyedeh Fatemeh Nabaei [1], Hossein Afzalimehr [2,*], Jueyi Sui [3,*], Bimlesh Kumar [4] and Seyed Hamidreza Nabaei [5]

1 Department of Water Engineering, Isfahan University of Technology, Isfahan 8415683111, Iran; f.nabaei@alumni.iut.ac.ir
2 Department of Civil Engineering, Iran University of Science and Technology, Tehran 1684613114, Iran
3 School of Engineering, University of Northern British Columbia, Prince George, BC V2N 4Z9, Canada
4 Department of Civil Engineering, Indian Institute of Technology Guwahati, Guwahati 781039, India; bimk@iitg.ac.in
5 Department of Mechanical Engineering, Tabriz University, Tabriz 5166616471, Iran; hamidreza.nabaei95@ms.tabrizu.ac.ir
* Correspondence: hafzali@iust.ac.ir (H.A.); jueyi.sui@unbc.ca (J.S.); Tel.: +1-250-960-6399 (J.S.)

Abstract: In the present experimental study, the effect of vegetation on flow structure and scour profile around a bridge abutment has been investigated. The vegetation in the channel bed significantly impacted the turbulent statistics and turbulence anisotropy. Interestingly, compared to the channel without vegetation, the presence of vegetation in the channel bed dramatically reduced the primary vortex, but less impacts the wake vortex. Moreover, the tangential and radial velocities decreased with the vegetation in the channel bed, while the vertical velocity (azimuthal angle > 90°) had large positive values near the scour hole bed. Results showed that the presence of the vegetation in the channel bed caused a noticeable decrease in the Reynolds shear stress. Analysis of the Reynolds stress anisotropy indicated that the flow had more tendency to be isotropic for the vegetated bed. Results have shown that the anisotropy profile changes from pancake-shaped to cigar-shaped in the un-vegetated channel. In contrast, it had the opposite reaction for the vegetated bed.

Keywords: local scour; semi-elliptical abutment; vegetated bed; Reynolds shear stress; turbulence intensity; Reynolds stress anisotropy

Citation: Nabaei, S.F.; Afzalimehr, H.; Sui, J.; Kumar, B.; Nabaei, S.H. Investigation of the Effect of Vegetation on Flow Structures and Turbulence Anisotropy around Semi-Elliptical Abutment. *Water* **2021**, *13*, 3108. https://doi.org/10.3390/w13213108

Academic Editors: Francesco Gentile and C. Jason Williams

Received: 31 August 2021
Accepted: 31 October 2021
Published: 4 November 2021

1. Introduction

Over the past several decades, researchers have emphasized the importance of the scour process for designing bridges and other hydraulic structures, e.g., [1–5]. One of the crucial causes of bridge failure is the scour around its abutment and the associated flow pattern around the structure [6]. The Federal Highway Administration of the USA investigated 383 cases of bridges damaged by catastrophic floods and estimated that 72% of the bridge failures were caused by abutment damage [7]. The flow pattern and scour mechanism around the bridge abutment are phenomena that result from the reaction between the three-dimensional turbulence flow field around the abutment foundation in erodible beds [8]. Around the abutments, the development of the boundary layer of the protrusion wall has created complexity in the flow field [9]. The flow field around an abutment involves a complex three-dimensional (3D) vortex flow, and this complexity is increased by the development of the scour hole, involving flow separation [10]. The scour hole around an abutment is developed by both primary vortices and the downflow, similar to a horseshoe vortex at piers. The downflow is the principal cause of the development of the scour hole. The secondary vortexes are created near the primary vortex, behind the abutment and at the separation zone, by limiting the power of the primary vortex in the scour hole development. Downstream of the abutment, the factor that causes the flow separation from the abutment creates the wake vortices [11]. In short, it can be said that the impact of the flow on the upstream face of the abutment and the separation of it downstream of the abutment is one of the most critical factors in the scouring process at the abutments [10]. By studying wing-wall abutment, Kw An and Melville [11] found that the wake vortices downstream of an abutment were caused by the flow separation at the abutment's corner. As such, the wake vortices form at the downstream region of the abutment. These vortices, with the vertical axis and low-pressure center, suck up sediment particles and move sediment particles downstream, after separation from the bed with the mainstream, creating an independent scour hole downstream of an abutment [10]. Readers can refer to figure 6.3 of Melville and Coleman [12], which illustrates the flow and scour patterns around a short abutment.

There have been numerous investigations for estimating the local scour rate around bridge abutments, e.g., [13–16]. Some researchers have also studied the flow field and characteristics of flow around bridge abutments within scour holes [3,17,18]. Most of these studies are concentrated on flow patterns around bridge abutments in an alluvial channel based on laboratory experiments.

Due to the complexity of the scouring process around an abutment, resolving the flow feature near the scour hole bed and turbulence characteristics is profoundly challenging [19]. The assumption of isotropic distribution of turbulent statistics in numerical models prohibits their application in scouring around the bed abutment. The purpose of conducting anisotropy analysis is to understand the turbulent flow characteristics better and determine the turbulence structure's sensitivity for different bed conditions [20–22]. By introducing the invariant functions, the turbulence anisotropy effectively reduces the complexity of a three-dimensional flow field to a two-dimensional flow that is simpler for analysis [23]. Thus, the Reynolds stress anisotropy study is an important research topic for developing turbulence theories and numerical simulations.

Lumley and Newman [20] proposed the technique of the anisotropy invariants, which provides a procedure to analyze turbulence. Their study used the normalized Reynolds stress anisotropy tensor invariants to establish the anisotropy invariant map (AIM, also referred to as the Lumley map). Mera et al. [24] analyzed the evolution of the magnitude and nature of anisotropy along a meandering river. AIM also characterized the spatial distribution of anisotropy invariants and the nature of turbulence anisotropy.

The riverine plants play a crucial role in river dynamics and hydraulic structure [25]. From a hydrological view, although it is of utmost importance to protect the ecosystem area of a river network dominated by vegetation [26], this issue has not been thoroughly studied and used effectively. Besides, some engineering measures and expensive solutions have been installed to prevent a local scour process around hydraulic structures (such as the collar, spur dike, and submerged vane). On the other hand, the application of vegetation for reducing scouring can provide an eco-friendly and economical outcome, as it can help with the protection of the river territory, and it is more compatible with the environment. Luhar et al. [27] used hydrodynamic concepts to explore how vegetation affects the flow and transmission and how the flow feedback can affect the spatial structure of vegetation. Some studies have been conducted to describe the flow in vegetated channels. Results indicated that the presence of vegetation in the channel bed modifies flow and sediment transport [28–30]. In these studies, it was found that vegetation in the channel increases flow resistance, modifies flow patterns, and provides additional drag while leading to a decrease in the bed shear stress, thus significantly decreasing sediment transport.

Afzalimehr et al. [31] stated that in channels with a submerged vegetated bed, its Reynolds stress distribution was nonlinear due to the presence of drag force. This drag force was caused by submerged plants in the water and had its maximum value slightly above the top of the vegetation canopy. Comparing Reynolds stress distribution and turbulence intensity graphs to those of velocity distribution in non-uniform decelerating flow with a vegetation bed, Keshavarz et al. [32] concluded that the zone for the maximum Reynolds stresses and turbulence intensity coincides with the region where the highest rate of change in velocity occurs. Afzalimehr et al. [33] asserted that vegetation on channel banks changes the shape of the scour hole compared to that without vegetation. They concluded that vegetated banks decreased Reynolds shear stress near the bed and played no role in the vortex structure downstream of the abutment. In such circumstances, understanding the effect of vegetation on flow conditions in river restoration projects is essential. Although some studies have been reported, knowledge gaps regarding the impact of vegetation around the hydraulic structure, such as the bridge abutment, remain.

The vegetation in a channel bed is one of the prominent factors in environmental hydraulics, which impacts the flow field and scouring process around the abutments. In this experimental study, a semi-elliptical abutment has been used because of its aerodynamical feature. The purpose of this study was to investigate the flow pattern around a semi-elliptical abutment in a channel with submerged vegetation (hereafter, we use the word vegetation to represent the submerged vegetation) and the impact of vegetation on the flow pattern. The effect of the anisotropic turbulence on the local scour has been introduced in the present work. In the present study, equivalent experimental conditions have been set to compare the effect of a vegetated channel bed on the local scour around an abutment to that without a vegetation cover. The following aspects have been investigated: the scouring mechanism, changes of the flow parameters, flow field, Reynolds shear stress distribution, Reynolds stress anisotropy, and evolution of the anisotropic invariant function around the semi-elliptical abutment.

2. Materials and Methods

Experiments have been conducted in a laboratory flume for two cases: channel bed without vegetation (first case) and channel bed with vegetation (second case). The flume is 16.0 m long, 0.9 m wide, and 0.6 m deep with a rectangular cross-sectional area. Experiments for each case have been repeated twice under the same hydraulic conditions. The flume has a glass floor with glass walls. The bed slope of this flume is 0.0003. A weir located at the end of the flume was used to adjust the water level in the flume. A pump was used to circulate the water with a maximum discharge capacity of 0.06 $m^3 s^{-1}$ from the sump. An electromagnetic flowmeter was installed in the supply conduit to measure the discharge passing through the flume continuously. A semi-elliptical abutment has been used in this experimental study. The semi-elliptical abutment has a length (along the streamwise direction) of l = 15 cm and a width (perpendicular to the streamwise direction) of b = 4 cm.

The abutment was made of Teflon (thermoplastic polymer), which has suitable properties such as lightness, high strength, and long-term durability in water. The selection of the abutment (b) width should be sufficient to avoid the influence of the channel banks on the scouring process around it (this means the sidewall has no effect on the scour in this case study). Chiew and Melville [34] stated that the distance ratio of the facing wall from the center of the abutment to the width of the abutment (b) is equal to at least 5. A semi-elliptical abutment was chosen in this study, since a few studies have been performed by using a semi-elliptical abutment [3,33]. The experimental section (sandbox) was 1.0 m long, 0.16 m deep, and 0.9 m wide, located at 10.3 m downstream of the flume entrance. In this testing section, the flow was developed. The abutment was embedded in this sandbox. Teflon panels were installed at 0.16 m above the flume bottom and covered with sediments. The one-meter-long sandbox between the upstream Teflon panel and the downstream Teflon panel was filled with sieved sand. According to Dey et al. [35], to eliminate the non-uniform effect of sediment particles that can significantly reduce the scour depth, the geometric standard deviation of the sediment particle size $\sigma_g = (d_{84}/d_{16})^{0.5}$ should be less than 1.4. In this equation, d_i is the grain size (mm), smaller than i percent of sediment particles. The sediment used in this study was uniform sand prepared by using mechanical sieve analysis tests. The median diameter of the sediment particles was d_{50} = 0.75 mm, and the geometric standard deviation of the particle size distribution σ_g was 1.22 (<1.4).

For this experimental study, the discharge was set up as $Q = 0.05\ m^3\ s^{-1}$ with a flow depth of 0.23 m in the flume, at a point where no movement of sediment was observed in the channel bed. As mentioned before, the flow depth in the flume was adjusted by the end slide gate. Moreover, under the condition of the same flow rate, water depth was gradually decreased, and velocity gradually increased until the motion of sediment particles was observed, namely, the incipient motion of sediment particles. The subcritical flow depth of 0.18 m for the incipient motion of sediment particles was obtained. Then, the approaching flow depth h was maintained at 19 cm. The experiment ran with an average approaching flow velocity of $U = 0.292\ m\ s^{-1}$ and Froude number of $Fr = 0.2138$ for each experiment, which satisfied the clear water scour condition of $u^*/u^*_c = 0.95$, where u^* is the shear velocity, which is calculated using the log law [36,37]. In this experimental study, u^*_c and u^* values were $0.0126\ m\ s^{-1}$ and $0.01197\ m\ s^{-1}$, respectively. The point gauge with an accuracy of ± 1.0 mm was used to measure flow depth.

In the first phase of this experimental study, the abutment was installed at 11.0 m downstream from the flume entrance. The flow in the testing section (sandbox) was fully developed turbulent flow, since it was confirmed by the longitudinal velocity profiles measured at the cross-sections of 7.5, 8, and 8.5 m from the flume entrance. At these three cross-sections upstream of the sandbox, the longitudinal velocity profiles had been matched to each other.

In the second phase of this experimental study, the effect of vegetation in channel bed on the flow characteristics around the semi-elliptical abutment was investigated. To the best of our knowledge, neither experimental nor numerical studies have been carried out in this field. In the first step, a glass panel with dimensions of 0.4 m $\times$ 0.5 m $\times$ 0.006 m (length $\times$ width $\times$ thickness) was placed on the bottom of the sandbox. Then, numerous bunches of vegetation were stuck on the glass bed. These vegetation are nearly uniformly distributed around the abutment and form a geometrical network in the form of rhombuses with equal dimensions and a side length of 14 cm (approximately) (Figure 1). Each vegetation bunch consists of about 55–65 pine needles.

Figure 1. (**a**) Vegetation bunches consisting of pine needles used in experiments; (**b**,**c**) vegetation bunches installation and arrangement.

In the present study, this type of geometrical distribution of vegetation has been selected to consider the effect of resistance created by the scattered vegetation spots in the flow and understand how it affects the scouring process and flow structure. Pine needles were used as vegetation because of their resistance, stability, and woodiness [38]. The height of the pine needles was 0.15 m. After installing the vegetation on the glass panel, the vegetation bed was filled with sieved sand with a depth of 0.11 m. The length of the vegetation above the sediment bed was only 0.04 m. With such a setup of vegetated bed around the semi-elliptical abutment, the experiment was run under clear water scour conditions with defined discharge.

According to Melville and Chiew [39], the scour equilibrium was achieved after 72 hours in the first experiment for channel bed without vegetation and 48 hours in the second experiment for channel bed with vegetation (Figure 2).

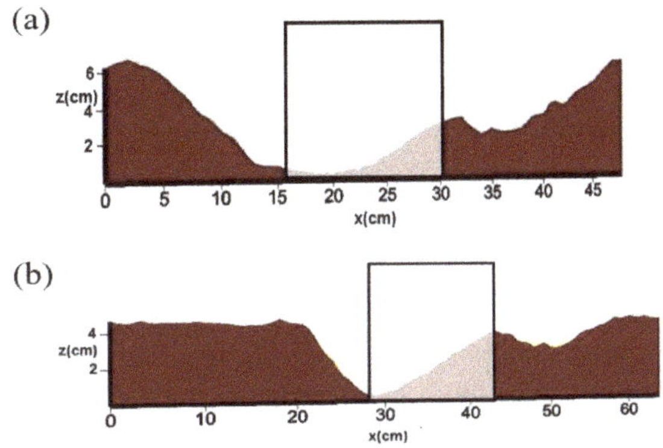

Figure 2. Profile of the scour hole after the equilibrium state was achieved (**a**) without vegetation and (**b**) with vegetation.

After the scour process reached the equilibrium state for each experimental run, the flow discharge was slowly reduced to zero, and the pump was switched off. After the flume water was completely drained, the scour hole was stabilized by spraying a thin layer of chemical material. A hypothetical network with a resolution of 1.0 cm was determined on the bed surface, and their intersection points were marked. Then, about 1500 points were measured using a mobile limnimeter (a point gauge). After that, by defining an origin mark (shown in Figure 3, with the coordinate (0,0,0)), the coordinates for the collected data were applied to each point, and the topography was drawn using Surfer software.

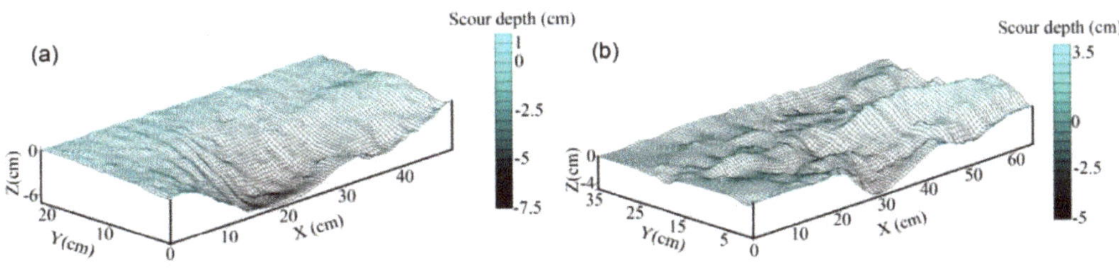

Figure 3. Topography of the final scouring hole around the abutment: (**a**) without vegetation (abutment coordinates: x = [16,31], y = [0,r], z= [0,40]); (**b**) with vegetation (abutment coordinates: x = [28,43], y = [0,r], z = [0,40]).

Figure 3 shows the topography of the flume bed after the scouring process reaches the equilibrium state. After measuring the scoured flume bed, the flume was gradually filled up with water again. Then, velocities at different azimuthal angles θ with specific radial distances r were measured. The instantaneous three-dimensional velocity components were measured at different sections using a down-looking acoustic Doppler velocimeter (ADV), Vectrino^{+} model, made by Nortek, with a duration of 120 s. The sampling frequency was set at 200 Hz [40]. The accuracy and quality of the collected data were controlled by two parameters, the correlation coefficient (COR) and the signal-to-noise ratio (SNR).

In this study, to investigate the flow patterns around the abutment, it was necessary to measure the velocities around the scour holes at different azimuthal sections. Therefore, we selected the measured points in cylindrical polar coordinates (θ, r, z) with azimuthal angles of 30°, 60°, 90°, 120°, and 160°. Additionally, using the ADV at each azimuthal angle, four profiles were measured at the radial axis with spacing distances of 4, 6, 8, and 10 cm from the abutment's surface. Along each vertical axis, data have been collected at 20–30 measuring points. Observations have been collected from 4 mm above the bed to the point 5 cm below the water surface.

Win ADV software [41] has been used to calculate turbulent flow statistics. The software also filters the inappropriate recorded data with SNR and COR less than 15 dB and 70%, respectively. In this study, we used this filter to obtain the desired data. Moreover, the filter provided by Goring and Nikora [40] for the phase-space threshold despiking has been used to detect and eliminate the spurious data.

3. Results and Discussions

The scouring process was initiated from the front face of the abutment in an unvegetated channel. The two-dimensional stream acceleration primarily causes scouring due to the effect of obstacles. Then, rill erosion was quickly formed around the abutment [42,43]. A few minutes later, a creeping motion of bed materials happened. The rill erosion reached the rear of the abutment, and the scouring process extended to the downstream region of the abutment. In other words, the shear stress in the vicinity of the abutment exceeds its critical value, hence causing bed erosion around the abutment. It was observed that sediment particles were transported further downstream of the abutment and progressively formed a deposition dune. This phenomenon can be explained as the washing out process of the bed particles in the front face of the abutment by the downward flow, which causes the development of the scour hole around the abutment. Besides, the presence of the wake vortices influences the movement of bed particles, which results in decreasing the dune's height. After the initial scouring process, the primary vortex (horseshoe vortex) structures seem to play a significant role in extending and developing the scour hole around the abutment. This process was observed in the case of the vegetated bed as well, but with less equilibrium scour depth. The maximum scour depth for the case without vegetation in the bed was located near the abutment nose (in the middle portion and upstream of the abutment) and reached 6.6 cm.

However, for the case with a vegetated bed, the maximum scour depth moved toward the front face of the abutment and reached a maximum depth of 4.3 cm. Results showed that the appearance of vegetation on the channel bed resulted in a noticeable reduction in the depth of the scour hole. Moreover, since the vegetation have significantly changed the flow velocity, the incipient motion process of sediment has been affected. The results of the incipient motion process around the abutment in a vegetated channel confirmed the findings of Shahmohammadi et al. [44,45], who studied the incipient motion of sediment in a vegetated channel without the presence of an abutment. Shahmohammadi et al. [44,45] claimed that the presence of vegetation patches resulted in a decrease in cross-averaged streamwise velocity by almost 20%. Moreover, the scouring process around multiple vegetation patches interacted with and affected the transport of particles from the vegetation patch to further downstream of the channel. In the present study, the eroded sediment particles were delivered downstream of the abutment to form the highest deposition dune at a 15 cm distance from the channel wall.

The instantaneous changes of the scour depth around the abutment for both the vegetated and un-vegetated bed are shown in Figure 4. One can see from Figure 4 that the presence of bed vegetation significantly reduced the depth of the scour hole. Up to $t = 120$ min into the scour process (the intersection of two trend lines for scour depth), the scour depth for the case with the vegetated bed was surprisingly higher than that of the un-vegetated bed. During the initial stage of the scouring process, in the vicinity of the upstream region of the abutment, the vegetation resulted in stronger primary vortices. After 2 h of the experimental run, the scour depth for the case without vegetation in the bed increased faster than for the case with the vegetated bed. For both cases, the scour hole depths increased gradually with time but at a decreasing rate.

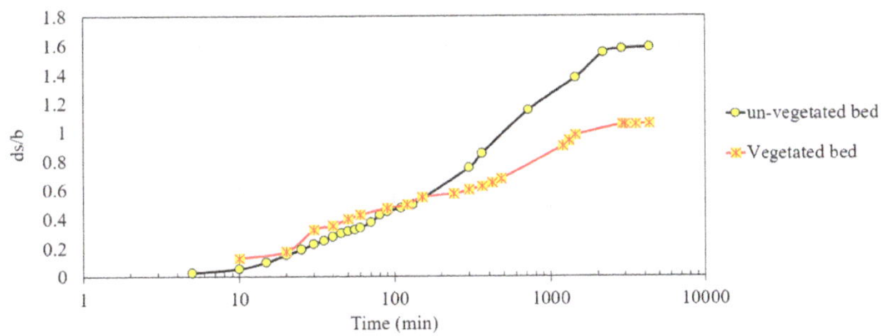

Figure 4. Variation of relative scour depth with time in the front face of the semi-elliptical abutment for both un-vegetated and vegetated beds.

3.1. Velocity Field around the Abutment

With the increase in the scour depth, the flow velocity decreased, since the flow cross-sectional area increased. Figure 5a,b show the streamlines for the un-vegetated bed and Figure 5c,d for the vegetated bed, after the scouring process reached the equilibrium state. The streamlines were drawn using a linear interpolation method at two heights of $d = 0.6$ cm and $d = 10.0$ cm. It was also observed that water moved slower with less momentum near the abutment wall and its base and then turned away more than the faster-moving water from the abutment and the bed. This phenomenon caused the skewing of velocity profiles, as explained by Dey [10]. Upstream of the abutment, close to the scour hole bed (mboxemphd = 0.6 cm), the circulation was strong and decreased with an increase in the azimuthal angles θ. The horizontal over the scour hole resulted in a reverse flow and wakes in the scour hole. Figure 5 shows that the appearance of vegetation around the abutment influenced the mean velocity field (U) patterns in both heights ($d = 0.6$ cm and $d = 10.0$ cm from the scour hole bed, respectively). It was observed that vegetation significantly reduces the mean velocity around the abutment at a lower height ($d = 0.6$ cm). This phenomenon was observed in the downstream area of the abutment (Figure 5c). Close to the scour hole bed ($d = 0.6$ cm), a significant decrease in the mean velocity was observed at the point (x, y) = (4 cm, 9 cm) for the case without vegetation in the bed and at (x, y) = (1 cm, 14 cm) for the case with the vegetated bed. The U values reached their minimum at the scour hole bed and increased with the distance from the scour hole bed.

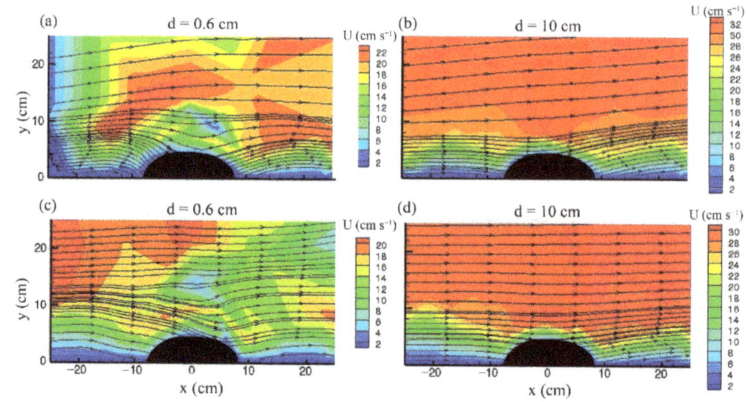

Figure 5. Streamlines around the abutment (**a**,**b**) for the un-vegetated bed and (**c**,**d**) for the vegetated bed at two heights of d = 0.6 cm and d = 10.0 cm from the stabilized scour bed. Note that the range of the color scale is not equalized for better illustration.

A cylindrical polar coordinate (Figures 6–11) was used to represent the flow patterns. The time-averaged velocity components in (θ, r, z) are respectively represented by (u, v, w), whose corresponding fluctuations are (u', v', w'). For this experimental study, u describes the tangential velocity, v is used to denote the radial velocity, and the vertical velocity is expressed as w. The positive directions of u, v, and w are counterclockwise, outward, and upward, respectively. The velocity distributions are plotted in the rz plane at different azimuthal angles θ of 30°, 60°, 90°, 120°, and 160°. The abutment wall represents r_0 = 0, and r_0 refers to Equation (1) for calculating the abscissa scale.

$$r_0 = r - r(\theta) = r - \frac{bl/2}{\sqrt{(b\cos\theta)^2 + (l/2\sin\theta)^2}} \tag{1}$$

For the case without vegetation in the bed, the contours of the tangential velocity u at different azimuthal planes (30°, 60°, 90°, 120°, and 160°) are shown in Figure 6. One can observe the features of the passage of the flow around the abutment. The tangential velocity u, among all 3D velocity components, played a crucial role in developing the scour hole and the turbulence structure. The magnitude of u without vegetation increased with the azimuthal degree θ from 0° to 120° and then decreased at θ = 160° (at the downstream zone of the abutment). The magnitude of u was more significant when the scour depth was smaller, while it decreased gradually (θ = 160°) with an increase in scour depth because of an increase in flow area. The strong circulation in the upstream region of the abutment caused a significant decrease in the tangential velocity to a great extent near the scour bed. The results clearly showed that this impact had been reduced with the distance from the abutment. In Figure 6, negative values have not been observed. However, by screening data obtained using the ADV, the negative instantaneous tangential velocities have been noticed. Either vortex or turbulence may cause these negative instantaneous tangential velocities. The magnitude of u increased in the vertical direction from the scour bed, with the exception of near the water surface for 160°. The results showed that the velocity gradient ($\partial u/\partial z$) inside the scour hole ($z < 0$) was more than that of the outside of the scour hole ($z \geq 0$). In the region downstream of the abutment, due to the effect of the wake vortex, u was declined near the water surface at θ = 160°. The contour lines of tangential velocity were concentrated near the scoured bed, especially at the edge of the scour hole, indicating the rapid change of u, namely a high-velocity gradient ($\partial u/\partial z$).

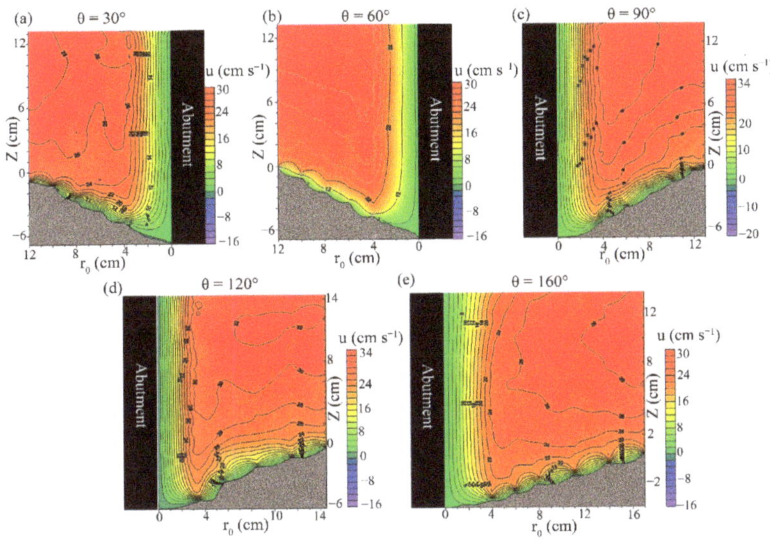

Figure 6. Velocity field for the case bed without vegetation in the tangential direction (u, cm/s) for different azimuthal sections. Note that the range of the color scale is not equalized for better illustratio.

The scour depth decreased significantly from 6.6 cm in the middle of the abutment to 4.3 cm in the front of the abutment, owing to less tangential velocity (Figure 7). In other words, vegetation cover in the channel bed resulted in different velocity patterns that generated different scouring processes. The tangential velocity decreased with the increase in the radial distances outside the scour hole, compared to the un-vegetated channel. This finding is similar to the observation of Yamasaki et al. [46]. The decrease in the streamwise velocity resulted from the adjacent vegetation's effect; a small deposition ridge was formed along the centerline. By comparing Figure 6 to Figure 7, regardless of whether there were vegetation bunchlets in the channel bed, the tangential velocity was positive. This finding confirmed the result reported by Jafari and Sui [47]. For the channel bed without vegetation, the tangential velocity component decreased rapidly near the scour hole bed, especially at the upstream face of the abutment. With the presence of vegetation, a region has been observed at the upstream face of the abutment. The velocity contour lines first concentrated with lower tangential velocity and then moved away from each other (Figure 7). This deceleration spot and its changes were firstly observed outside the scour hole ($\theta = 30°$) and then, with the increase in azimuthal angle, it moved inside the scour hole ($\theta = 60°, 90°$) and approached the abutment wall. Although the tangential velocity increased with the azimuthal angle, the deceleration rate at $\theta = 90°$ was more significant in the marked spot (black circle) than those at $\theta = 30°, 60°$.

Consequently, the maximum scour depth moved from the abutment nose to the abutment tip (corner of the abutment attached to the wall). In other words, the presence of vegetation caused part of the deceleration to occur above the original bed level. Thus, the scouring depth was reduced.

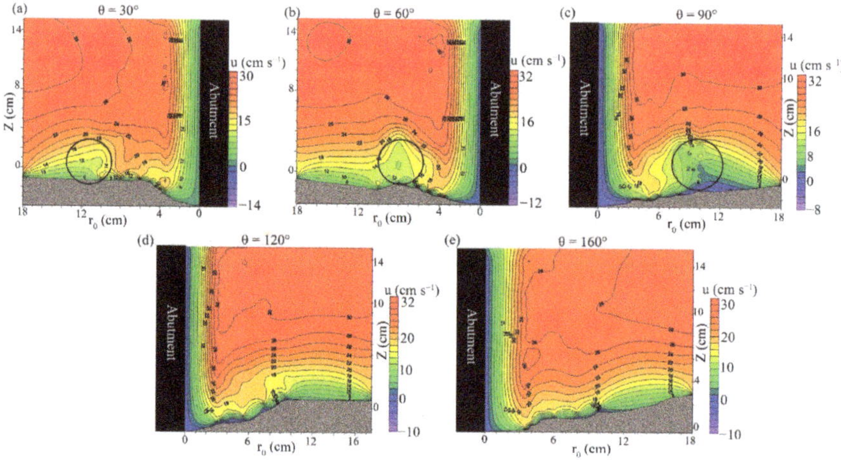

Figure 7. Velocity field for the case with a vegetated bed in the tangential direction (u, cm/s) at different azimuthal sections. Note that the range of the color scale is not equalized for better illustration.

Figure 8 shows a comparison between tangential velocity profiles for the vegetated bed and un-vegetated bed at azimuthal planes (30°, 60°, 90°, 120°, and 160°) at different distances from the abutment surface (4, 6, 8, and 10 cm). From Figure 8, it can be seen:

(1) The larger the distance from the abutment in the radial direction, the more the skewness in the distribution pattern of tangential velocity, especially at the upstream face of the abutment in an un-vegetated channel. Additionally, with the increase in the azimuthal angle (moving downstream), the skewness decreased. In other words, by moving downstream, the tangential velocity near the scoured bed tended to increase.

(2) For the vegetated bed case, the tangential velocity profile distribution had an "S" shape (changed near the scour bed), and this can be seen at all azimuthal angles, except the 160°.

(3) With the presence of vegetation in the channel bed, the tangential velocity profile has two turning points caused by the interplay between the drag force, the gravitational potential, and the momentum flux gradient. Such a result has been reported by Huai et al. [48] and Shahmohammadi et al. [44].

(4) For the case with a vegetated bed, the velocity gradient near the channel bed became negative with the increase in the radial distance from the abutment, especially between 5 cm and 12 cm at all azimuthal angles, except the 160°. With the increase in the azimuth angle, the tangential velocity increased slower than the case without vegetation in bed (Mostly at $z < 0.3h$).

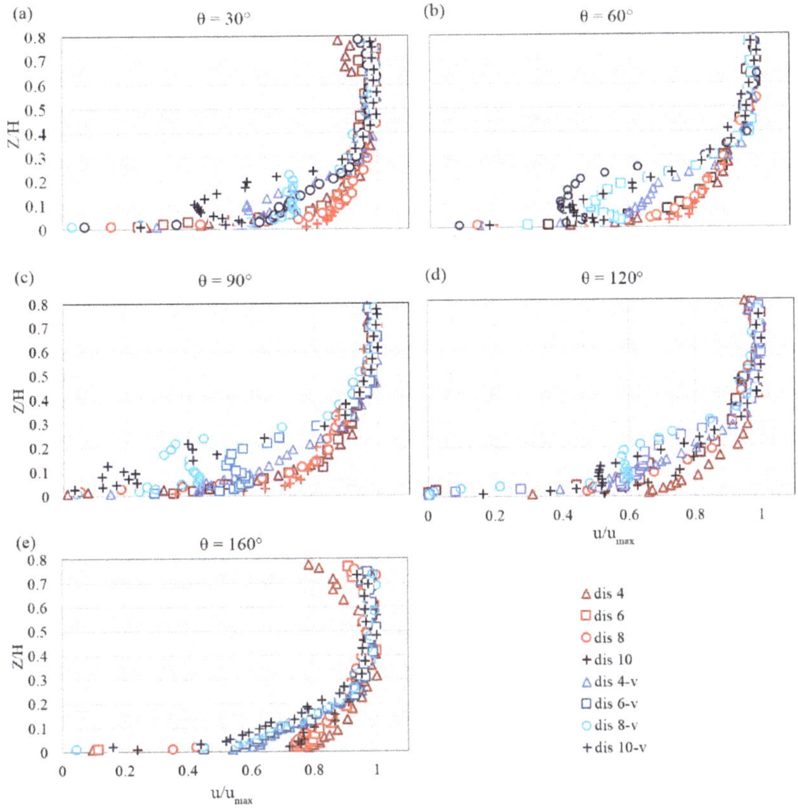

Figure 8. Profiles of the normalized tangential velocity u at different azimuthal planes, with specific distances from the abutment surface (4, 6, 8, and 10 cm) for the cases of un-vegetated bed (marked by red symbols) and vegetated bed (marked by blue symbols).

For the case with a vegetated bed, it is found that the presence of strong circulation around the abutment led to the reduction of the tangential velocity near the scoured bed. Figure 9 shows the maximum difference in tangential velocity between vegetated and un-vegetated cases. One can see that the difference in tangential velocity depended on the distance from the abutment and the azimuthal angle. The maximum difference at all angles was observed at a distance of either 8 or 10 cm from the abutment. This circulation was stronger for the case without vegetation in the bed rather than the vegetated one. It can be seen that the maximum difference occurred at 90° and the minimum difference at 160°. Further downstream, the exited flow from the scour hole joined to the main flow.

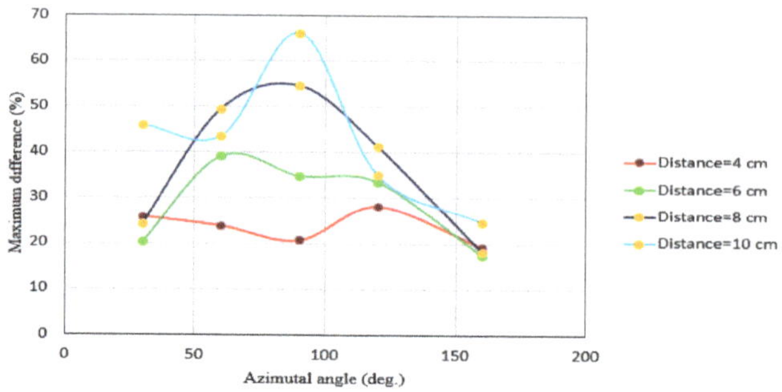

Figure 9. The maximum difference in tangential velocity between vegetated and un-vegetated cases of Figure 8 vs. the azimuthal angle at different distances from the abutment.

As showed in Figure 10, the radial-velocity component changed direction toward the contour line $v = 0$. This variation of the radial-velocity component at the upstream face of the abutment proved the existence of the primary vortices. Before entering the scour hole, the radial velocity was always positive; inside the hole and near the abutment base, the radial velocity had a negative and great absolute value. The flow zone with negative radial velocity created circulation due to the existence of the primary vortices. By moving from $\theta = 30°$ to $90°$, there were large negative values inside the scour hole and near the abutment. This phenomenon of reducing the negative values of v with an increase in θ was noticeable from the azimuthal angle of $0°$ to $90°$ as a result of primary vortex attenuation. This decrease resulted from the vortex changes, so that at $\theta = 90°$ it covers only a small part of the scour hole, which caused a small helicoidal flow close to the abutment. The reversal in the velocity during the development of a scour hole was related to the exposure of a larger flow area. By increasing the angle from $30°$ to $90°$, the positive values were first observed in the scour hole with a distance of $r_0/3$ from the abutment and close to the edge of the scour hole. For larger angles such as $120°$ and $160°$ (moving toward the downstream region), positive radial velocities have been observed at all points. Moreover, for larger angles (moving toward the downstream region), in addition to having positive radial velocities (v), which is a significant factor for creating the wake vortices, the radial velocity (v) became stronger and had greater absolute value rather than those upstream.

The observation from the Figure 11 confirmed that the vortex intensity in regions both upstream and downstream of the abutment could be reduced by placing vegetation in the channel bed. It was noticed from the radial velocity field (v) that the values upstream of the abutment (the azimuthal angle ranges from $30°$ to $90°$) were negative within the scour hole ($z \leq 0$) and outside of it ($z \geq 0$) for the case with a vegetated bed, different from those of the un-vegetated bed. For the case with a vegetated bed, the radial velocity at most of the measuring points near the scour bed was less than that of the un-vegetated bed. For the case with a vegetated bed (Figure 11), the radial velocity at $120°$ also had a negative value in the upper portion of the scour hole near the abutment. At an angle of $\theta = 90°$, radial velocities had negative values for both vegetated and un-vegetated cases. The changes of signs of radial velocities showed a weak reverse flow due to the flow separation. A similar observation of the negative value of the radial velocity at a wing-wall abutment was reported by Dey and Barbhuiya [49] and Afzalimehr et al. [33]. With the increase in the radial distance from the abutment surface, the radial velocity changed to a positive and larger value. At an angle of $160°$, the radial velocity at all points was positive. The maximum positive values of the radial velocity at $160°$ for both cases (vegetated and un-vegetated) were observed in a limited region around the abutment, with a distance from 3.5 cm to 6 cm.

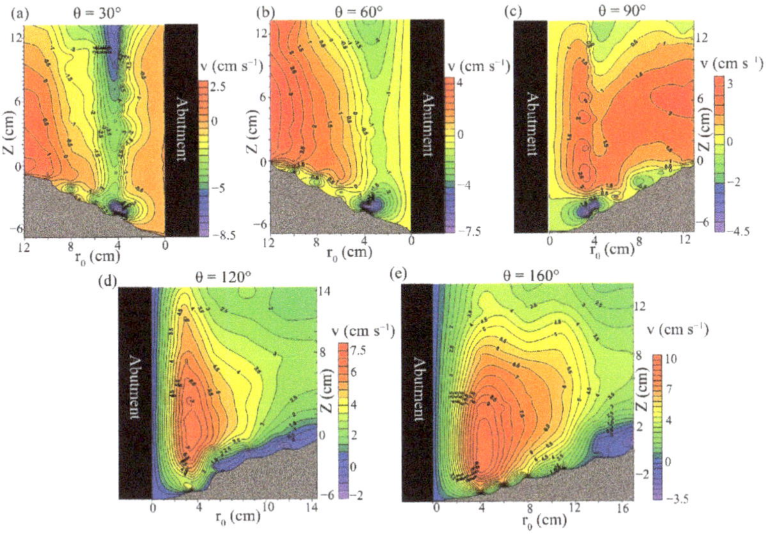

Figure 10. Velocity field in the radial direction (v, cm/s) for the un-vegetated case at different azimuthal sections of θ. Note that the range of the color scale is not equalized for better illustration.

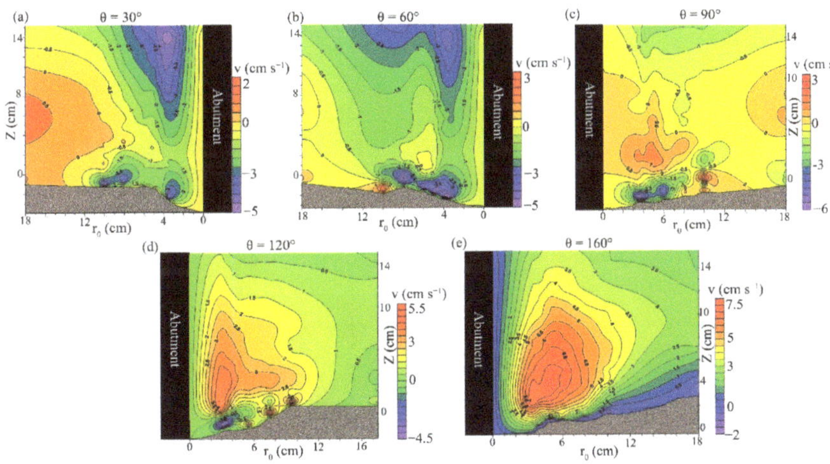

Figure 11. Velocity field in the radial direction (v, cm/s) for the case of a vegetated bed at different azimuthal sections of θ. Note that the range of the color scale is not equalized for better illustration.

For the case without vegetation in the bed, results presented in Figure 12 revealed that the magnitude (absolute value) of the vertical velocity (w) in the region upstream of the abutment increased significantly toward the scour hole bed. The results confirmed that there existed a negative pressure gradient and strong vortices in this region. The reason can be attributed to the obstruction of the flow by the abutment. On the other hand, in the downstream region (i.e., θ = 120° and 160°), the vertical velocity (w) near the scour bed became positive (with upward direction), indicating the suction process adjacent to the abutment, which is caused by separation flow. The direction of the vertical velocity (w) in most of the flow zone (both upstream and downstream) was downward. The maximum vertical velocity for the case without vegetation in the bed was observed at θ = 120°. Dey and Barbhuiya [50] reported that the positive vertical velocities occurred at the downstream region of the abutment.

The magnitude of the negative vertical velocity was diminished upstream of the abutment and inside the scour hole (Figure 13), which is different from that for the case without vegetation in the bed, indicating that the power of the downflow leads to the primary vortices being reduced. In other words, the downflow plays an essential role during the scouring process, because the downward velocity increases and strengthens the primary vortex system, which results in a deeper scour hole [47]. For both cases of un-vegetated and vegetated beds, the reversal nature of the vertical velocity (w) near the scour hole bed in the upstream zone of the abutment was not distinct. However, the negative values of the vertical velocity declined with the increase in the azimuthal angle, implying a weak downward velocity, which led to the attenuation of the primary vortex toward the downstream region. For the case with a vegetated bed, at each angle, there is a spot of the maximum absolute vertical velocity (w) that occurred near the abutment at a depth of approximately 4 cm. Its location at the upstream region of the abutment was inside the scour hole, and with the increase of the azimuthal angle (toward downstream), the spot of the maximum absolute vertical velocity (w) goes to a higher depth ($z > 0$) and out of the scour hole. However, for the case with a vegetated bed, there were negative values inside the scour hole; this spot of the minimum vertical velocity is located outside of the scour hole on both upstream and downstream sides of the abutment. With the increasing azimuthal angle, the spot of the minimum vertical velocity reached close to the water's surface.

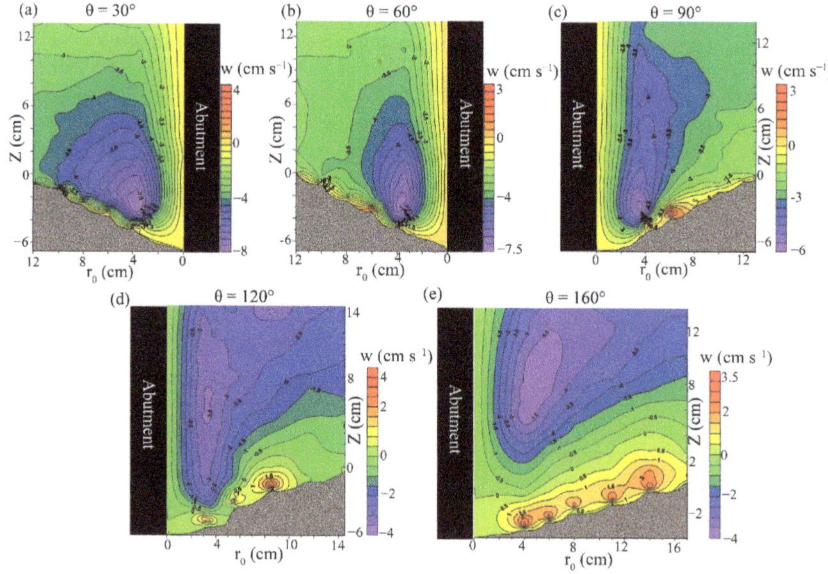

Figure 12. Velocity field in the vertical direction (w, cm/s) for the case of the un-vegetated bed at different azimuthal angles of θ. Note that the range of the color scale is not equalized for better illustration.

Figure 14 displays the vertical distributions of the normalized vertical velocity component at different vertical sections. On either side of the abutment, the direction of the vertical velocity changes, showing the existence of a helicoidal flow upstream and a wake vortex near the scour hole bed. With the increase in the azimuthal angle, there will be more and more positive vertical velocities near the bed. Moreover, with the increase in the azimuthal angle, these positive values will be located further away from the scour hole bed.

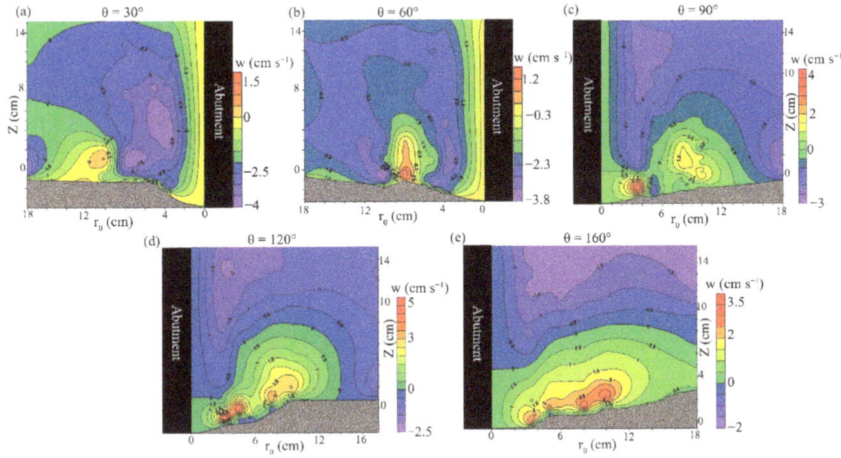

Figure 13. Velocity field in the vertical direction (w, cm/s) for the case of the vegetated bed at different azimuthal angles of θ. Note that the range of the color scale is not equalized for better illustration.

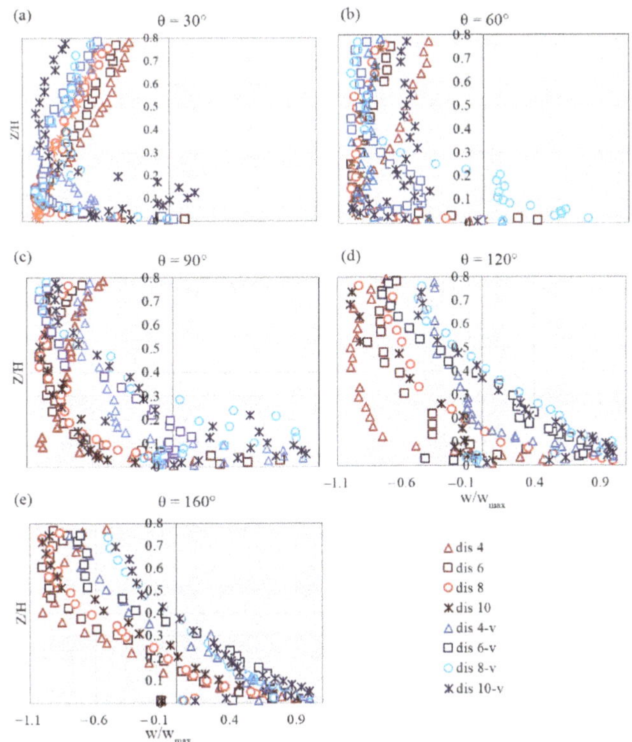

Figure 14. Comparison of the normalized vertical velocity (w) at the azimuthal planes with specific distances from the abutment surface for the un-vegetated bed (marked by red symbols) and the vegetated bed (marked by blue symbols).

For the case without vegetation in the bed, the maximum values of the absolute vertical velocity in the upstream region of the abutment occurred around the initial level of the sand bed (the original bed level), then they moved toward the water surface. This trend changes if $\theta > 90°$. In such a way, near the scour bed, the values of w become positive. With the increase in the distance from the scour hole bed, the vertical velocity around the original bed level approached zero. When the distance increased, the absolute of the negative values increased. Such a trend is observed for the case with a vegetated bed with an azimuthal angle > 90°. For the case with a vegetated bed, positive values near the scour hole and the original bed level were slightly larger than those without vegetation in the bed.

3.2. Reynolds Shear Stress Distribution

The vertical distribution of the normalized Reynolds shear stress ($-u'w'$) for an un-vegetated case was defined as the stress at a plane that is parallel to the flume's wall, and it is more important for us in analyzing than other Reynolds shear stresses. The normalized Reynolds shear stress distribution in Figure 15a shows the pattern of Reynolds stress at different azimuthal angles for the case without vegetation in the bed and an equal distance of 4 cm from the abutment.

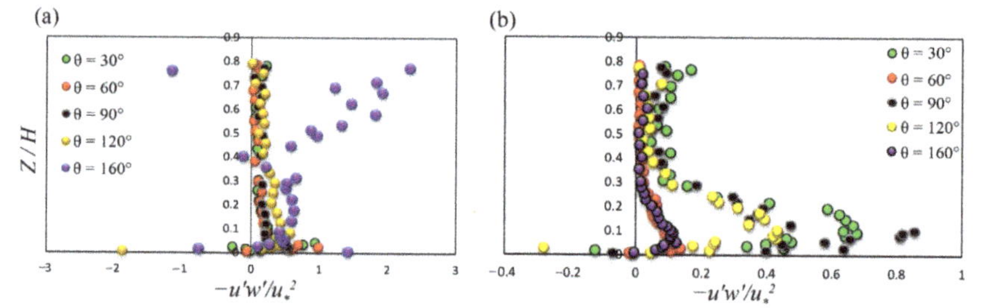

Figure 15. Vertical distribution of $u'w'$ at the azimuthal planes in the (**a**) un-vegetated bed and the (**b**) vegetated bed at equivalent distances from the abutment (4 cm).

Outside of the scour hole, the Reynolds stress had positive and linear distribution at all azimuthal angles (with the exception of 160°) near the original bed, which then decreased toward the water surface. However, near the scour bed it became negative, except at 90°, and was positive at other points, which could be recognized from the bulge shape in the Reynolds stress profile. In other words, the Reynolds stress had a negative value near the scour hole bed, the absolute value of the Reynolds stress declined with the increase in the (dimensionless) depth from the scour hole bed, reached zero, and then increased up to a depth around $z = 0.0255h$. Afterward, the reduction started again in the third step. This means that the maximum stress occurred at a depth around $0.0255h$. Note that this depth was not identical for different profiles. The reason for the sign-changing near the bed was a significant momentum transfer due to convective acceleration.

It should be noted that the negative Reynolds shear stress indicated the reverse flow, and the point where the sign changed from the positive to the negative indicated the separation point. The maximum value of the normalized Reynolds shear stress ($-u'w'$) near the scour bed was $1.43u^{*2}$, located at $\theta = 160°$ in the radial distance from the abutment and with $z = 0.2$ cm. The high values of Reynolds shear stress could be attributed to the high-pressure gradient and the effort of the flow for separation. Both stress and the high-pressure gradient above the scour bed were partially responsible for developing the primary vortex and the scour hole. In the downstream region (at $\theta = 160°$), the Reynolds stresses did not follow a continuous trend. The shear stress increased dramatically at this angle and reached its maximum value near the water surface at a depth of $z = 0.775h$.

Dey and Barbhuiya [50] pointed out that Reynolds stress had uneven distribution and an unknown pattern in the region downstream of the abutment. They attributed this phenomenon to the flow separation and vortex shedding in this region. In the present study, this point was observed at $\theta = 160°$ (Figure 15a), and there was no specified collocation at this azimuthal angle. For the case with a vegetated bed (Figure 15b), a special pattern has been observed at most of the azimuthal angles (except $\theta = 90°$). One can see from Figure 15b that, inside the scour hole, the vertical distribution of the normalized Reynolds shear stress has a bulges shape.

For the case with a vegetated bed, the vertical distribution of the normalized Reynolds shear stress above the scour hole had an approximately linear trend and decreased toward the water surface, although at some angles it increased near the water surface. For the case with a vegetated bed, the magnitude of $-u'w'$ near the scour bed and above it decreased (except in $\theta = 90°$) with the increase in the azimuthal angle, indicating the decline of the primary vortex and pressure gradient around the abutment. In general, by comparing the Reynolds shear stress ($-u'w'$) for the case without vegetation in the bed to that for the case with a vegetated bed, one could say that the maximum stress near the bed occurred at $\theta = 160°$ for the case without vegetation in the bed (which also had a significant value), while the maximum stress was observed at an azimuthal angle of 90° for the case with the vegetated bed.

The distributions of Reynolds shear stresses ($-u'w'$) for 40 measuring positions were presented in Figure 16. Toward the scour hole, the velocity decreases with the increase in the flow depth, leading to $dp/d\theta > 0$. Therefore, the trend of the pressure gradient and shear stress near the scour hole bed is positive, presented by a convex distribution curve. A convex distribution indicates a decelerating flow due to the generation of turbulence. For the case without vegetation in bed, the convex distribution of Reynolds stress was observed at all angles from 0° to 120°, except at a distance of 10 cm from the abutment surface. At the angle of 160°, it is observed that at certain distances away from the abutment ($d = 8$ and 10 cm), the distributions of Reynolds shear stresses ($-u'w'$) are similar to the convex shape. While for the case with the vegetated bed, the distributions of Reynolds shear stresses ($-u'w'$) have a convex shape at all azimuthal angles and follow a certain trend. For the case with the vegetated bed, a convex distribution of Reynolds stress at 160° has an increasing value compared to those at other angles. The comparison of Reynolds shear stress for the vegetated case to that for the un-vegetated case indicates that vegetation in the channel bed dramatically reduces the Reynolds shear stress component. The Reynolds shear stresses ($-u'w'$) for the case with the vegetated bed, at all angles except at ($\theta = 160°$, $r_0 = 4$ cm), are less than those for the case with the un-vegetated bed. Moreover, the Reynolds shear stress ($--u'w'$) shows a decreasing trend in the vertical direction toward the water surface from a depth of $Z/H > 0.05$ at the upstream surface of the abutment and from a depth of $Z/H > 0.14$ at the downstream surface of the abutment. In the scour hole, however, the Reynolds shear stress ($-u'w'$) displays an increasing trend in the vertical direction from the scour bed toward a depth of Z/H, as mentioned above. Siniscalchi et al. [51] stated that a negative Reynolds shear stress ($-u'w'$) indicates an upward vertical momentum transport with negative velocity gradients. The magnitude of negative Reynolds shear stress ($-u'w'$) can be seen in the downstream side of the abutment ($\theta > 90°$), and their values increase with the increase in the azimuthal angle (θ). Such Reynolds stress $-u'w'$ near the scour bed confirms the development of the wake zones downstream of the abutment, which occurs at 120° for the case with the vegetated bed and 160° for the case without vegetation in the bed. These wake zones lead to the generation of strong turbulence.

3.3. Turbulence Intensity

The vertical distributions of the normalized tangential, radial, and vertical turbulent intensity components are shown in Figure 17.

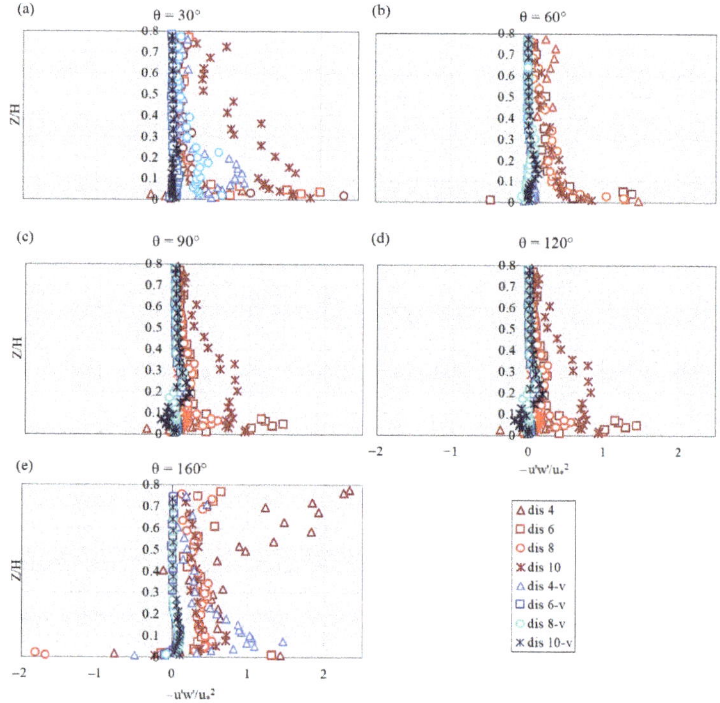

Figure 16. Comparison of the Reynolds shear stress distribution at the azimuthal planes with specific distances from the abutment surface (4, 6, 8, and 10 cm) for the case without vegetation in the bed (marked by red symbols) and with vegetation in the bed (marked by blue symbols).

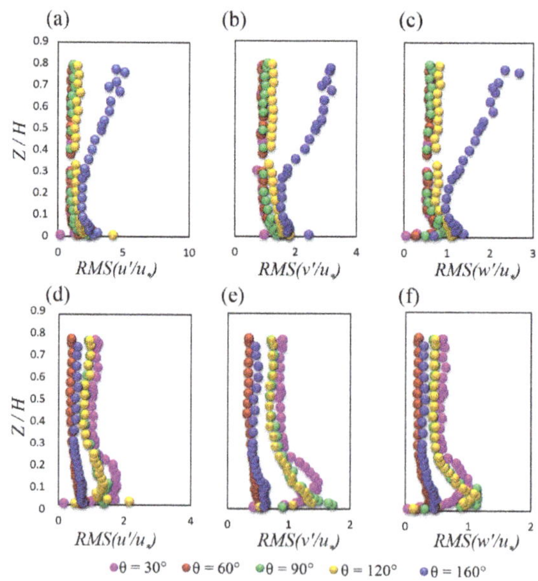

Figure 17. Turbulence intensity distributions in 3D at different azimuthal sections in the (**a–c**) channel bed without vegetation and (**d–f**) the channel bed with vegetation, at distances of 4 cm from the abutment.

The turbulence intensity values in both tangential and radial directions were greater than those in the vertical direction. For the case with vegetation in the bed, turbulence intensities were lower than those for the case without vegetation in the bed, with the exception of inside the scour hole at the azimuthal angles of θ = 30° and θ = 90°. The turbulence intensity values decreased with the increase in the distance from the scour bed and changed to a linear trend (except at θ = 160° for the case without vegetation in the bed). The maximum value of turbulence intensity was observed at the angle of 160° for the case without vegetation in the bed.

3.4. Reynolds Stress Anisotropy

The Reynolds averaged Navier–Stokes equation is used to define the normalized anisotropy tensor b_{ij}. The difference between the ratio of Reynolds stress tensor terms to turbulence kinetic energy (TKE) and its isotropic equivalent quantity gives the Reynolds stress anisotropy tensor b_{ij}, which provides an estimate of the degree of departure from the idealized isotropic turbulence [20,52]. The b_{ij} is given by:

$$b_{ij} = \frac{\overline{\acute{u}_i \acute{w}_j}}{2k} - \frac{1}{3}\delta_{ij} \tag{2}$$

where $\acute{u}_i$ is the instantaneous velocity fluctuation in the direction i, δ_{ij} is the Kroecker delta function, which is $\delta_{ij}(i \neq j) = 0$ and $\delta_{ij}(i = j) = 1$, k is the turbulent kinetic energy of the flow defined as k = average TKE $= 0.5(\overline{\acute{u}\acute{u}} + \overline{\acute{v}\acute{v}} + \overline{\acute{w}\acute{w}})$, and i, j = 1, 2, 3 are the spatial components. The normalized anisotropy tensor b_{ij} has a zero trace as a consequence of its formulation, which two independent invariants can represent. These invariants are defined as follows:

$$\begin{aligned} I_2 &= -b_{ij}b_{ji}/2 \\ I_3 &= -b_{ij}b_{jk}b_{ki}/3 \end{aligned} \tag{3}$$

The anisotropy invariant map (AIM) of Lumley and Newman [20], also called the "Lumley triangle", uses this second I_2 and third principal I_3 components of turbulence anisotropy to create the coordinate system (I_3, I_2). Another method that can construct a nonlinear anisotropy invariant map (AIM) is based on the invariants I_2 and I_3 called the "Turbulence triangle", which in general is also termed the Lumley triangle and uses the coordinate system (ξ, η), where:

$$\begin{aligned} \xi &= (I_3/2)^{1/3} \\ \eta &= (-I_2/3)^{1/2} \end{aligned} \tag{4}$$

When plotting ξ vs. η, the domain of both invariants confined by three lines is reduced to the interior of a triangle [52]. The triangle boundaries define several characteristics of turbulence states classified based on the shape of the eddies. In the AIM, it is observed that three vertices and edges of the triangle correspond to isotropic turbulence (denoted by 3D) where the three normal stresses are equal, one-component isotropic turbulence (denoted by 1D), and two-component isotropic turbulence (denoted by 2D), respectively (Figure 18). The limitation of turbulent structures to two distinct types transfers turbulence from the 3D to 2D and/or 1D: (1) the left side of the triangle corresponds to pancake-shaped turbulence, where the fluctuations of turbulence exist along with two directions with equal magnitude. These two equal components (which show with σ_i as Reynolds normal stress) have a considerably higher amplitude than the small one ($\sigma_1 = \sigma_2 > \sigma_3$), and this state is known by 2D turbulence; (2) the right side of the triangle corresponds to a rod-like or cigar-shaped turbulence where the turbulent fluctuations only exist along one direction (1D), in which one principal component is larger than the other two equal components ($\sigma_1 = \sigma_2 < \sigma_3$). The upper boundary curve of the triangle, defined by ($\eta^2 = (1/27 + 2\xi^3)$), is used to represent the isotropy of the two-component turbulence. Briefly, as it has been said, the turbulence triangle consists of bottom, right, and left vertices (3D, 1D, and 2D, respectively) bounded by three boundaries (two linear boundaries known as axisymmetric turbulence and one nonlinear boundary). The region within this triangle and far from the specified limit indicates the general tridimensional turbulence condition. Any data points that lie within this limitation of the turbulence triangle refer to a specific turbulence state.

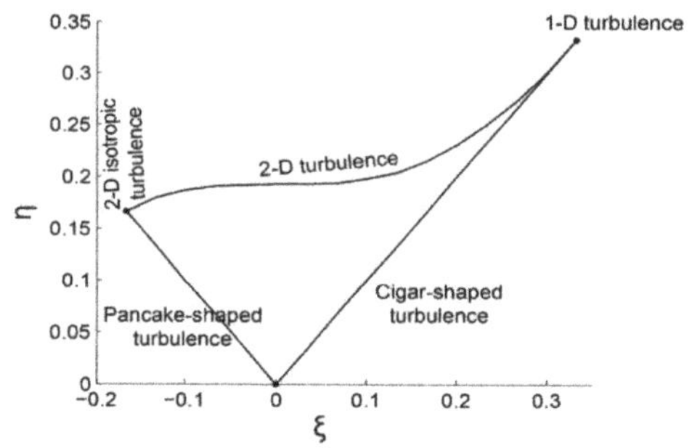

Figure 18. A schematic shape of a Lumley triangle in the η (2nd invariant) and ξ (3rd invariant).

For deeper insight into the anisotropy level, an invariant function (F) is expressed as the following [53]:

$$F = 1 + 9I_2 + 27I_3$$
$$\rightarrow If: \begin{cases} F = 0 \rightarrow \text{Two dimensional (top side of the triangle)} \\ F = 1 \rightarrow \text{Isotropic turbulence (bottom vertex of the triangle)} \end{cases} \quad (5)$$

The invariants ξ and η have been plotted on the Lumley triangle for both cases with vegetation and without vegetation in channel the bed (Figure 19). The evolution of the invariants η and ξ across the z-direction has been analyzed at different azimuthal angles (marked by different symbols) and radial distances from the abutment (4, 6 (or) 8, and 10 cm, highlighted by different colors). By comparing Figure 19a to Figure 19b, it is realized that the invariants for the case with vegetation in the bed showed the tendency to be closer to the origin than that of the case without vegetation. This result indicated that it is slightly closer to isotropic for the vegetated bed than that without vegetation, because the placement of the vegetation on the bed decreased τ_{uw} around the abutment, especially inside the scour hole. One can also see from Figure 19 that there was both pancake-shaped and cigar-shaped turbulence, based on the data measured in the laboratory and according to the spread of points in the respective turbulence triangles. As can be seen in Figure 19, most of the anisotropy invariants for both cases (with and without vegetation in the bed) have been gathered between the 2D and the cigar-shaped limit. In the triangle for the un-vegetated case, it is observed that the level of anisotropy increases with an increase in the distance from the abutment.

The majority of the values of ξ were less than zero, and a trend toward pancake-shaped turbulence has been observed (Figure 19) (even by increasing the radial distance). At the azimuthal angle of θ = 90°, the interaction region appeared, which showed a tendency toward the cigar-shaped boundary. In the region downstream of the abutment, there was no trend at the angles of 120° and 160° (except at the distance of 4 cm) caused by the separation of flow. Note that the turbulence anisotropy at the angle of 160° and close to the abutment (with a distance of 4 cm from the abutment) was in the opposite direction of the trend for the angle of 30° (with the same distance). This state at this angle (θ = 160°) is also called rod-like or cigar-shaped turbulence.

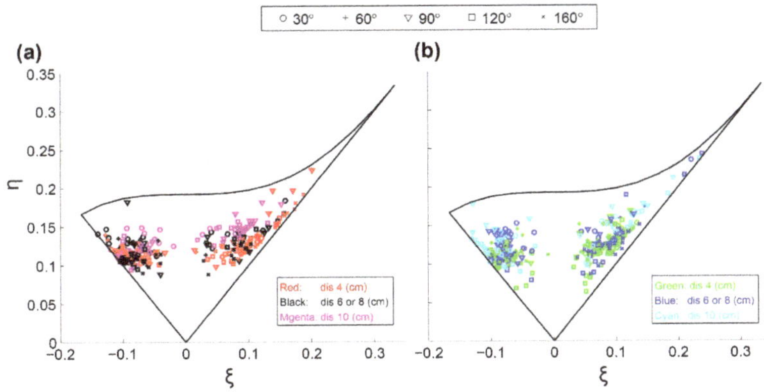

Figure 19. Turbulence triangle plotted in the $\xi - \eta$ plane at specified azimuthal sections in the case (**a**) without vegetation in the bed and (**b**) with vegetation in the bed, at radial distances from the abutment.

For the vegetated channel bed (Figure 19b) in the upstream region of the abutment (i.e., θ = 30° and 60°), an opposite trend has been noticed compared to the case without vegetation in bed. For the majority of the values of ξ that were greater than zero and at an azimuthal angle of 30°, there was a tendency to be cigar-shaped. At an azimuthal angle of 90°, similar to the case of an un-vegetated bed, the pattern of the turbulence anisotropy changed, data points lay on the left line curve ($\xi < 0$). In the region downstream of the abutment, a similar trend toward that for the un-vegetated bed occurred. Note that for the angle of 160° and with a distance 4 cm from the abutment, the data points lay close to the left line curve ($\xi < 0$) for the point near the scour bed.

As showed in Figure 20, the turbulence anisotropy across the vertical direction has been evaluated at a constant distance from the abutment (4 cm) at different azimuthal planes. This figure uses two different symbols (circles and squares) for un-vegetated and vegetated beds, respectively. The movement of the anisotropy vs. depth, fitted with two curves of black and gray color with arrowheads, is illustrated.

For the case with an un-vegetated bed, with the increase in vertical distance, the turbulence anisotropy invariants change from pancake-shaped to cigar-shaped in the front face of the abutment. Namely, upstream of the abutment near the scour bed, the anisotropy tends to be pancake-shaped, and its values increase with the increase in the azimuthal angle. Eventually, near the water surface, the turbulence anisotropy reaches the right boundary at which one of the components of TKE is larger than the other two. At the azimuthal angle of 90°, the anisotropic turbulence state and kinetic energy act inversely. Thus, there is a trend toward the cigar-shaped boundary with a high anisotropic level near the scour bed. At the downstream side of the abutment with an azimuth angle of 160°, the turbulence anisotropy is more uniform, which shows a tendency toward the cigar-shaped boundary. Moreover, the level of anisotropy increases with the depth increase and follows a path almost parallel to the right boundary. Therefore, as presented in Figure 20, it is clear that the data points of the surface zone at all angles for the case without vegetation in the bed approach the cigar-shaped limit.

The degree of turbulence anisotropy for the case with vegetation in the bed, close to the abutment (4 cm) at the upstream region and inside the scour hole, is higher than that for the case without vegetation in the bed. In addition, unlike the turbulence anisotropy for the case without vegetation in the bed, the turbulence anisotropy for the case with vegetation in the bed starts from the cigar shape, and then the turbulence anisotropy moves to the opposite direction as the vertical distance increases. The tendency to the cigar-type structure near the scour bed in the region upstream of the abutment indicates the presence of a dominant direction of velocity fluctuations. At an azimuthal angle of θ = 90°, similar to a trend for the un-vegetated case, an opposite trend toward the upstream region of the abutment is observed near the scour bed, which shows a tendency toward the pancake anisotropy limit. In the region downstream of the abutment (θ = 120°, 160°), different from the results for the case with the un-vegetated bed, no definite relation between the spatial distribution of the turbulence and the vertical position is observed either near the scour hole or at the water surface. Moreover, the turbulence anisotropy pattern rotates from the pancake shape to the cigar shape from the bottom towards the surface.

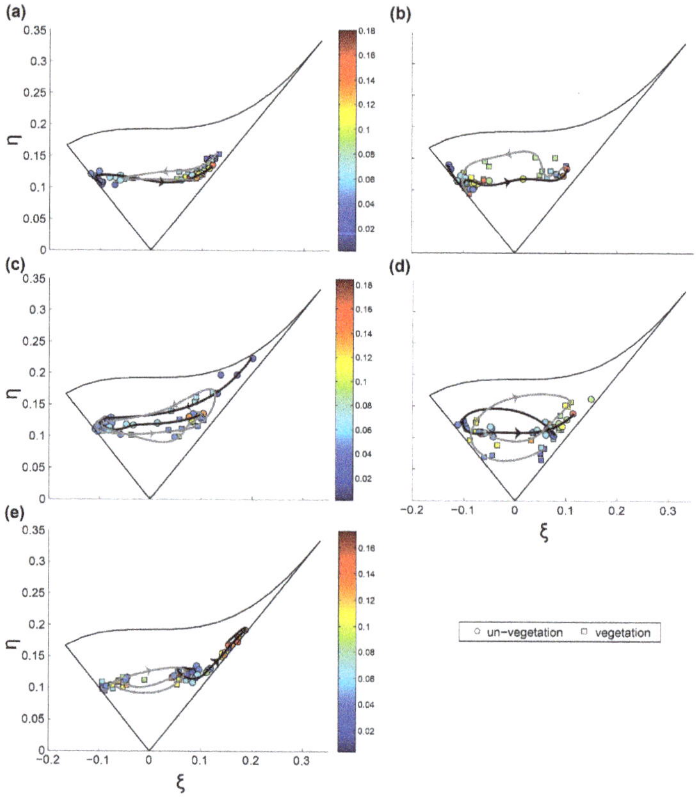

Figure 20. Maps of the anisotropic invariant for the cases of the un-vegetated bed (labeled by circular shapes) and vegetated bed (labeled by rectangular shapes) at the azimuthal sections of θ = (**a**) 30°, (**b**) 60°, (**c**) 90°, (**d**) 120°, (**e**) 160° with a constant distance from the abutment (4 cm). The depth of the measurement is color-coded in meters. The distribution of the turbulence anisotropy for both cases of un-vegetated and vegetated bed throughout the water column is denoted by black and gray curves, respectively. The direction from the bottom to the water surface is indicated by the arrowheads.

On the other hand, the degree of the turbulence anisotropy downstream of the abutment near the scour hole has its lowest value. For all angles (except for θ = 30°), it has been observed that if data points near the scour bed approach one of the anisotropy types, it tends to the opposite region near the water surface. For instance, near the scour hole at θ = 160°, data points are confined to the left side of the Lumley, whereas the approach to the right side of the Lumley is near the surface zone. The highest degree of anisotropy for the un-vegetated bed usually occurs at the points near the scour hole and/or at the points near the scour hole or the water surface. This result was also reported by Mera et al. [24]. Note that the highest degrees of anisotropy occur near the water surface in the region downstream of the abutment. While for the case with vegetation in the bed, at θ =

30°, the highest degree of anisotropy is observed near the scour bed. With the increases in the angle, the highest degree of anisotropy is located out of the scour hole. In the region downstream of the abutment, the lowest degree of anisotropy occurs near the scour bed.

To preciously assess the turbulence anisotropy changes at specific angles with a constant distance from the abutment (4 cm) along the tangential direction, a parameter named anisotropic invariant function (*F*) was defined here. Figure 21 shows the anisotropic invariant function (*F*) for the flow depth (*Z/H*). At the upstream face of the abutment, the value of the invariant function (F) near the scour hole for the case of the un-vegetated bed is more than that for the case with vegetation in the bed. At all angles, inside the scour hole and near the scour bed, the value of F is close to 0.7 for both vegetated and un-vegetated cases. However, at an angle of $\theta = 90°$ and with a very close distance to the scour bed, the value of F for an un-vegetated bed is 0.08, which corresponds to two-dimensional isotropy. Downstream of the abutment, the opposite condition happens, and the value of F for the case of the vegetated bed near the scour hole is greater than that for the case of the un-vegetated bed. Moreover, at an angle of $\theta = 120°$ inside the scour hole and near the scour bed, the value of F indicates three-dimensional isotropy.

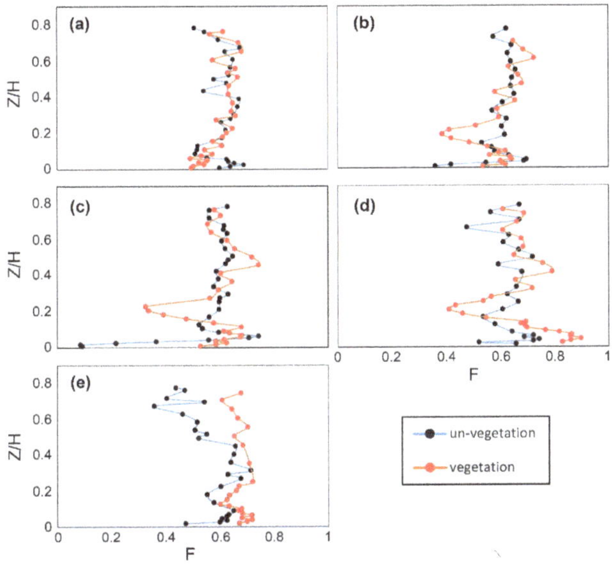

Figure 21. Evolution of the invariant function (*F*) at the azimuthal sections of θ = (**a**) 30°, (**b**) 60°, (**c**) 90°, (**d**) 120°, and (**e**) 160° for the cases of the un-vegetated bed and the vegetated bed at an equal distance of 4 cm from the abutment.

As the vertical distance approaches very close to the scour bed at all azimuthal angles, the turbulence anisotropy tends to return to the isotropic limit. Then, the turbulence anisotropy moves away from the isotropic limit with a gradual increase in vertical distance, and the invariant function (*F*) decreases. The changes of the invariant function (*F*) were observed between 0.5 and 0.7, and the invariant function (*F*) increased with the angle (except for 160°) for the case of the vegetated bed.

4. Conclusions

The following conclusions can be drawn from this study:

(1) Compared to the case of an un-vegetated bed, with vegetation around the abutment, the maximum scour depth occurred at the abutment tip instead of the upstream nose of the abutment. The presence of vegetation in the channel decreases the maximum

values of the 3D velocity components (except for the positive values of w downstream and near the scour bed) and reduces the turbulence intensities inside the scour hole. For this experimental study, the presence of vegetation in the channel bed caused a reduction of 34.8% of the scour depth. The time required for achieving the equilibrium condition also decreased. Interestingly, within the first 120 min of the experiments, the scour depth for the case with vegetation in the bed was slightly higher than that for the case without vegetation.

(2) The circulation and vortices have been weakened due to the presence of vegetation in the channel bed. In other words, the primary vortex for the case of an un-vegetated bed was stronger than that for the vegetated bed. The turbulence kinetic energy inside the scour hole is reduced.

(3) The tangential velocity component was always positive and decreased rapidly near the scour bed, especially around the upstream face of the abutment. Additionally, the deceleration process of flow has proceeded slowly with vegetation in the channel bed. The tangential velocity increased with the increase in the azimuthal angle for both vegetated and un-vegetated cases, but the rate of deceleration at $\theta = 90°$ for the vegetated case was greater than those of $\theta = 30°, 60°$. The maximum tangential velocity difference between two cases was observed at an angle of 90°. With vegetation's presence, the tangential velocity profile (at all angles except the 160°) near the bed had changed to an "S" shape, which showed two turning points. Near the vegetated bed, the tangential velocity gradient with radial distance from the abutment became negative, especially between 5 cm to 12 cm (except at 160°).

(4) The contours of radial velocity (v) upstream of the abutment (30° to 90°) had negative values for the case with vegetation in the bed within the scour hole ($z \leq 0$) and above it ($z \geq 0$), different from that for the case of un-vegetated bed. Downstream of the abutment, the sign change in radial velocity was observed for the case of the un-vegetated bed only at 90° and for the case of the vegetated bed at $\theta = 90°, 120°$. This indicated a week flow reversal as a result of backflow. For the vegetated bed, variation in radial velocity inside the scour hole was less than that for the un-vegetated one (except 90° near the scour bed), indicating reduction in the primary vortex by vegetation.

(5) For both vegetated and un-vegetated cases, the magnitude of the vertical velocity (w) toward the scour bed increased significantly, resulting in the powerful downflow around the abutment. For the case with vegetation in the bed inside the scour hole upstream of the abutment, the magnitude of the negative vertical velocity is diminished in this region compared to that for the un-vegetated bed, indicating that the power of downflow leads to a reduction of the primary vortices. The extent of the positive vertical velocities for the case with vegetation in the bed was increased and covered broader if the azimuthal angle was more than 90°, compared to the un-vegetated case. The maximum positive vertical velocities were observed at 160° for the un-vegetated bed and 120° for vegetation. It is found that vegetation could not significantly reduce the effect of upward flow in the downstream region adjacent to the abutment due to the suction that takes place.

(6) Although radial and vertical velocities played an utmost important role in the flow field, the effect of tangential velocity was the highest, implying that the tangential velocity (u) contributed more to the turbulence intensities, Reynolds shear stress, and consequently the development of the scour holes.

(7) The presence of vegetation in the bed dramatically reduces the Reynolds shear stress component, except at the distance of 4 cm and 160° near the scour bed. For the case of the un-vegetated bed, the convex distribution of Reynolds shear stress ($-u'w'$) was observed at all angles from 0° to 120°, except at a distance of 10 cm. With vegetation, Reynolds shear stress profiles were convex at all azimuthal angles and had a certain trend. The magnitude of negative Reynolds shear stress ($-u'w'$) confirms the development of wake zones downstream of the abutment, which occur at 120° for the

vegetated bed and 160° for the case of the un-vegetated bed. These wake zones lead to the generation of strong turbulence. At a constant distance of 4 cm from the abutment, inside the scour hole, the vertical distribution of Reynolds shear stress and turbulence intensity possessed a convex shape. Near the scour bed, with the increase in the azimuthal angle (except 90°), both Reynolds shear stress and turbulence intensity decreased for the case of the vegetated bed. For the case of an un-vegetated bed, the maximum stress occurred near the bed at an angle of $\theta = 160°$.

(8) By placing vegetation in the bed, for the vegetated bed, the flow tends to be more isotropic than that for the un-vegetated case. For the un-vegetated bed, the anisotropy level increases with the radial distance from the abutment due to the turbulence triangle. Moreover, at the upstream face of the abutment, most of the data points in the Lumley triangle tend to be pancake-shaped turbulence. However, for the case of a vegetated bed, there was a tendency to be cigar-shaped. Downstream of the abutment, due to flow separation, there was no definite trend. For the un-vegetated bed, upstream of the abutment, the anisotropy at a close distance of 4 cm from the abutment changes from pancake-shaped to cigar-shaped with the increase in the vertical distance (from the scour bed). For the case of a vegetated bed, the anisotropy starts from the cigar shape then changes to the opposite direction. Inside the scour hole in this zone, the degree of anisotropy for the case of the vegetated bed is larger than that for the un-vegetated bed. Downstream of the abutment, for the case of the un-vegetated bed, the anisotropy of turbulence is more uniform at $\theta = 160°$ and presents a tendency toward the cigar-shaped boundary. The highest degree of anisotropy for the case of an un-vegetated bed occurs near the scour hole or near the water surface. The vegetated bed was observed near the scour bed at $\theta = 30°$, but it was located outside of the scour hole by increasing the angle.

(9) For the case of an un-vegetated bed, inside the hole upstream of the abutment, the invariant function (F) value was greater than that for the vegetated bed. Downstream of the abutment, however, it was the opposite. Moreover, the F value for the case of the un-vegetated bed near the scour hole at $\theta = 90°$ corresponds to two-dimensional isotropy. For the vegetated bed, however, inside the hole at $\theta = 120°$, F leads to three-dimensional isotropy, with the highest amount of all the azimuthal angles for both cases.

Author Contributions: Conceptualization, S.F.N., H.A., J.S., B.K. and S.H.N.; methodology, S.F.N., H.A. and B.K.; validation, S.F.N., H.A., J.S., B.K. and S.H.N.; formal analysis, S.F.N., H.A. and S.H.N.; investigation, S.F.N., J.S. and B.K.; resources, S.F.N., H.A., J.S., B.K.; data curation, S.F.N. and S.H.N.; writing—original draft preparation, S.F.N. and S.H.N.; writing—review and editing, H.A., J.S. and B.K.; visualization, S.F.N.; supervision, H.A., J.S. and B.K.; funding acquisition, H.A. and J.S. All authors have read and agreed to the published version of the manuscript.

Funding: This research received no external funding.

Institutional Review Board Statement: The study does not involve the use of humans or animals.

Informed Consent Statement: Not applicable.

Data Availability Statement: The data presented in this study are available on request from the first author.

Conflicts of Interest: The authors declare no conflict of interest.

References

1. Sui, J.; Fang, D.; Karney, B.W. An experimental study into local scour in a channel caused by a 90 bend. *Can. J. Civ. Eng.* **2006**, *33*, 902–911. [CrossRef]
2. Sui, J.; Faruque, M.; Balachandar, R. Influence of channel width and tailwater depth on local scour caused by square jets. *J. Hydro-Environ. Res.* **2008**, *2*, 39–45. [CrossRef]
3. Afzalimehr, H.; Moradian, M.; Singh, V.P. Flow field around semi-elliptical abutments. *J. Hydrol. Eng.* **2018**, *23*, 04017057. [CrossRef]

4. Sui, J.; Faruque, M.; Balachandar, R. Local scour caused by submerged square jets under model ice cover. *J. Hydraul. Eng.* **2009**, *135*, 316–319. [CrossRef]
5. Wu, P.; Hirshfield, F.; Sui, J. Armour layer analysis of local scour around bridge abutments under ice cover. *River Res. Appl.* **2015**, *31*, 736–746. [CrossRef]
6. Arneson, L.; Zevenbergen, L.; Lagasse, P.; Clopper, P. *Evaluating Scour at Bridges–Federal Highway Administration Hydraulic Engineering Circular No. 18*, 5th ed.; FHWA-HIF-12-003; FHWA: Washington, DC, USA, 2012.
7. Richardson, J.; Richardson, E. The fallacy of local abutment scour equations. In Proceedings of the Hydraulic Engineering, San Francisco, CA, USA, 25–30 July 1993; pp. 749–754.
8. Melville, B.W. Pier and abutment scour: Integrated approach. *J. Hydraul. Eng.* **1997**, *123*, 125–136. [CrossRef]
9. Kothyari, U.; Ranga Raju, K. Scour around spur dikes and bridge abutments. *J. Hydraul. Res.* **2001**, *39*, 367–374. [CrossRef]
10. Dey, S. *Fluvial Hydrodynamics*; Springer: Berlin/Heidelberg, Germany, 2014.
11. Kw An, R.T.; Melville, B.W. Local scour and flow measurements at bridge abutments. *J. Hydraul. Res.* **1994**, *32*, 661–673. [CrossRef]
12. Melville, B.W.; Coleman, S.E. *Bridge Scour*; Water Resources Publication: Littleton, CO, USA, 2000.
13. Wu, P.; Hirshfield, F.; Sui, J.; Wang, J.; Chen, P.-P. Impacts of ice cover on local scour around semi-circular bridge abutment. *J. Hydrodyn.* **2014**, *26*, 10–18. [CrossRef]
14. Wu, P.; Balachandar, R.; Sui, J. Local scour around bridge piers under ice-covered conditions. *J. Hydraul. Eng.* **2016**, *142*, 04015038. [CrossRef]
15. Sui, J.; Afzalimehr, H.; Samani, A.K.; Maherani, M. Clear-water scour around semi-elliptical abutments with armored beds. *Int. J. Sediment Res.* **2010**, *25*, 233–245. [CrossRef]
16. Yang, Y.; Xiong, X.; Melville, B.W.; Sturm, T.W. Dynamic morphology in a bridge-contracted compound channel during extreme floods: Effects of abutments, bed-forms and scour countermeasures. *J. Hydrol.* **2021**, *594*, 125930. [CrossRef]
17. Bressan, F.; Ballio, F.; Armenio, V. Turbulence around a scoured bridge abutment. *J. Turbul.* **2011**, N3. [CrossRef]
18. Koken, M. Coherent structures at different contraction ratios caused by two spill-through abutments. *J. Hydraul. Res.* **2018**, *56*, 324–332. [CrossRef]
19. Namaee, M.R.; Sui, J. Velocity profiles and turbulence intensities around side-by-side bridge piers under ice-covered flow condition. *J. Hydrol. Hydromech.* **2020**, *68*, 70–82. [CrossRef]
20. Lumley, J.L.; Newman, G.R. The return to isotropy of homogeneous turbulence. *J. Fluid Mech.* **1977**, *82*, 161–178. [CrossRef]
21. Sarkar, S.; Dey, S. Turbulent length scales and anisotropy downstream of a wall mounted sphere. *J. Hydraul. Res.* **2015**, *53*, 649–658. [CrossRef]
22. Sharma, A.; Kumar, B. Anisotropy Properties of Turbulence in Flow Over Seepage Bed. *J. Fluids Eng.* **2021**, *144*, 021501. [CrossRef]
23. Ali, N.; Hamilton, N.; Calaf, M.; Cal, R.B. Classification of the Reynolds stress anisotropy tensor in very large thermally stratified wind farms using colormap image segmentation. *J. Renew. Sustain. Energy* **2019**, *11*, 063305. [CrossRef]
24. Mera, I.; Franca, M.J.; Anta, J.; Peña, E. Turbulence anisotropy in a compound meandering channel with different submergence conditions. *Adv. Water Resour.* **2015**, *81*, 142–151. [CrossRef]
25. Velasco, D.; Bateman, A.; Redondo, J.M.; DeMedina, V. An open channel flow experimental and theoretical study of resistance and turbulent characterization over flexible vegetated linings. *Flow Turbul. Combust.* **2003**, *70*, 69–88. [CrossRef]
26. Piedelobo, L.; Taramelli, A.; Schiavon, E.; Valentini, E.; Molina, J.-L.; Nguyen Xuan, A.; González-Aguilera, D. Assessment of green infrastructure in Riparian zones using copernicus programme. *Remote Sens.* **2019**, *11*, 2967. [CrossRef]
27. Luhar, M.; Rominger, J.; Nepf, H. Interaction between flow, transport and vegetation spatial structure. *Environ. Fluid Mech.* **2008**, *8*, 423–439. [CrossRef]
28. Nepf, H.M. Hydrodynamics of vegetated channels. *J. Hydraul. Res.* **2012**, *50*, 262–279. [CrossRef]
29. Kabiri, F.; Afzalimehr, H.; Sui, J. Flow structure over a wavy bed with vegetation cover. *Int. J. Sediment Res.* **2017**, *32*, 186–194. [CrossRef]
30. D'Ippolito, A.; Calomino, F.; Alfonsi, G.; Lauria, A. Drag coefficient of in-line emergent vegetation in open channel flow. *Int. J. River Basin Manag.* **2021**, 1–11. [CrossRef]
31. Afzalimehr, H.; Moghbel, R.; Gallichand, J.; Sui, J. Investigation of turbulence characteristics in channel with dense vegetation. *Int. J. Sediment Res.* **2011**, *26*, 269–282. [CrossRef]
32. Keshavarz, A.; Afzalimehr, H.; Singh, V.P. Impact of unfavourable pressure gradient and vegetation bed on flow. *J. Hydraul. Eng.* **2015**, *4*, 10–16.
33. Afzalimehr, H.; Bakhshi, S.; Gallichand, J.; Sui, J. Effect of vegetated-banks on local scour around a wing-wall abutment with circular edges. *J. Hydrodyn.* **2014**, *26*, 447–457. [CrossRef]
34. Chiew, Y.-M.; Melville, B.W. Local scour around bridge piers. *J. Hydraul. Res.* **1987**, *25*, 15–26. [CrossRef]
35. Dey, S.; Bose, S.K.; Sastry, G.L. Clear water scour at circular piers: A model. *J. Hydraul. Eng.* **1995**, *121*, 869–876. [CrossRef]
36. Afzalimehr, H.; Moradian, M.; Gallichand, J.; Sui, J. Effect of adverse pressure gradient and different vegetated banks on flow. *River Res. Appl.* **2016**, *32*, 1059–1070. [CrossRef]
37. Bolhassani, R.; Afzalimehr, H.; Dey, S. Effects of relative submergence and bed slope on sediment incipient motion under decelerating flows. *J. Hydrol. Hydromech.* **2015**, *63*, 295–302. [CrossRef]
38. Bhandari, L.; Upadhyay, C.; Mohite, P.; Gupta, A.; Sarkar, S.; Panwar, V.S. Tensile test on pine needles and crack analysis of pine needles short fiber reinforced composites. *IOSR J. Mech. Civ. Eng. (IOSRJMCE) E-ISSN* **2015**, *12*, 2278-1684.

39. Melville, B.W.; Chiew, Y.-M. Time scale for local scour at bridge piers. *J. Hydraul. Eng.* **1999**, *125*, 59–65. [CrossRef]
40. Goring, D.G.; Nikora, V.I. Despiking acoustic Doppler velocimeter data. *J. Hydraul. Eng.* **2002**, *128*, 117–126. [CrossRef]
41. Wahl, T. Analyzing ADV data using WinADV. In Proceedings of the Joint Conference on Water Resources Engineering and Water Resources Planning and Management, Minneapolis, MN, USA, 30 July–2 August 2000; pp. 1–10.
42. Lança, R.M.; Fael, C.S.; Maia, R.J.; Pêgo, J.P.; Cardoso, A.H. Clear-water scour at comparatively large cylindrical piers. *J. Hydraul. Eng.* **2013**, *139*, 1117–1125. [CrossRef]
43. Lee, S.O.; Sturm, T.W. Effect of sediment size scaling on physical modeling of bridge pier scour. *J. Hydraul. Eng.* **2009**, *135*, 793–802. [CrossRef]
44. Shahmohammadi, R.; Afzalimehr, H.; Sui, J. Impacts of turbulent flow over a channel bed with a vegetation patch on the incipient motion of sediment. *Can. J. Civ. Eng.* **2018**, *45*, 803–816. [CrossRef]
45. Shahmohammadi, R.; Afzalimehr, H.; Sui, J. Assessment of Critical Shear Stress and Threshold Velocity in Shallow Flow with Sand Particles. *Water* **2021**, *13*, 994. [CrossRef]
46. Yamasaki, T.N.; Jiang, B.; Janzen, J.G.; Nepf, H.M. Feedback between vegetation, flow, and deposition: A study of artificial vegetation patch development. *J. Hydrol.* **2021**, *598*, 126232. [CrossRef]
47. Jafari, R.; Sui, J. Velocity Field and Turbulence Structure around Spur Dikes with Different Angles of Orientation under Ice Covered Flow Conditions. *Water* **2021**, *13*, 1844. [CrossRef]
48. Huai, W.-x.; Zhang, J.; Katul, G.G.; Cheng, Y.-g.; Tang, X.; Wang, W.-j. The structure of turbulent flow through submerged flexible vegetation. *J. Hydrodyn.* **2019**, *31*, 274–292. [CrossRef]
49. Dey, S.; Barbhuiya, A.K. 3D flow field in a scour hole at a wing-wall abutment. *J. Hydraul. Res.* **2006**, *44*, 33–50. [CrossRef]
50. Dey, S.; Barbhuiya, A.K. Turbulent flow field in a scour hole at a semicircular abutment. *Can. J. Civ. Eng.* **2005**, *32*, 213–232. [CrossRef]
51. Siniscalchi, F.; Nikora, V.I.; Aberle, J. Plant patch hydrodynamics in streams: Mean flow, turbulence, and drag forces. *Water Resour. Res.* **2012**, *48*. [CrossRef]
52. Pope, S.B. *Turbulent Flows*; Cambridge University Press: Cambridge, UK, 2000; Available online: https://www.cambridge.org/ir/academic/subjects/physics/nonlinear-science-and-fluid-dynamics/turbulent-flows?format=PB&isbn=9780521598866 (accessed on 20 October 2021).
53. Lumley, J.L. Computational modeling of turbulent flows. In *Advances in Applied Mechanics*; Elsevier: Amsterdam, The Netherlands, 1979; Volume 18, pp. 123–176.

water

MDPI

Article

Assessment of Critical Shear Stress and Threshold Velocity in Shallow Flow with Sand Particles

Reza Shahmohammadi [1], Hossein Afzalimehr [2] and Jueyi Sui [3,*]

[1] Department of Water Engineering, Isfahan University of Technology, Isfahan 8415683111, Iran; Reza.shahmohammadi@gmail.com
[2] Department of Civil Engineering, Iran University of Science and Technology, Tehran 1684613114, Iran; hafzali@iust.ac.ir
[3] School of Engineering Program, University of Northern British Columbia, Prince George, BC V2N 4Z9, Canada
* Correspondence: jueyi.sui@unbc.ca; Tel.: +1-250-960-6399

Abstract: In this study, the incipient motion of four groups of sand, ranging from medium to very coarse particles, was experimentally examined using an acoustic Doppler velocimeter (ADV) in different water depths under the hydraulically transitional flow condition. The transport criterion of the Kramer visual observation method was used to determine threshold conditions. Some equations for calculating threshold average and near-bed velocities were derived. Results showed that the threshold velocity was directly proportional to both sediment particle size and water depth. The vertical distributions of the Reynolds shear stress showed an increase from the bed to about 0.1 of the water's depth, after performing a damping area, then a decrease toward the water surface. By extending the linear portion of the Reynolds shear stress in the upper zone of the damping area to the bed, the critical shear stress, particle shear Reynolds number, and critical Shields parameter were calculated. Results showed that the critical Shields parameter was located under the Shields curve, showing no sediment motion. This indicates that the incipient motion of sediment particles occurred with smaller bed shear stress than that estimated using the Shields diagram in the hydraulically transitional flow region. The reason could be related to differences between the features of the present experiment and those of the experiments used in the development of the Shields diagram, including the approaches to determine and define threshold conditions, the accuracy of experimental tools to estimate critical shear stress, and sediment particle characteristics. Therefore, the change in the specifications of experiments from those on which the Shields diagram has been based led to the deviation between the estimation using the Shields diagram and that of real threshold conditions, at least in the hydraulically transitional flow region with sand particles.

Keywords: incipient motion; threshold condition; critical shear stress; threshold velocity; hydraulically transitional flow

Citation: Shahmohammadi, R.; Afzalimehr, H.; Sui, J. Assessment of Critical Shear Stress and Threshold Velocity in Shallow Flow with Sand Particles. *Water* **2021**, *13*, 994. https://doi.org/10.3390/w13070994

Academic Editors: Jihn-Sung Lai and David Dunkerley

Received: 15 February 2021
Accepted: 29 March 2021
Published: 4 April 2021

1. Introduction

Reservoir dams for flood control, power generation, and water supply for irrigation and municipalities and industries involve huge costs. The service life of reservoirs depends on the amount of sediment delivered by rivers. On a sediment bed, if the flow velocity increases so much that the hydrodynamic forces, including drag and lift, exceed the stabilizing forces resulting from the particles' submerged weight, the sediment particle motion is intermittently and randomly initiated. The state of flow that is just sufficient to start sediment particle motion is called the threshold or critical condition [1]. The threshold of sediment particle motion plays an important role in many river engineering issues and some special problems, e.g., [2–4].

In field or laboratory experiments, the threshold condition can be determined by two methods—the bed-load extrapolation method and the visual observation method [1]. In the bed-load extrapolation method, the critical shear stress is defined by extrapolating paired measurements of bed shear stress and bed-load transport rate to zero or low reference transport rate of sediment flux [5]. This method is sensitive to the way of extrapolation [6] and the determined reference transport value [7]. Abbott and Francis [8] classified grains movement into three different types, namely, (1) rolling at which shear stress is only a little more than the critical value, (2) grains ballistic jumps or saltation influenced by bed mean shear stress and roughness, and (3) suspended motion known by generally longer trajectories. Against suspended movement, rolling and saltation are limited to near the bed. They named the ratio of the shear velocity to critical shear velocity as the transport stage. In the transport stage, it is not expected the value is less than 1.0, but with values larger than 1.0 rolling, saltation and suspension could be detected. The visual observation method is based on monitoring the sediment particle movement. Kramer [9] defined four levels of sediment movement [5], i.e., the first stage (no sediment transport), with no movement of sediment particles; the second stage (weak sediment transport), with the movement of a small number of the smallest particles in the isolated parts of the bed; the third stage (medium sediment transport), with the motion of a large number of medium-sized particles, considering bed-surface configuration is not affected; and the fourth stage (general sediment transport), with the motion of all sizes of particles in all parts of the bed, considering it is strong enough to change the bed-surface configuration. Different definitions of threshold conditions in various investigations have led to conflicting results and have made it difficult to compare [5,10,11].

Some research studies show that the sediment motion is influenced by the near-bed turbulence, indicating the nature of hydrodynamic forces acting on the grain particles [12]. Bialik [13] applied a Lagrangian perspective to study numerically the role of the coherent structure in the incipient motion of sediment particles. He used a 3D relevant model of grains, in which a special procedure has been designed to generate coherent structures. The numerical results showed that the sweeps and outward events play a generally dominant role in the initiation of particles saltating. Dey et al. [12] attempted to quantify the turbulence characteristics of near-bed flows in threshold conditions of non-cohesive sediments. Their analysis of experimental data measured in flows over immobile and threshold condition beds showed the changes in the turbulence characteristics due to differences in bed conditions. They applied quadrant analysis of the data of velocity fluctuations and concluded that sweep events are the dominant mechanism toward sediment movement, and ejection events are prevalent at the top of the wall–shear layer. In this condition, the turbulent dissipation exceeds the turbulent generation. Nikora et al. [14] illustrated a physically based explanation of the dispersion relation, introducing two types of sand movement in the form of sand waves related to the region of small and large wavenumbers. They explained that the formation of small sand waves is a result of the individual sand particles' motion, while larger sand waves form due to the motion of smaller waves.

Threshold average and near-bed velocities are equal to the average flow velocity and near-bed (at the sediment particle level) velocity under the defined threshold conditions, respectively [1]. Numerous studies have provided equations for estimating the threshold average and near-bed velocities, corresponding to water depth and sediment particle characteristics [15–17]. However, the precise size of sediment particles and hydraulic flow regime have usually not been clearly demonstrated, and the determination of the threshold average and near-bed velocities is still a challenging issue [1], which requires more in-depth studies, especially in the presence of non-cohesive sand particles at the hydraulically transitional flow condition. On the other hand, Einstein [18], Velikanov [19], Yalin [20], and Ling [21] investigated the effect of lift force on sediment motion. However, the effect of drag force, along with that of lift force, on the threshold conditions for sediment motion should also be considered [1]. A theoretical sediment threshold condition model should consider the effect of both lift and drag forces against the particle stabilizing force resulting from the submerged particle's weight. Regarding the effect of bed-shear stress on the threshold of sediment particle motion, all studies have only been based on laboratory measurements to yield empirical equations with different and approximate results [1].

The semi-theoretical method proposed by Shields [22] had phenomenally improved the estimation accuracy of the threshold condition [1]. Even today, it is the most commonly used method to estimate the threshold condition of non-cohesive sediments. The data points located on the Shields curve represent the threshold conditions and the regions above and below the curve represent the regions with and without sediment motion, respectively. With regard to hydrodynamic conditions, the Shields diagram is divided into three different flow regions [1], i.e., the hydraulically smooth flow with the particle shear Reynolds number less than 2, the hydraulically transitional flow with the particle shear Reynolds number between 2 and 500, and the hydraulically rough flow with the particle shear Reynolds number more than 500. In these regions, the viscous sublayer thickness is larger, almost equal to, and smaller than sediment particle diameter, respectively. The critical Shields parameter has a minimum value (about 0.032) when the particle shear Reynolds number ranges from 9 to 20 and a constant value of about 0.056 in the hydraulically rough flow region.

Despite extensive applications of the Shields diagram, many researchers have challenged its validity [5,11,23–27]. Regarding the rough turbulent flow, it is reported that the results of the Shields diagram for the incipient motion of coarse materials are not appropriate [28]. In this case, Neil [29] claimed the critical Shields parameter equal to 0.03 for the particle shear Reynolds number more than 500, while Gessler [30] obtained 0.046 for a similar condition. Unreliability of the Shields diagram makes it problematic as a recommendation for engineering applications with coarse materials, which expresses results should be divided by the number of two [28]. Some studies such as Miller et al. [23] and Yalin and Karahan [31] reported a wide range of the Shield parameter from 0.02 to 0.065. Other studies have been attempted to improve Shields' results [32–36], but the issue has remained challenging, particularly about the hydraulically transitional flow region of the Shields diagram linked to sand particles, which is related to the curved zone of the Shields diagram. Considering that most studies concentrate on the incipient motion of the coarse materials including hydraulically rough flow, it must be valuable to study the threshold condition of the sand particles in the hydraulically transitional flow regime experimentally.

Most of the fundamental studies for developing practical equations to determine the threshold average and near-bed velocities and non-dimensional critical shear stress trace back to the time when advanced experimental equipment for collecting data were not available, and the main focus was on the hydraulically rough flow. Nowadays, advancement in experimental tools provides new opportunities to improve the results of the previous experimental studies. In this study, to estimate the sand particles threshold condition in the hydraulically transitional flow accurately, the vertical distribution of velocity and Reynolds shear stress were acquired more accurately using an ADV instrument. Additionally, the exact flow discharge was measured using an electromagnetic flowmeter. Thereafter, compared to the results of previous studies, more accurate experimental results about threshold conditions have been obtained.

2. Theoretical Background

2.1. Threshold Average Velocity Equations

The following equation was developed by Goncharov [15] to calculate the threshold average velocity U_{cr}:

$$U_{cr} = 1.07\sqrt{\Delta g d}\log\left(8.8\frac{h}{d_{95}}\right), \tag{1}$$

where h is the water depth, d is the median sediment grain size, d_{95} is the size for which 95% of the sediment particles are smaller, g is the gravitational acceleration, Δ is the submerged relative mass density of sediment particles, and $\sqrt{\Delta g d}$ has a dimension of velocity. This equation was derived for turbulent flow around the sedimentary bed, indicating a hydraulically rough flow regime. Goncharov [15] claim that the equation does not vary for cases that lower part of the grain particles is located inside the laminar boundary layer, which indicate equation maybe could be considered for hydraulically transitional flow regime. Neill [16] developed the following equation for estimating the threshold average velocity for coarse gravel particles for hydraulically rough flow condition:

$$U_{cr} = 1.41\sqrt{\Delta g d}\left(\frac{h}{d}\right)^{1/6}. \tag{2}$$

Based on a large amount of data in hydraulically rough flow under the threshold condition, Garde [17] proposed the following equation:

$$U_{cr} = \sqrt{\Delta g d}\left(0.5log\frac{h}{d} + 1.63\right). \tag{3}$$

2.2. Threshold Near-Bed Velocity Equation

Garde [17] proposed the following equation for calculating the threshold near-bed velocity in hydraulically rough flow:

$$u_{cr} = 1.51\sqrt{\Delta g d}. \tag{4}$$

2.3. Shields Approach

Shields claimed that nondimensional critical bed shear stress should be a function of shear Reynolds number [28]. In the Shields diagram, the below-mentioned particle shear Reynolds number R_* and critical Shields parameter Θ_c have been considered on the horizontal and vertical axis, respectively:

$$R_* = \frac{k_S\sqrt{\tau_{OC}/\rho}}{\upsilon} \overset{k_S=d}{\Rightarrow} R_* = \frac{d\sqrt{\tau_{OC}/\rho}}{\upsilon} \tag{5}$$

$$\Theta_c = \frac{\tau_{OC}}{\Delta\rho g d}, \tag{6}$$

where k_s is the Nikuradse's equivalent roughness, which is usually assumed to be equal to the sediment median grain size, ρ is the mass density of water, and τ_{oc} is the critical shear stress that, divided by $\Delta\rho g d$, is converted into the non-dimensional critical bed shear stress.

3. Materials and Methods

Experiments were conducted in a rectangular flume that is 15 m long, 0.9 m wide, and 0.6 m deep (Figure 1). In order to observe the movement of sediment particles, sidewalls of the flume were made of transparent glass. A slide-gate was located at the end of the flume, which allowed the water to spill over into a downstream reservoir. This reservoir was equipped with a sediment trap and was connected to the suction pipe of a centrifugal pump. The pumping system circulated the water between the flume downstream reservoir and the water tank located upstream of the flume. In order to monitor the water temperature during the experiment, a floating electrical thermometer was put in the upstream water tank.

In the upstream tank, a multi-layer grid and a secondary stilling basin were installed to dissipate water energy and reduce water flow oscillation at the flume entrance. The pumping system consisted of an electromotor pump with a discharge capacity of 50 lit/s, a piping system, an electromagnetic flowmeter, a three-phase switchboard, and a variable frequency drive. The pump discharge was controlled by adjusting the input frequency of the electromotor pump using the variable frequency drive. The electromagnetic flowmeter was installed on the outlet pipe of the pump and measured the discharge within the maximum relative percentage error of 0.5%. In addition, a liminimeter (Laboratory of The Isfahan University of Technology, Isfahan, Iran) with a measuring resolution of 0.5 mm was used to acquire the water depth. By changing the flow discharge and adjusting the end slide gate, it was possible to reach the desired flume water depth and velocity.

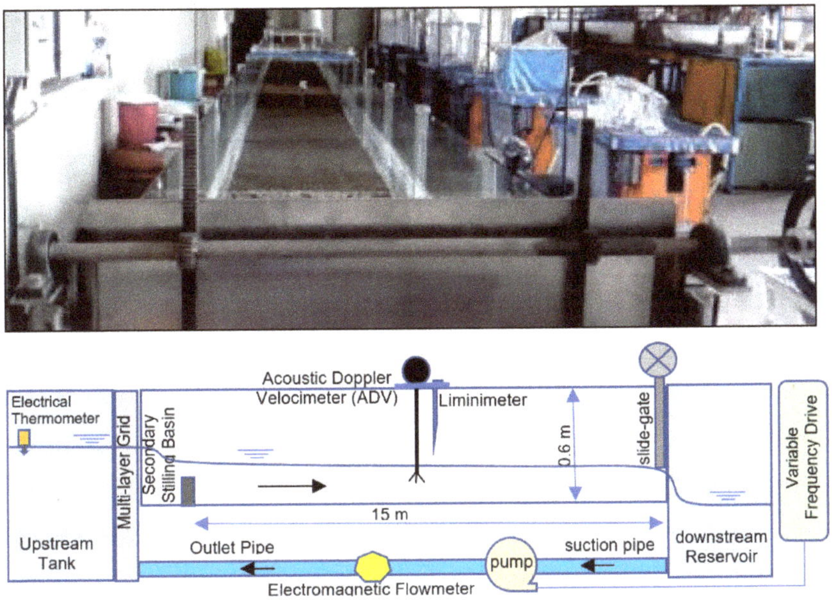

Figure 1. Flume used for experiments (dimensions: 15 m long, 0.9 m wide, and 0.6 m deep).

In this experimental study, an acoustic Doppler velocimeter (ADV) was used to record velocity time series data and determine velocity fluctuations. This ADV is manufactured by the Nortek Corporation with a maximum 0.5% relative percentage error and acoustic frequency of 10 MHz according to instrument instruction available at www.nortek-as.com (accessed on 1 October 2004). The ADV acquires three-dimensional velocity data including u (streamwise), v (spanwise to the left side), and w (vertical toward water surface). The negative values of w indicate a downward flow from the water surface toward the bed. The duration for data collection lasted 2 min at each point with a sampling frequency of 200 Hz, according to the latest version of the Vectrino Plus interface software (version 1.22, Nortek Corporation, Vangkroken, Norway). In this way, 24,000 data were recorded at each point along verticals for velocity measurements. The ADV was set up at each desired position using a three-dimensional moveable device. Flow velocities along all verticals along the centerline of the flume were measured. Along each vertical, approximately 20-point velocities were recorded from the sand bed to the water surface so that 50 percent of the points were in the inner layer of the velocity profile (20 percent of the water depth near the bed).

The nominal distance of the ADV transmit transducer to the focal point of sampling volume was about 50 mm. On the other hand, in order to prevent it from interference with air bubbles of the water surface, it required at least 10 mm submergence of the transmit transducer and its four receiving transducers. As a result, due to the ADV inherent limitation, the first point for velocity measurement from the water surface was at least 60 mm below the water surface. Similarly, due to the effect of the sand bed on the sampling volume, it was impossible to measure flow velocities in a zone of about 3–4 mm near the bed. Therefore, measurements of velocity along verticals were limited to a range from 3–4 mm above the sand bed to 60 mm below the water surface. It is noticeable that ADV is affected by Doppler noise and spikes caused by aliasing of the Doppler signal due to the shifting phase between the outgoing and incoming pulse [37]. The velocity data were filtered using WinADV software (version 2.024, Bureau of Reclamation, Washington, DC, USA) and aliases. Spikes were removed using the phase–space threshold despiking filter, developed by Goring and Nikora [38] and modified by Wahl [39], together with a minimum acceptable correlation coefficient of 70 and signal-to-noise ratio of 15. On average, 18% of the data were ignored and the rest of the data were verified for analysis. Using those valid data, velocity and Reynolds shear stress profiles were analyzed and plotted by means of a program that was developed using the Excel visual basic for application (VBA) programming language.

Natural quartz sand with a mass density (ρ_s) of 2.65 g/cm^3 was used as sediment particles in this experimental study. Therefore, the relative mass density of sediment particles (S) and the submerged relative mass density of sediment particles (Δ) were equal to 2.65 and 1.65, respectively. As shown in Table 1, after screening and grading, four groups of sediment particles numbered as I, II, III, and IV, were obtained with median grain sizes of 0.43, 0.83, 1.38, and 1.94 mm, respectively. According to the criterion for sediment size classification [40], group I is medium sand, group II is coarse sand, and groups III and IV are very coarse sand. On the other hand, the median diameter of all four material groups is less than 2 mm, indicating the range size of sand material [41]. In Table 1, d_i is the size (mm), which is smaller than i percent of sediment particles, and σ_g is the geometric standard deviation of sediment particles, σ_g = $(d_{84}/d_{16})^{0.5}$. The values of σ_g less than about 1.4 indicated a uniform distribution of sediment particles [42]. According to those values in Table 1, sediment particles have an acceptable uniform distribution. In Figure 2, the gradation (particle-size distribution) curves of sediments are shown in a semi-logarithmic graph. As observed, the gradation curve patterns indicated a uniform gradation.

Table 1. Grain size characteristics of four groups of screened and graded sediments—mm.

Sediment	Class	d_{16}	d_{35}	d_{50}	d_{84}	σ_g
I	Medium Sand	0.27	0.36	0.43	0.56	1.45
II	Coarse Sand	0.72	0.78	0.83	1.11	1.24
III	Very Coarse Sand	1.22	1.31	1.38	1.58	1.14
IV	Very Coarse Sand	1.71	1.84	1.94	2.28	1.16

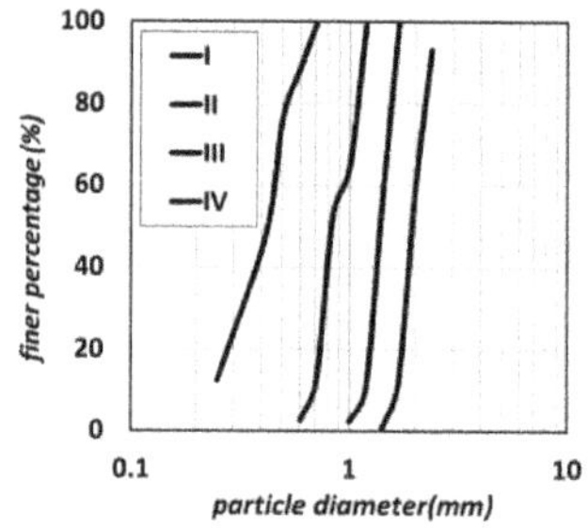

Figure 2. Gradation curves of sediment particles (from left to right are I, II, III, and IV, respectively).

The upstream edge of the sand bed was 8 m downstream from the flume entrance and the downstream boundary of the sand bed was 12 m downstream from the flume entrance. Namely, this 4 m long flume section was covered by sand with a thickness of 3 cm. Along this sand-bed section, the turbulent flow was fully developed, and the influence of the tailgate was negligible. As shown in Figure 3, for all sediment groups, the velocity distributions have a similar pattern in successive cross sections P1, P2, and P3, which were 9 m, 10 m, and 11 m downstream from the flume entrance, respectively. The legends in Figure 3 are explained as follows: for example, "II-H3-P3" describes a velocity profile over a sand bed with sediment group "II" and water depth of "H3" at the location of "P3". One can infer from Figure 3 that all velocity profiles at P1, P2, and P3 have similar shapes; in particular, the vertical velocity distributions at P2 and P3 are nearly the same. In this study, velocity profiles at P3 were used in the hydraulic analysis.

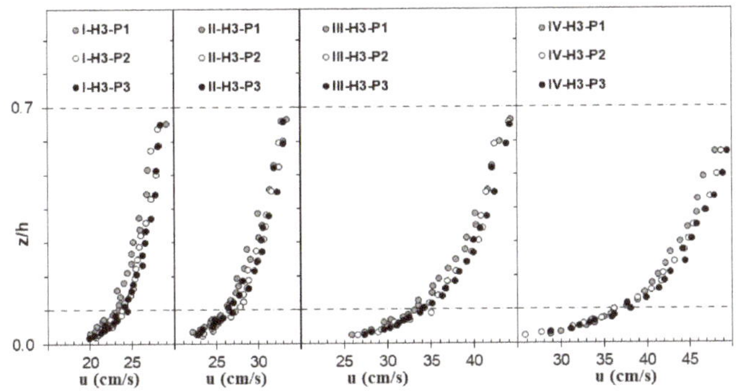

Figure 3. Flow velocity profiles at cross sections P1, P2, and P3 within the study reach (cm/s).

The velocity verticals were acquired using the ADV along this 4 m long sand-bed flume section. As shown in Figure 4, the upstream and downstream of the study sand-bed reach were covered by coarse particles that were not mobilized during experimental runs. The coarse-grained material upstream of the sand-bed facilitates a fully turbulent flow condition. In order to eliminate the effect of the bed roughness change on flow characteristics, the first (P1) and the last (P3) positions were located at a distance of one meter from the course materials. The position of P2 was located between P1 and P3, namely, in the middle of the study sand bed.

Figure 4. A view of the sand-bed section with coarse particles upstream and downstream.

The criterion for the incipient motion of sediment particles was determined based on the medium transport of the Kramer visual observation method that defined the movement of a large quantity of medium-size particles. To achieve the threshold conditions, firstly, the liminimeter was set up at the appropriate level. During experiments, the desired water depth is acquired by touching the pinpoint of the liminimeter on the water surface. In order to prevent sediment particles from being washed away at the beginning of the experiment run, the pump was started with a low flow rate (of about 5 lit/s), while the downstream slide gate was closed. In this way, water was gradually spilled into the flume from the tank upstream of the flume. Considering the closure of the end slide gate, the water level in the flume gradually increased. Once the water level touched the pinpoint of the liminimeter, the flume end slide gate was opened so slightly as to create a low-velocity flow for the intended water depth by creating an equilibrium state, namely, the discharge spilled into the flume equals to the flow rate out of the end slide gate. At the desired water depth, to achieve threshold conditions, the flow velocity should be increased. Thus, the pump discharge and the end slide gate opening were increased step by step, maintaining the balance between the inflow into the flume and the outflow from the flume by maintaining the intended water depth. After reaching the threshold conditions, to make sure that the motion of the sediment particles was stable, the continuous movement of sediment particles was kept for 10–15 min. One can infer from Figure 5 that after finishing the ADV data acquisition, a number of sediment particles were eroded from the studied sand-bed section and deposited downstream reach, implying that the threshold condition was determined appropriately.

Figure 5. Transportation of a number of sediment particles out of the studied sand-bed reach.

4. Results

4.1. Characteristics of Flow over the Sand Bed

Table 2 shows the results and characteristics of experiments for the incipient motion of sediment particles. The experiments for each sediment group I, II, and III were conducted with three water depths of H1 = 100, H2 = 120, and H3 = 140 mm, respectively. For sediment group IV, experiments were carried out for water depths of H1 = 91, H2 = 104, and H3 = 120 mm, respectively.

Table 2. Results and characteristics of experiments for incipient motion of sediment.

Experiment Name:	I (d = 0.43 mm)			II (d = 0.83 mm)			III (d = 1.38 mm)			IV (d = 1.94 mm)		
	H1	H2	H3	H1	H2	H3	H1	H2	H3	H1	H2	H3
Water Depth (h)—mm	100	120	140	100	120	140	100	120	140	91	104	120
Relative Submergence (h/d)	235	282	329	120	144	168	72	87	101	47	54	62
Relative Roughness (d/h) × 10^{-3}	4	4	3	8	7	6	14	11	10	21	19	16
Aspect Ratio (B/h)	9.0	7.5	6.4	9.0	7.5	6.4	9.0	7.5	6.4	9.9	8.7	7.5
Water Temperature—°C	17	15	19	18	20	19	21	22	17	24	24	21
Discharge (Q)—L/s	21.3	27.3	33.4	27.1	33.0	38.5	33.2	41.6	49.3	35.7	41.6	49.3
U_{cr} = Q/(Bh)—cm/s	23.6	25.3	26.5	30.1	30.6	30.6	36.9	38.5	39.1	43.6	44.4	45.7
Re = Uh/ν × 10^3	24	30	37	30	37	43	37	46	55	40	46	55
Fr = U/$(gh)^{0.5}$	0.24	0.21	0.19	0.31	0.26	0.22	0.38	0.33	0.28	0.49	0.44	0.39

4.2. Threshold Average Velocity

Based on experiments data, the relationship between two dimensionless parameters of $\frac{h}{d}$ and $\frac{U_{cr}}{\sqrt{\Delta g d}}$ is plotted in Figure 6, and the following equation was obtained:

$$U_{cr} = \sqrt{\Delta g d}\left(0.0024\left(\frac{h}{d}\right) + 2.34\right);\ R^2 = 0.93 \quad (7)$$

where the threshold average velocity (U_{cr}) is in cm/s, d and h are in cm, g is equal to 981 cm/s^2, and Δ is equal to 1.65. Table 3 summarizes the values of critical average velocities determined by Equation (7) and those estimated using equations developed by other researchers. Figure 7 illustrates flow velocity distribution under threshold conditions for the four-group sediments under different water depths.

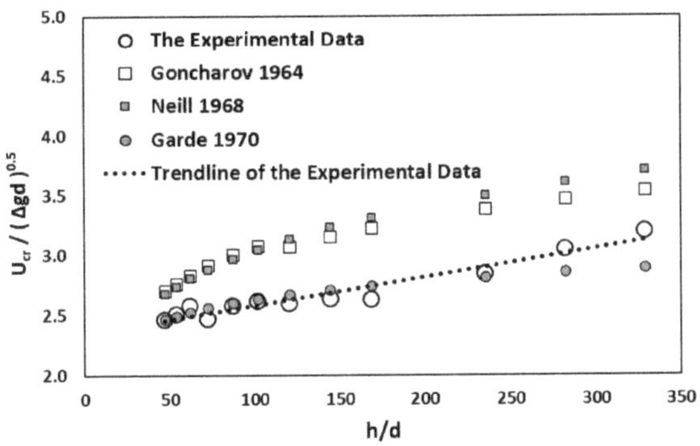

Figure 6. Relationship between $\frac{h}{d}$ and $\frac{U_{cr}}{\sqrt{\Delta g d}}$ in this study, compared to results of other studies.

Table 3. Estimation of critical average velocities obtained using Equation (7), compared to results using other equations (cm/s).

Experiment:		I (d = 0.43)			II (d = 0.83)			III (d = 1.38)			IV (d = 1.94)		
		H1	H2	H3	H1	H2	H3	H1	H2	H3	H1	H2	H3
Experimental Data		23.6	25.3	26.5	30.1	30.6	30.6	36.9	38.5	39.1	43.6	44.4	45.7
Extracted Equation (Equation (7))		24.1	25.0	26.0	30.5	31.2	31.8	37.6	38.1	38.6	43.5	43.7	44.1
Goncharov	Value	28.1	28.8	29.4	35.6	36.6	37.4	43.6	44.9	46.0	47.9	49.0	50.2
	Difference with Equation (7)	−14%	−13%	−12%	−14%	−15%	−15%	−14%	−15%	−16%	−9%	−11%	−12%
Neill	Value	29.1	30.0	30.8	36.3	37.5	38.4	43.0	44.4	45.5	47.5	48.5	49.7
	Difference with Equation (7)	−17%	−16%	−16%	−16%	−17%	−17%	−13%	−14%	−15%	−8%	−10%	−11%
Garde	Value	23.4	23.7	24.0	31.0	31.4	31.8	38.3	38.8	39.3	43.7	44.2	44.8
	Difference with Equation (7)	3%	6%	8%	−2%	−1%	0%	−2%	−2%	−2%	−1%	−1%	−1%

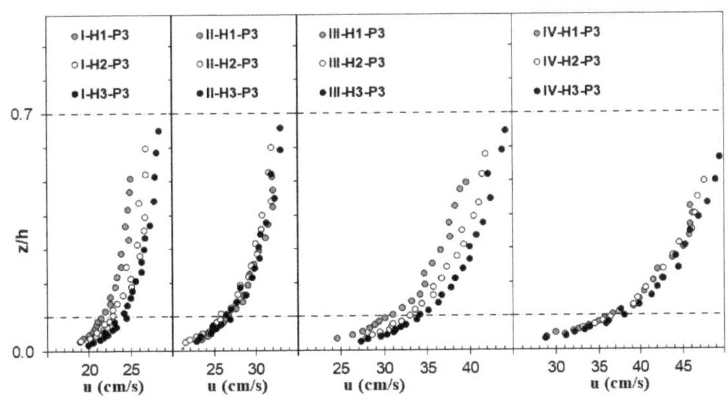

Figure 7. Flow velocity distribution under threshold conditions (cm/s).

4.3. Threshold Near-Bed Velocity

The average velocity under threshold conditions is often calculated by dividing the flow rate by the flow cross-sectional area. In reality, the flow velocity near the sand bed plays a more important role in the movement of sediment particles [1]. Since it is difficult to measure the threshold near-bed velocity precisely at the particle level, one can indirectly obtain it by extrapolating the time-averaged streamwise velocity verticals up to the particle level [1]. By extrapolating the presented flow velocity verticals in Figure 7 toward the bed, the threshold near-bed velocity was obtained, as presented in Table 4. Considering the validity of the law of the wall in the inner layer (z/h < 0.2), the extrapolation is performed using an exponential function in the inner layer toward the adjacent of the bed.

Table 4. Threshold near-bed velocity (cm/s).

Experiment	I (d = 0.43 mm)			II (d = 0.83 mm)			III (d = 1.38 mm)			IV (d = 1.94 mm)		
	H1	H2	H3	H1	H2	H3	H1	H2	H3	H1	H2	H3
u_{cr} (cm/s)	18.1	18.2	19.0	21.5	20.2	19.6	22.2	23.5	24.1	25.3	24.9	24.8

Figure 8 shows the relation between the threshold near-bed velocity and sediment particle size and can be described as the following equation:

$$U_{cr} = 44.2\, d + 16.7; R^2 = 0.99 \tag{8}$$

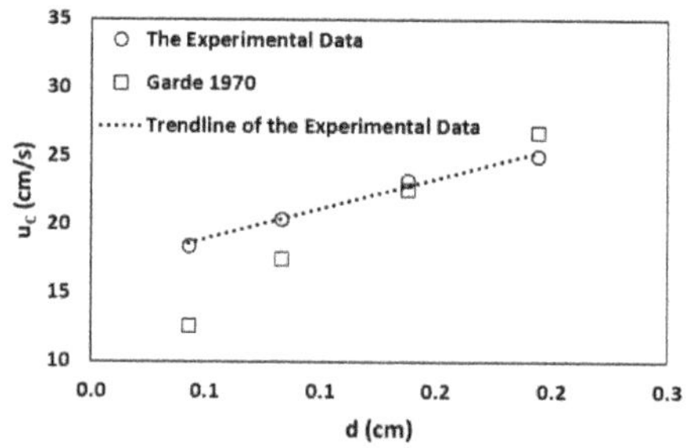

Figure 8. Threshold near-bed velocity vs. sediment particle size in comparison with results using equations developed by other researchers.

In Table 5, the threshold near-bed velocities, calculated by Equation (8) and equations proposed by Garde [17] (Equation (4)), are presented.

Table 5. Estimation of the threshold near-bed velocity using Equation (8) and those proposed by other researchers (cm/s).

Sediment Group:		I (d = 0.43)	II (d = 0.83)	III (d = 1.38)	IV (d = 1.94)
Averaged Experimental Data		18.4	20.4	23.2	25.0
Extracted Equation (Equatiuon (8))		18.6	20.4	22.8	25.3
Garde [17] (Equatiuon (4))	Value	12.6	17.5	22.6	26.8
	Difference with Equatiuon (8)	48%	16%	1%	−6%

4.4. Shields Diagram

To calculate the parameters of the Shields diagram (particle shear Reynolds number and critical Shields parameter), it is necessary to calculate the critical shear stress. As shown in Figure 9, the vertical distribution pattern of the Reynolds shear stress $-\overline{u'w'}$ profiles were convex, including a damping zone when the dimensionless water depth was about z/h = 0.1–0.2. The Reynolds shear stress increased from the sand bed to the water depth of z/h = 0.1 and then decreased toward the water surface. The most dominant method to estimate bed shear stress was the Reynolds shear stress distribution. This method can be used in the case of steady, uniform flow [1,28]. In this case, one should notice that there is a near-bed damping zone, and extrapolation should be taken in the linear portion of the Reynolds shear stress distribution above the bed [1]. According to the widely accepted approach, it was extended the linear portion of the Reynolds shear stress in the upper zone of the damping zone to the bed. Then, particle shear Reynolds number and critical Shields parameter were calculated (Table 6).

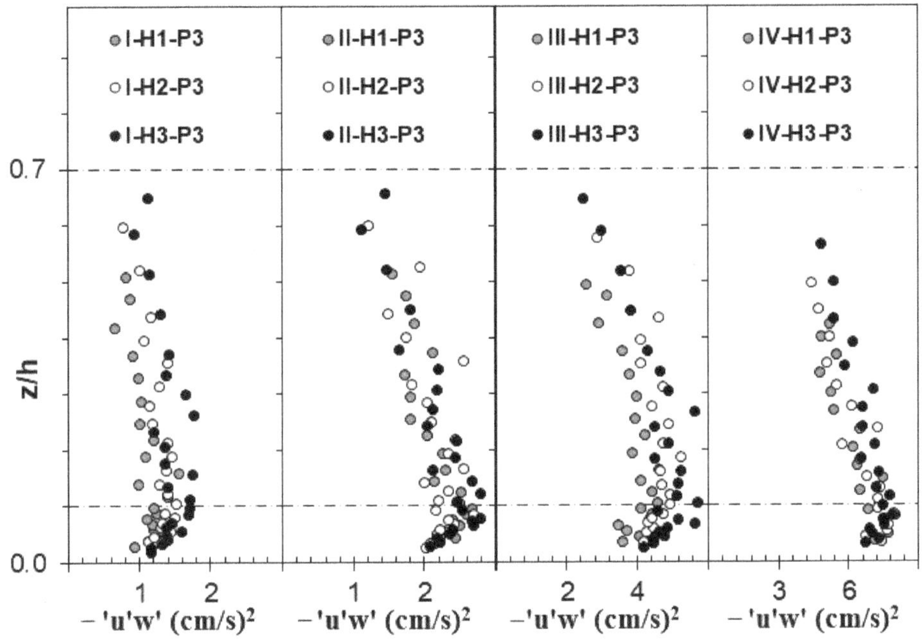

Figure 9. Reynolds shear stress $-\overline{u'w'}$ distributions under threshold condition $(cm/s)^2$.

Table 6. The critical shear stress, particle shear Reynolds number, and critical Shields parameter of the experiments.

Experiment	I (d = 0.43 mm)			II (d = 0.83 mm)			III (d = 1.38 mm)			IV (d = 1.94 mm)		
	H1	H2	H3	H1	H2	H3	H1	H2	H3	H1	H2	H3
τ_{oc} $(cm/s)^2$	1.5	1.8	2.2	2.9	2.9	3.0	5.7	6.4	6.9	8.2	8.4	9.0
R_*	5.3	5.8	6.4	14.1	14.1	14.4	32.9	34.9	36.2	55.6	56.2	58.2
Θ_c	0.022	0.026	0.032	0.022	0.022	0.022	0.026	0.029	0.031	0.026	0.027	0.029

5. Discussion

5.1. Characteristics of Flow over the Sand Bed

Considering the values of Reynolds and Froude numbers, flow conditions in all experiments were turbulent and subcritical. The threshold average velocities for the incipient motion of sediment particles varied from 23.6 to 45.7 cm/s. The ratios of flume width B to flow depth h, which is termed aspect ratio, varied from 6.4 to 9.9. Considering that the velocity data is acquired in the middle line of the flume with 45 cm distance from the sidewalls, and with the high enough aspect ratio, it is expected that sidewall effects and secondary currents be negligible and velocity profiles are influenced only by bed materials, including maximum water velocity at the water surface. This could be an important factor for gaining more reliable results at the threshold conditions.

5.2. Threshold Average Velocity

As shown in Figure 6 and Table 3, the calculated threshold average velocities using Equation (7) are in good agreement with those of experiments (presented in Table 2), in which the maximum difference between them was 2 cm/s (less than 4%). The differences between the calculated threshold average velocities using Equation (7) and equations of other researchers are less than 20%, confirming the desired performance of Equation (7). The results indicate well agreement between derived and Garde's equations.

The main reason for the difference between results using Equation (7) and those obtained by others should be related to the differences between flow regimes at the threshold condition. As already mentioned, Goncharov [15], Neill [16], and Garde [17] conducted experiments using coarse materials (gravel bed) under a rough turbulent flow regime of the threshold condition, whereas current experiments performed using sand bed under the hydraulically transitional flow regime. Additionally, it is not deniable that differences in the definition of threshold conditions and the accuracy of the velocity estimation could be other reasons for the difference. For example, in this experiment, an electromagnetic flowmeter was used to determine water discharge and velocity, which is more precise than equipment used by previous researchers.

According to Figure 7 and Equation (7), the threshold average velocity is directly proportional to the water depth. This means that by increasing the water depth, the threshold condition occurs at a higher velocity value. An increase in the sediment particle size and its submerged weight tends to increase in required hydrodynamic forces to move the particle. Thus, the threshold average velocity for the incipient motion of sediment increases with the sediment particle size.

5.3. Threshold Near-Bed Velocity

According to the values presented in Table 4, it can be concluded that similar to the equation proposed by Garde [17], for a specified sediment particle, changes in water depth do not significantly affect the threshold near-bed velocity. One can also infer from Tables 4 and 5 that results obtained using Equation (8) are in good agreement with the experimental data, and the difference between calculation results and those of experiments is less than 0.4 cm/s (less than 2%). According to Table 5, except for sediment group I, results using Equation (8) were in good agreement with those using equation proposed by Garde [17]. The reason for the differences in the results should be attributed to threshold average velocity because with increasing the sand particle size and increasing the particle shear Reynolds number, differences between the current experiment and Gard's equation are reduced.

5.4. Shields Diagram

Regarding the range of the particle shear Reynolds number, it was observed that all experimental data lay in the range of the hydraulically transitional flow. Iwagaki [33] pointed out that in the range $6.83 < R_* < 51.1$, the bed particles size is in the order of the viscous sublayer thickness. The flow is in the hydraulically transitional flow regime, and both viscous and turbulence are effective in the bed particles. Obviously, the range of the particle shear Reynolds number discussed by Iwagaki [33] is the same as this experiment condition, as presented in Table 6. As shown in the table, for a sediment group, an increase in the water depth leads to some extent larger critical shear stress. However, the effect of the water depth on critical shear stress is not noticeable, maybe due to the non-obvious differences in water depths. It is also noticed that, despite an increase in critical shear stress with the increase in sediment particle size, the variations in the critical Shields parameters for all sediment groups were not considerable, merely ranging from 0.022 to 0.032.

Figure 10 shows the relation between critical Shields parameter and relative roughness d/h of the present study, compared to that presented by Buffington and Montgomery [5]. Buffington and Montgomery [5] collected experimental data using different methods to determine the incipient motion mainly through visual observation and bed-load extrapolation (reference transport rate). According to Figure 10, they claimed that there was an overall positive correlation between critical Shields parameter and sediment relative roughness for $d/h \geq 0.01$ (groups III and IV of the particles), and an inverse correlation for $d/h \leq 0.01$ (groups I and II of the particles).

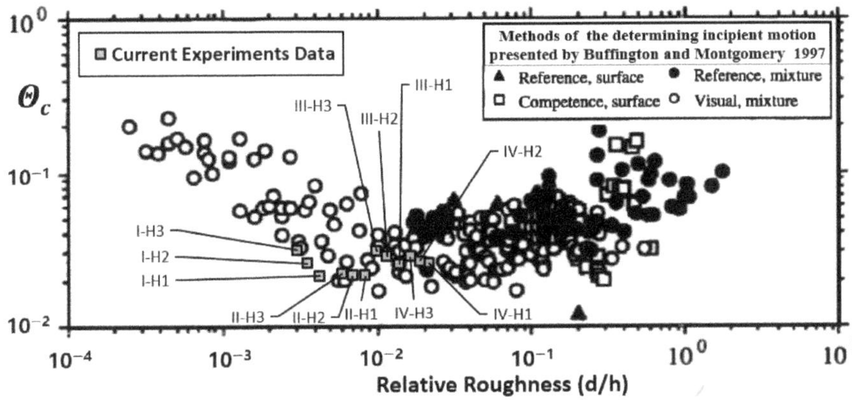

Figure 10. Relation between critical Shields parameter and the relative roughness of the present study, compared to those using different methods for determining the incipient motion presented by Buffington and Montgomery [5].

Experimental results are presented in the Shields diagram, as shown in Figure 11 and Table 6. It is expected that the points related to the experiments should lie on the Shields diagram curve, indicating threshold conditions. However, all data points of experiment results lay below the Shields diagram curve, where it was supposed to indicate no sediment motion. The following reasons may be attributed to this phenomenon:

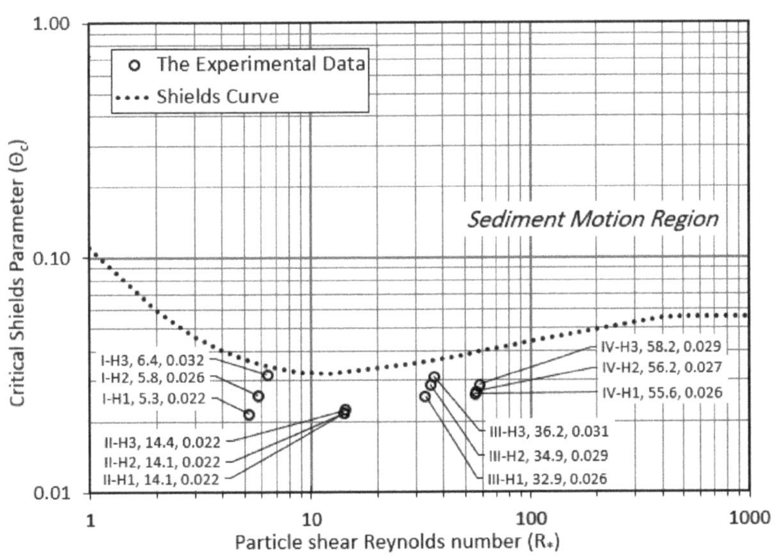

Figure 11. Current experimental results presented in the Shields diagram.

- Method for the definition of the threshold condition: Although Shields did not clearly explain the method, he used to determine the threshold conditions in his experiments, Kennedy [43] claimed that he probably applied Kramer's general motion criteria. Generally, there is an agreement that he used the bed-load extrapolation method to estimate the critical shear stress [23,44,45]. In the present experiments, the medium transport criterion of the Kramer visual observation method without any bed form was used. Different methods for threshold definition could yield different values of critical shear stress. Based on a lot of different studies, Buffington and Montgomery [5] clearly showed that the bed-load extrapolation method overestimated the critical Shields parameter, compared to the visual observation method. Comparing the results of the critical Shields parameters in the present experimental study to those presented by Buffington and Montgomery [5], it clearly showed in Figure 12 that although the critical Shields parameters of the present experiments are a little smaller than those of the bed-load extrapolation method (Figure 12a), they are in agreement with those of experiments based on the visual observation method (Figure 12b). Therefore, an overestimation of the critical Shields parameters using the bed-load extrapolation method could be expected. The main reasons for the overestimation of the critical Shields parameters based on the bed-load extrapolation could be related to more probable bedform creation in the bed-load extrapolation experiments because of the method's inherent condition [46]. Bedforms influence both the bed shear stress and sediment transport by exerting more drag force and turbulence [47]. It can dissipate bed shear stress [48–50], causing significant overestimation of the bed shear stress. In addition, the negligible change of water depth owing to bedform can lead to some overestimation [47];

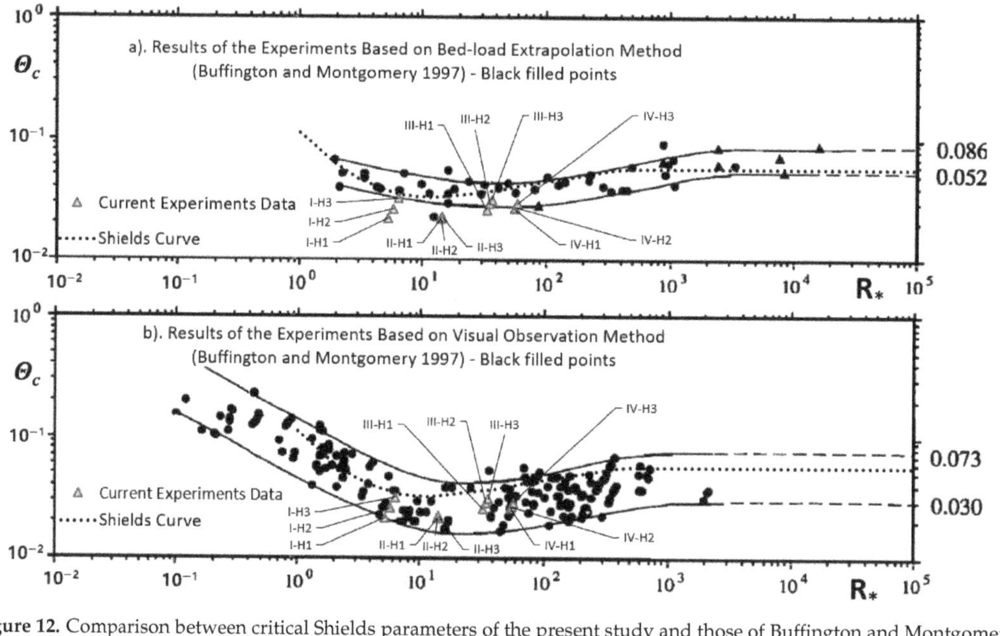

Figure 12. Comparison between critical Shields parameters of the present study and those of Buffington and Montgomery obtained via Bed-load Extrapolation Method (**a**) and Visual Observation Method (**b**) [5].

- Experimental measuring tools: Nowadays, the tools for acquiring vertical distributions of velocities and Reynolds shear stress in the hydraulic laboratory are more advanced, compared to those used by Shields [22] a long time ago, as also reported by Buffington and Montgomery [5]. Using an ADV in the present experimental study, it was possible to determine the critical shear stress more precisely, leading to more precise and reliable results with some expected differences from those reported by Shields [22] and Buffington and Montgomery [5];

- Sediment characteristics: The differences in sediment characteristics, such as size, shape, roundness, sorting, packing, and mass density, can affect the results of incipient motion [47,51–53]. Shields used four types of sediment including nearly uniform grains of brown coal, amber cuttings, crushed barite, and crushed granite, with the mass density ranging from 1060 to 4300 kg/m^3 and the median grain size ranging from 0.36 to 3.44 mm, which were sub-angular to very angular [47]. In this experimental study, naturally rounded quartz particles were used. Obviously, both crushed and angular particles cause more resistance to incipient motion due to higher friction of the particles [11], and this leads to an increase in the critical shear stress.

6. Conclusions

In this study, the incipient motion of four groups of sand particles has been investigated. The medium transport criterion of the Kramer visual observation method was used to determine the threshold conditions for the incipient motion. An electromagnetic flowmeter was used to measure discharge. To have reliable results, an acoustic frequency ADV was used to acquire velocity time series data. Therefore, profiles of the mean point flow velocity and Reynolds shear stress were obtained. Then, by extrapolation toward the sand bed, the near-bed velocity and critical shear stress were determined. Appropriate equations were derived to calculate the threshold average and near-bed velocities. Comparison between results using derived equations and those using equations developed by other researchers showed that there were some differences between our results and those of others. The main reason for the differences should be related to differences between bed materials and hydraulic flow regimes under the threshold condition. Additionally, the differences in the definition of threshold conditions and the accuracy of the water velocity estimation should not be ignored.

Experimental results of the current study showed that the points were located below the curve of the Shields diagram, namely, in the region for no sediment motion. Results of the present study indicate that the initiation of sediment particles occurred with bed shear stress, which is less than that estimated by the Shields diagram. This could be attributed to some differences between the current experiment and those of the Shields experiments, including approaches for determining threshold conditions, measurement accuracy of experiment tools, and sediments used in experiments. Comparison between the critical Shields parameters of the present study and those of Buffington and Montgomery [5] clearly showed that the critical Shields parameters of the present study were smaller than values related to the bed-load extrapolation method. However, these results were in satisfactory agreement with those of experiments based on the visual observation method. Results of this study support the claims of researchers who questioned the accuracy of the Shields diagram and showed that sediments start to move with smaller shear stress than that estimated by the Shields diagram.

Author Contributions: Conceptualization, R.S., H.A. and J.S.; methodology, R.S., H.A.; software, R.S.; validation, R.S., H.A. and J.S.; formal analysis, R.S. and H.A.; investigation, R.S., H.A. and J.S.; resources, R.S., H.A. and J.S.; data curation, R.S.; writing—original draft preparation, R.S.; writing—review and editing, R.S., H.A. and J.S.; visualization, R.S.; supervision, H.A. and J.S.; project administration, H.A. and J.S. All authors have read and agreed to the published version of the manuscript.

Funding: This research received no external funding.

Institutional Review Board Statement: Not applicable.

Informed Consent Statement: Not applicable.

Data Availability Statement: The data used in this manuscript are available by writing to the first author.

Conflicts of Interest: The authors declare no conflict of interest.

Notation

$-\overline{u'w'}$	Reynolds shear stress
τ_{oc}	Critical shear stress
U	Average velocity
U_{cr}	Threshold average velocity (under threshold condition)
u_{cr}	Threshold near-bed velocity (at the sediment particles level)
d_{16}	The size of which is finer than 16% of the sediment particles
d_{35}	The size of which is finer than 35% of the sediment particles
d_{50} or d	Sediment median grain size (the size of which is finer than 50% of the particles)
d_{86}	The size of which is finer than 86% of the sediment particles
D_{95}	The size of which is finer than 95% of the sediment particles
σ_g	Geometric standard deviation of sediment particles, $\sigma_g = (d_{84}/d_{16})^{0.5}$
g	Gravitational acceleration (981 cm^3/s)
v	Coefficient of water kinematic viscosity (equal to 0.01 cm^2/s)
ρ	Mass density of water (equal to 1.0 g/cm^3)
ρ_s	Mass density of sediment particles (equal to 2.65 g/cm^3)
S	Relative mass density of sediment particles ($S = \rho_s/\rho$, is equal to 2.65)
Δ	Submerged relative mass density of sediment particles ($\Delta = S - 1 = 1.65$ for quarts)
k_s	Nikuradse's equivalent roughness
R_*	Particle shear Reynolds number
Θ_c	Critical Shields parameter (non-dimensional critical shear stress)
ADV	Acoustic Doppler velocimeter
h	Water depth
h/d	Relative submergence of sediment particles
d/h	Relative roughness of sediment particles
B	Flume width (is equal to 90 cm)
B/h	Aspect ratio (the ratio between the flow width to the water depth)
Q	Flow rate (water discharge measured by electromagnetic flowmeter)
Re	Reynolds number
Fr	Froude number

References

1. Dey, S. *Fluvial Hydrodynamics: Hydrodynamic and Sediment Transport Phenomena*; Springer: Berlin/Heidelberg, Germany, 2014.
2. Wang, J.; Sui, J.; Karney, B.W. Incipient motion of non-cohesive sediment under ice cover-an experimental study. *J. Hydrodyn.* **2008**, *20*, 117–124. [CrossRef]
3. Sui, J.; Afzalimehr, H.; Kabiri Samani, A.; Maherani, M. Clear-water scour around semi-elliptical abutments with armored beds. *Int. J. Sediment Res.* **2010**, *25*, 233–245. [CrossRef]
4. Shahmohammadi, R.; Afzalimehr, H.; Sui, J. Impacts of turbulent flow over channel bed with vegetation patch on the incipient motion of sediment. *Can. J. Civ. Eng.* **2018**, *45*, 803–816. [CrossRef]
5. Buffington, J.M.; Montgomery, D.R. A systematic analysis of eight decades of incipient motion studies, with special reference to gravel-bedded rivers. *Water Resour. Res.* **1997**, *33*, 1993–2029. [CrossRef]
6. Diplas, P. Bed load transport in gravel-bed streams. *J. Hydraul. Eng.* **1987**, *113*, 277–292. [CrossRef]
7. Wilcock, P.R. Methods for estimating the critical shear stress of individual fractions in mixed-size sediment. *Water Resour. Res.* **1988**, *24*, 1127–1135. [CrossRef]
8. Abbott, J.E.; Francis, J.R.D. Saltation and suspension trajectories of solid grains in a water stream. *Philos. Trans. R. Soc. Lond. Ser. A Math. Phys. Sci.* **1977**, *284*, 225–254.
9. Kramer, H. Sand mixtures and sand movement in fluvial model. *Trans. Am. Soc. Civ. Eng.* **1935**, *100*, 798–838. [CrossRef]
10. Paintal, A. Concept of critical shear stress in loose boundary open channels. *J. Hydraul. Res.* **1971**, *9*, 91–113. [CrossRef]
11. Afzalimehr, H.; Dey, S.; Rasoulianfar, P. Influence of decelerating flow on incipient motion of gravel-bed streams. *Sadhana Acad. Proc. Eng. Sci.* **2007**, *32*, 545–559. [CrossRef]
12. Dey, S.; Sarkar, S.; Solari, L. Near-bed turbulence characteristics at the entrainment threshold of sediment beds. *J. Hydraul. Eng.* **2011**, *137*, 945–958. [CrossRef]
13. Bialik, R.J. Numerical study of near-bed turbulence structures influence on the initiation of saltating grains movement. *J. Hydrol. Hydromech.* **2013**, *61*, 202–207. [CrossRef]
14. Nikora, V.I.; Sukhodolov, A.N.; Rowinski, P.M. Statistical sand wave dynamics in one-directional water flows. *J. Fluid Mech.* **1997**, *351*, 17–39. [CrossRef]
15. Goncharov, V.N. *Dynamics of Channel Flow*; Israel Programme for Scientific Translation: Moscow, Russia, 1964.

16. Neill, C.R. Note on initial movement of coarse uniform bed-material. *J. Hydraul. Res.* **1968**, *6*, 173–176. [CrossRef]
17. Garde, R.J. Initiation of motion on a hydrodynamically rough surface, critical water velocity approach. *J. Irrig. Power* **1970**, *27*, 271–282.
18. Einstein, H.A. *The Bed-Load Function for Sediment Transportation in Open Channel Flows*; Technical bulletin number 1026; Department of Agriculture, Soil Conservation Service: Washington, DC, USA, 1950.
19. Velikanov, M.A. *Dynamics of Alluvial Stream*; State Publishing House of Theoretical and Technical Literature: Moscow, Russia, 1955; Volume 2.
20. Yalin, M.S. An expression for bed-load transportation. *J. Hydraul. Div.* **1963**, *89*, 221–250. [CrossRef]
21. Ling, C.H. Criteria for incipient motion of spherical sediment particles. *J. Hydraul. Eng.* **1995**, *121*, 472–478. [CrossRef]
22. Shields, A.F. *Application of Similarity Principles and Turbulence Research to Bed-Load Movement*; Mitteilungen der Preussischen Versuchsanstalt für Wasserbau und Schiffbau: Berlin, Germany, 1936; Volume 26, pp. 5–24.
23. Miller, M.C.; McCave, I.N.; Komar, P.D. Threshold of sediment motion under unidirectional currents. *Sedimentology* **1977**, *24*, 507–527. [CrossRef]
24. Paphitis, D. Sediment movement under unidirectional flows: An assessment of empirical threshold curves. *Coast Eng.* **2001**, *43*, 227–245. [CrossRef]
25. Dey, S.; Papanicolaou, A. Sediment threshold under stream flow: A state-of-the-art review. *KSCE J. Civ. Eng.* **2008**, *12*, 45–60. [CrossRef]
26. Emadzadeh, A.; Chiew, Y.M.; Afzalimehr, H. Effect of accelerating flow on incipient motion in a gravel-bed stream. *Adv. Water Resour.* **2010**, *33*, 1094–1104. [CrossRef]
27. Bolhassani, R.; Afzalimehr, H.; Dey, S. Effects of relative submergence and bed slope on sediment incipient motion under decelerating flows. *J. Hydrol. Hydromech.* **2015**, *63*, 295–302. [CrossRef]
28. García, M.H. *Sedimentation Engineering: Processes, Measurements, Modeling, and Practice. ASCE Manuals and Reports on Engineering Practice, no. 110*; ASCE: Reston, VA, USA, 2008.
29. Neill, C.R. Note on initial movement of coarse uniform material. *J. Hydraul. Res.* **1968**, *6*, 157–184.
30. Gessler, J. Beginning and ceasing of sediment motion. In *River Mechanics*; Shen, H.W., Ed.; Water Resources Publications: Littleton, CO, USA, 1971; Chapter 7.
31. Yalin, M.S.; Karahan, E. Inception of sediment transport. *J. Hydraul. Div.* **1979**, *105*, 1433–1443. [CrossRef]
32. White, C.M. The equilibrium of grains on the bed of a stream. *Philos. Trans. R. Soc. Lond.* **1940**, *174*, 322–338.
33. Iwagaki, Y. Fundamental study on critical tractive force. *Trans. Jpn. Soc. Civ. Eng.* **1956**, *41*, 1–21.
34. Wiberg, P.L.; Smith, J.D. Calculations of the critical shear stress for motion of uniform and heterogeneous sediments. *Water Resour. Res.* **1987**, *23*, 1471–1480.
35. Miller, R.L.; Byrne, R.J. The angle of repose for a single grain on a fixed rough bed. *Sedimentology* **1966**, *6*, 303–314. [CrossRef]
36. Dey, S. Sediment threshold. *Appl. Math. Model.* **1999**, *23*, 399–417. [CrossRef]
37. Nikora, V.I.; Goring, D.G.; Biggs, B.J.F. Silverstream eco-hydraulics flume: Hydraulic design and tests. *N. Z. J. Mar. Freshwater Res.* **1998**, *32*, 607–620. [CrossRef]
38. Goring, D.G.; Nikora, V.I. Despiking acoustic doppler velocimeter data. *J. Hydraul. Eng.* **2002**, *128*, 117–126. [CrossRef]
39. Wahl, T.L. Discussion of "Despiking acoustic doppler velocimeter data" by Derek G. Goring and Vladimir I. Nikora. *J. Hydraul. Eng.* **2003**, *129*, 484–487. [CrossRef]
40. Lane, E.W. Report of the subcommittee on sediment terminology. *Trans. Am. Geophys. Union* **1947**, *28*, 936–938. [CrossRef]
41. Török, G.T.; Józsa, J.; Baranya, S. A shear reynolds number-based classification method of the nonuniform bed load transport. *Water* **2019**, *11*, 73. [CrossRef]
42. Dey, S.; Sarkar, A. Scour downstream of an apron due to submerged horizontal jets. *J. Hydraul. Eng.* **2006**, *132*, 246–257. [CrossRef]
43. Kennedy, J.F. The Albert Shields story. *J. Hydraul. Eng.* **1995**, *121*, 766–772. [CrossRef]
44. Mantz, P.A. Incipient transport of fine grains are flakes by fluids—Extended Shields diagram. *J. Hydraul. Div.* **1977**, *103*, 601–615. [CrossRef]
45. Lavelle, J.W.; Mofjeld, H.O. Do critical stresses for incipient motion and erosion really exist? *J. Hydraul. Eng.* **1987**, *113*, 370–385. [CrossRef]
46. Wilcock, P.R.; Southard, J.B. Experimental study of incipient motion in mixed-size sediment. *Water Resour. Res.* **1988**, *24*, 1137–1151. [CrossRef]
47. Buffington, J.M. The legend of A.F. Shields. *J. Hydraul. Eng.* **1999**, *125*, 376–387. [CrossRef]
48. Parker, G.; Peterson, A.W. Bar resistance of gravel-bed streams. *J. Hydraul. Div.* **1980**, *106*, 1559–1575. [CrossRef]
49. Prestegaard, K.L. Bar resistance in gravel bed streams at bankfull stage. *Water Resour. Res.* **1983**, *19*, 473–476. [CrossRef]
50. Hey, R.D. Bar form resistance in gravel-bed rivers. *J. Hydraul. Eng.* **1988**, *114*, 1498–1508. [CrossRef]
51. Li, Z.; Komar, P.D. Laboratory measurements of pivoting angles for applications to selective entrainment of gravel in a current. *Sedimentology* **1986**, *33*, 413–423. [CrossRef]
52. Kirchner, J.W.; Dietrich, W.E.; Iseya, F.; Ikeda, H. The variability of critical shear stress, friction angle, and grain protrusion in water worked sediments. *Sedimentology* **1990**, *37*, 647–672. [CrossRef]
53. Buffington, J.M.; Dietrich, W.E.; Kirchner, J.W. Friction angle measurements on a naturally formed gravel streambed: Implications for critical boundary shear stress. *Water Resour. Res.* **1992**, *28*, 411–425. [CrossRef]

water

MDPI

Article

A Numerical Study of the Flow and Sediment Interaction in the Middle Reach of the Huai River

Jin Ni, Bangyi Yu and Peng Wu *,†

Key Laboratory of Water Conservancy and Water Resources in Anhui Province, Anhui & Huai River Institute of Hydraulic Research, Bengbu 233000, China; gpsnj@126.com (J.N.); yubangyi@vip.163.com (B.Y.)
* Correspondence: Peng.Wu@uregina.ca
† Current address: Environmental Systems Engineering, University of Regina, Regina, SK S4S 0A2, Canada.

Abstract: In recent years, the incoming sediments from upstream of the Huai River have continuously decreased. The relationship between flow and sediment has significantly changed. Therefore, the erosion and deposition characteristics of the river could be affected. To investigate this interaction between flow and sediment, the present study was conducted using the Wanglin section in the middle reach of the Huai River as the study site. A 1D hydrodynamic model was developed and validated using field data. Data from 1985–2014 were used as a continuous series while data from 2004–2014 were used as a repetitive series. The sediment variation and distribution processes at different locations were discussed. It was found that the river channel displayed several notable characteristics. In the flow direction, the channel had frontal erosion and backward deposition. The variation rate was relatively slow. With reduced sediment, the overall deposition at the Wanglin section was significantly mitigated. Future recommendations are provided based on the present simulation for flood mitigation along the Huai River.

Keywords: Huai River; river channel variation; erosion; deposition; numerical simulation

Citation: Ni, J.; Yu, B.; Wu, P. A Numerical Study of the Flow and Sediment Interaction in the Middle Reach of the Huai River. *Water* **2021**, *13*, 2041. https://doi.org/10.3390/w13152041

Academic Editor: Jueyi Sui

Received: 28 May 2021
Accepted: 22 July 2021
Published: 27 July 2021

1. Introduction

Predicting the response of natural or man-made channels to imposed supplies of flow and sediment is one of the most difficult practical problems faced by hydraulic engineers. The flow of the river moves the sediment and the channel bed is altered by erosion and the deposition of sediment. The shape of the river channel further modifies the flow. The interaction between flow and sediment is one of the primary dynamic factors in shaping riverbed channels. This complex interaction has been a continuous interest for river researchers and engineers. It plays an important role in determining erosion and deposition characteristics in natural river channels [1]. Much work has been conducted globally in the past to study the interaction between vegetation, flow and sediment [2–4], over grassed areas [5], and through numerical simulations [6,7]. However, few studies can be found which investigate the interaction of flow and sediment in China. A research gap was noted. In recent years, significant changes have been noted in flow and sediment interaction due to the combined effects of climate change and human activities. Some changes are noted along the middle reach of Yangtze River and the estuary of the Yellow River. The riverine ecosystems have been affected by the changes in the flow and sediment relationship. Additionally, as a major river in China, the Huai River is also notoriously vulnerable to flooding. The interaction between flow and sediment along the Huai River has been significantly changed. A significant reduction in sediment has been noted in the middle reach of the Huai River [8]. The main reasons for the reduction in sediment include: the construction of hydraulics structures, such as dams and sluices; soil conservation practice; and high-intensity sand mining. As shown in Figure 1, the average annual sediment runoff in the mainstream station of Lutaizi has decreased from 18 million tons in the 1950s to 3 million tons in 2000. However, the flow rate has remained relatively constant over the years. The decrease in sediment can have a significant impact to river channel development in the Huai River, which has received wide attention. Yu and Yu et al. [9–11] analyzed the erosion and deposition characteristics of the mainstream channel based on field data. It was noted that the deposition from Wangjiaba to Lutaizi was reduced, while from Lutaizi to Bengbu, the sediment movement changed from deposition to erosion. Between Bengbu and Fushan, further erosion was noted. Liu and Yang et al. [12,13], used topographic data of river cross-sections in the mainstream of the Huai River gathered over the past 70 years. It was found that the river channel erosion had been significantly increased due to sediment mining activities, which caused an uneven channel cutting in many sections of the Huai River. Moreover, with a deep cut and steep slope, river banks have a high potential of collapse, which poses a significant threat to flood control, navigation and shoreline utilizations. The behavior of sediment transport could potentially increase the flooding risk and pose a significant threat to public safety [14]. From the 1990s onwards, flood disasters in the Huai River have shown an increasing trend [15]. Investigation into the flow and water interaction in the Huai River is thus urgently needed.

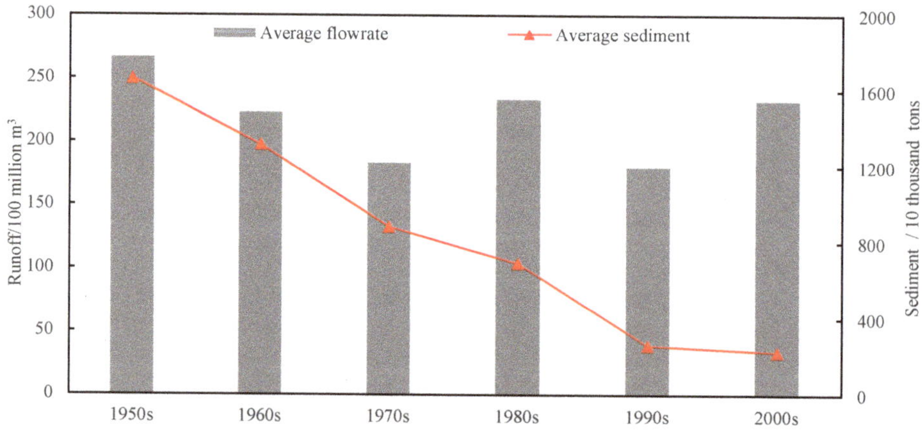

Figure 1. Changes of hydro-sediment in the mainstream of Huai River at Lutaizi station.

Previous studies focused on the relationship between flow and sediment along the Huai River are mainly based on field data analysis or the temporal evolution of sediment movement. It is difficult to predict the long-term equilibrium sediment variation considering the many influencing parameters. The detailed characteristics of the river channel variation are not well understood. To date, many numerical models of river sedimentation have been developed. Some of the models, such as Deflt3D and EFDC, have been well applied to rivers. However, none of these models have been applied to the middle reach of the Huai River. It is also noted that in many cases, it is not reasonable to apply a 3D model to a river with a width of several hundred channels and very limited cross-sections. A 1D flow model is more appropriate [16]. In the present study, a 1D hydrodynamic model was used to investigate the interaction of flow and sediment along the Huai River using scenario analysis. The impacts of flow variation on sediment were numerical simulated, which could provide a direct insight to allow decision makers to better manage the Huai River.

2. Study Site

The middle reach of the Huai River, from Wangjiaba to Linhuagang (Wanglin section), was used as the study site for this simulation. This section is located at the transition zone between the upstream and middle reach of the river, in a region comprising mountains and plains. A higher concentration of sediment was noted in this region due to the changes in the river channel slope and velocity [13]. Additionally, a basin-wide flood occurred in this section in July 2020, with overland flooding reaching heights of 0.22–0.46 m. The flooding had a significant impact on the local economy and communities. The current flow capacity from Wangjiaba to Nanzhaoji is around 6000 m^3/s, which is below the designed flow rate of 7400 m^3/s. Urgent action is required to investigate the relationship between flow and sediment in order to mitigate flooding risk.

The location of the study site can be found in Figure 2. In total, the Wanglin section is 110.98 km in length, with high elevations on both sides and low elevations in the middle. The tributaries include the Hong River, the Hong River Flood Diversion Channel, the Gu River, the Run River, the Bailu River and the Shiguan River. Most of these branches are relatively short in length, but high in velocities. Both the Hong River and the Shiguan River have large watersheds compared to the other branches. In the study area, the section between Wangjiaba to Nanzhaoji is treated as a flooding basin, as shown in Figure 2. From Nanzhaoji to Linhuagang, the river channel is relatively straight. Linhuagang station is the largest water control station along the middle reach of the Huai River.

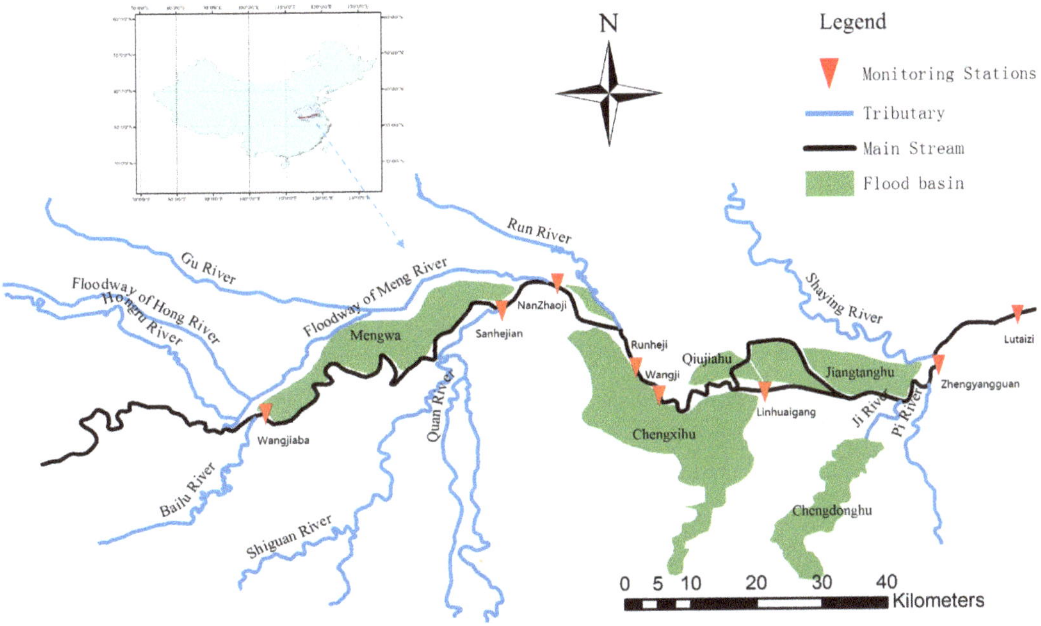

Figure 2. The branches of the Huai River and flood basins in the present study.

To understand the relationship between flow and incoming sediment, data from between 1985 and 2014 were collected from the following stations: Wangjiaba, Shangang, Dili and Jiangjiaji. It was found that the multiyear average flow rate at Wangjiaba is 318 m^3/s, with 2.99 million tons of sediment. The flood season is from June to September, which accounts for 61% of the annual flow and 75% of the annual sediment [17,18]. To further present the relationship between flow rate and sediment yield in the study site, the K-mean method was used, which can be seen in Table 1. A relationship between flow rate and sediment can be observed. The rich/medium/poor vs. rich/medium/poor is used to indicate the relationship between flow rate and sediment. A high flow rate normally has a higher sediment yield. The numerical study was conducted in order to understand the relationship between flow rate and sediment in the study site.

Table 1. The flow and sediment at Wanglin section from 1985–2014.

Year	Annual Flow Rate/m^3·s^{-1}	Annual Sediment/10,000 Tons	Flow–Sediment Type	Year	Annual Flow Rate/m^3·s^{-1}	Annual Sediment/10,000 Tons	Flow–Sediment Type
1985	323.25	349.53	both medium	2000	512.11	525.85	rich–medium
1986	190.64	256.21	poor–medium	2001	117.10	34.15	both poor
1987	613.71	1028.33	both high	2002	359.43	314.32	both medium
1988	149.86	196.76	both poor	2003	768.90	477.67	rich–medium
1989	402.21	700.72	medium–rich	2004	290.81	184.51	medium–poor
1990	300.77	281.38	both medium	2005	637.24	759.56	both rich
1991	670.35	939.31	both rich	2006	518.74	167.98	rich–poor
1992	125.03	69.84	both poor	2007	419.33	391.54	both medium
1993	206.02	119.25	both poor	2008	376.84	354.99	both medium
1994	140.06	66.49	both poor	2009	176.53	98.30	both poor
1995	167.90	301.27	poor–medium	2010	312.79	150.73	medium–poor
1996	478.98	419.31	rich–medium	2011	81.62	12.36	both poor
1997	185.38	112.77	both poor	2012	104.55	76.97	both poor
1998	543.51	408.26	rich–medium	2013	85.20	26.59	both poor
1999	76.71	21.44	both poor	2014	206.59	109.60	both poor

3. Numerical Models

3.1. Governing Equations

Many branches have been noted at the Wanglin section, as seen in Figure 2. The unsteady flow model was selected, as it achieved better accuracy in the flooding simulation and sediment calculation. Additionally, the flow in this section was dominated by suspended sediment transport. Therefore, the unsteady flow and sediment model was applied. The governing equations can be found below.

$$\text{Continuity equation:} \frac{\partial Q}{\partial x} + B\frac{\partial Z}{\partial t} = q \tag{1}$$

$$\text{Momentum equation:} \frac{\partial Q}{\partial t} + \frac{\partial}{\partial x}\left(\frac{\alpha Q^2}{A}\right) + gA\frac{\partial Z}{\partial x} + \frac{g}{A}\frac{n^2 Q|Q|}{R^{4/3}} = qV_x \tag{2}$$

$$\text{Suspended sediment advection equation:} \frac{\partial (AS_k)}{\partial t} + \frac{\partial (QS_k)}{\partial x} = \alpha_s \omega_k B(S_k^* - S_k) \tag{3}$$

$$\text{Riverbed deformation equation:} \gamma' \frac{\partial A_d}{\partial t} = \sum_{k=1}^{N} \alpha_s \omega_k B(S_k^* - S_k) \tag{4}$$

$$\text{Sediment transport formula:} S_k^* = k_s\left(\frac{V^3}{gh\omega_k}\right)^m \tag{5}$$

$$\text{Sediment transport equilibrium equation: } \sum_{i=1}^{L(n)} Q_i S_{i,k} = \sum_{j=1}^{M(n)} Q_j S_{j,k} \tag{6}$$

where Q is the flow rate (m^3/s); Z is the water level (m); B is the water surface width (m); x is the flow path (m); t is the time (s); q is the side inflow flow rate, positive for inflow and negative for outflow (m^3/s); A is the cross-sectional area (m^2); n is roughness factor; α is the coefficient of momentum correction; V_x is the component of side inlet velocity on the mainstream (m/s); S_k and S_k^* are the sediment concentration and sediment holding capacity of the group k particle size (kg/m^3); ω_k is settling velocity of the group k particle size(m/s);

k_s and m are calibration parameters of sediment transport formula—gain and exponent; a_s. is coefficient of saturation recovery; γ' is the dry capacity of sediment (kg/m^3); A_d. is the area of deposition (m^2); $L(n)$ and $M(n)$ are the total numbers of inlet and outlet reaches connected with branching point; and n. $S_{i,k}$. and $S_{j,k}$ are the sediment concentration of the group k sediment grain size (kg/m^3).

The model uses Visual C#2013 as the platform. The SQL Server 2012 is used as the database. The two modules used are a flow calculation module and sediment transport module. The two modules are connected using a non-coupling approach. The flow is calculated first to provide hydraulic parameters for the sediment module. The discrete solution of each module can be found in the references [19–23].

3.2. Model Calibration and Validation

The model was calibrated using field data from Honghekou to Lutaizi in the mainstream of the Huai River, which covers the study area of the Wanglin section. The flow rate, water level and sediment concentration data from 1999 to 2008 were used. In the present study, the aerial survey map of 1999 was adopted for model calculation.

The application of erosion and deposition in the reach of the mainstream of the Huai River below Bengbu showed that the values of some parameters in the model were relatively stable. The sediment carrying capacity index m had a value of around 0.92; the average value of a_s is between 0.5 and 0.6. Empirical values were used for such parameters to reduce uncertainty. From the debugging and calculation, the roughness of the main channel was 0.0215 and the roughness of the bank regions was 0.034. The sediment carrying capacity coefficient ks changed with the erosion and deposition of the river channel. When sand content S was near the sand carrying capacity S^*, a value of ks = 0.025 was used. Under other conditions, $k_s = 0.025\Delta k$, in which Δk is the adjustment coefficient, which can be expressed as follows:

$$k_s = 0.025 \times \begin{pmatrix} 1+(\frac{S}{S^*}-1.25)\times 0.6 & \frac{S}{S^*} > 1.25 \\ 1 & 1.25 \geq \frac{S}{S^*} \geq 0.75 \\ 1-(0.75-\frac{S}{S^*})\times 0.6 & \frac{S}{S^*} < 0.75 \end{pmatrix} \tag{7}$$

The perpendicular method was used to allocate the erosion and deposition area, which can be seen below [20].

$$\Delta B_i \frac{\delta Z_i}{\delta A} = \frac{\eta_i^c \Delta B_i}{\sum\limits_{i=1}^{n} \eta_i^c \Delta B_i} \tag{8}$$

In which $\delta Z_i \Delta B_i$ are the thickness and width of erosion and deposition in the perpendicular direction; δA is the calculated erosion and deposition are of the cross-section; η_i is the perpendicular flow depth; and c is the adjustment index, which equals to 1.5 for scouring and 0.5 for deposition.

The comparison between the field data and simulation can be found in Table 2. It should be noted that for both calibration and validation, the difference between the field data and simulation is below 15%, with the exception of 2001. The 2005 data are used in Figure 3 to demonstrate the simulation results. A good agreement was reached between the simulation and field data regarding both water level and flow rate. The present model was then applied to the Wanglin section for erosion and deposition simulation.

Table 2. Comparison of annual sediment transport rate at Lutaizi station (unit: 10,000 t/year).

Calibration					Validation				
Year	Measured	Simulation	Absolute Error	Relative Error/%	Year	Measured	Simulation	Absolute Error	Relative Error/%
1999	48.6	57.6	9.0	18.4	2004	248.7	276.8	28.1	11.3
2000	509.0	565.9	56.8	11.2	2005	340.4	385.8	45.4	13.3
2001	23.2	19.0	−4.2	−18.1	2006	129.6	108.3	−21.3	−16.4
2002	191.1	212.5	21.4	11.2	2007	235.1	269.2	34.1	14.5
2003	449.8	485.1	35.3	7.8	2008	235.0	266.9	32.0	13.6

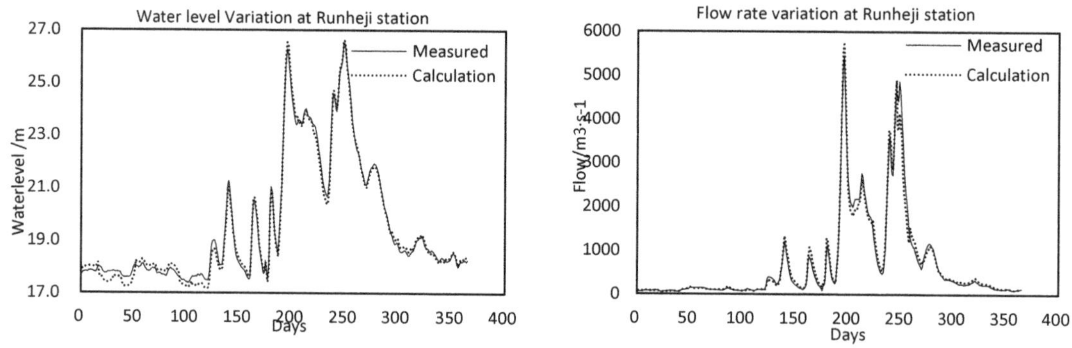

Figure 3. Comparison of water level and flow process at Run River station in 2005.

3.3. Boundary Conditions

The computational boundary conditions were determined using a typical flow and sediment relationship for the analysis of long-term river channel variation. It was noted that the incoming sediment to the Huai River continuously decreased from 1950 onwards. After 2000, this decreasing trend stabilized. It was further analyzed, using statistical analysis, that the incoming sediment to the Huai River would be low in the next few years [24].

In the present study, the following two scenarios of flow and sediment transport rate were conducted, as shown in Table 3. Scenario 1 was a continuous series considering the flow–sediment interaction from 1985 to 2014. In this scenario, 30 years of data were used to investigate the impacts of flow–sediment interaction to river channel morphology. Scenario 1 can be used as a baseline study. Scenario 2 was a repetitive series using field data from 2000 to 2014. The collected data were repeated once to form a 30-year flow–sediment interaction. This set of data were based on the field data which showed a good indication of current situations in the Huai River, with a 17% reduction in the incoming sediment. Using this scenario, the future river channel morphology can be further predicted.

Table 3. Statistics of hydro-sediment volume in the river under different flow–sediment scenarios.

Name	Series Composition	Runoff Flow Rate		Sediment	
		Annual Flow Rate/$m^3 \cdot s^{-1}$	Percentage of Flood Season/%	Annual Average Value/10,000 Tons	Percentage of Flood Season/%
Continuous series	1985–2014	318.0	61.7	298.5	76.1
Repetitive series	2000–2014	331.1	62.8	245.7	78.7

The modeling region from Wangjiaba to Linhuaigang is shown in Figure 1. From Wangji to the Meng River station, the initial topography was constructed based on the most recent measurement from 2018 with cross-sections 500 m apart. From Wangji to the Linhuagang station, the cross-section measurements from 2011 300 m apart were used. The proposed two scenarios were used regarding upstream boundary conditions, while downstream boundary condition can be analyzed using the stage-discharge relation curve at Linhuaigang station. The sediment diameter was selected from a field measurement in 2008, with d_{50} equal to 0.04 mm.

To ensure the stability of calculation, 300 s was used as time step. For each time step, daily data were used to calculate the flow and sediment boundary. It is noted that using the time period of 30 years was very time consuming in regard to the calculation. A threshold was used for erosion and deposition. The file was only updated when the cumulative elevation change exceeded the threshold. In the present research, a threshold of 0.05 m was used.

4. Results and Analysis

4.1. Erosion and Deposition Process

The sediment deposition under the simulated scenarios was numerically simulated. From the simulation, it was found that the deposition at the Wanglin station could be significantly reduced at Scenario 2, as shown in Figure 4. At the end of 10 years, the sediment deposition under Scenario 2 was 2.01 million m^3, which is 63.7% less compared to that of Scenario 1. At the end of the 30-year calculation period, the sediment deposited at the Wanglin station was calculated to be 6.19 million m^3 under Scenario 2, which is 34.6% less compared to that of Scenario 1. The main difference between the two scenarios is in the first 15 years. Due to the increased incoming sediment from 1987 to 1991, the sediment concentration was accumulated rapidly in Scenario 1. The largest difference was regarding the value at the end of 12 years, with an amount of 4.23 million m^3.

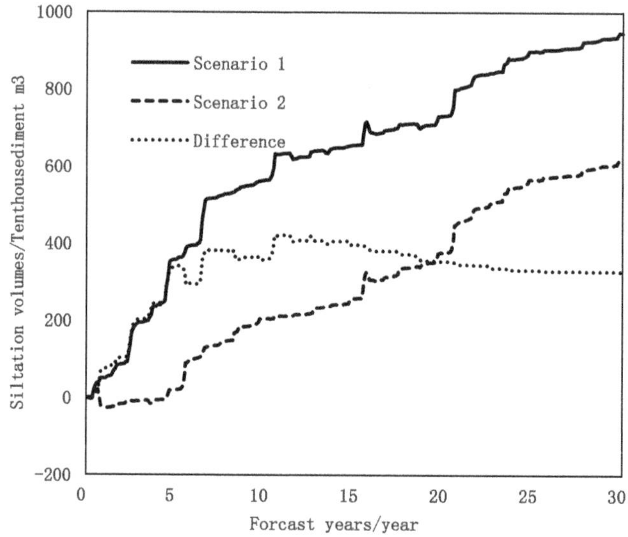

Figure 4. Cumulative deposition process in Wanglin section under two flow–sediment series.

From the simulation, Wanglin station shows a continuous accumulation of sediment under both scenarios. After 30 years, no equilibrium state was reached. It is extremely hard to transform the river channel to stabilized conditions considering the many human activities undertaken in this section. Sediment mining should be prohibited at this section. It may take hundreds of years for the river channel to reach an equilibrium state under natural flow conditions.

Table 4 is used to show the relationship between flow rate and sediment concentration in the past 30 years. The relationships between deposition and erosion at different years are presented using the maximum value and average value. It is found that a higher flow rate can result in a relative larger deposition value. Overall, the erosion is more favorable at a larger flow rate and lower incoming sediment, which can avoid deposition. From the table, the average ratio of incoming sediment to transported sediment is around 0.97 for Data Number 2, which indicates a state of relative equilibrium. In certain years—for example, Data Number 5 in Table 4—with limited flow and large amounts of incoming sediment, the sediment deposition could become more obvious. In the repetitive series, it may take a shorter period of time to reach a state of equilibrium.

Table 4. Deposition and sediment delivery ratio of different flow–sediment year types.

Flow–Sediment Type	Deposition Volume /10,000 $m^3 \cdot year^{-1}$			Incoming/Outgoing Sediment Ratio		
	Maximum	Minimum	Average	Maximum	Minimum	Average
both rich	122.04	73.23	101.09	0.84	0.79	0.82
rich- medium	33.23	−9.04	7.07	1.04	0.90	0.97
both medium	50.1	7.87	28.49	0.97	0.76	0.86
Medium–poor	24.22	4.16	14.19	0.95	0.78	0.86
Poor–medium	69.7	36.19	52.94	0.77	0.62	0.69
both poor	45.12	4.48	15.54	0.79	0.40	0.67

4.2. Deposition Distribution

The simulated deposition at the end of the 30 years is presented in Table 5. From the numerical modeling, it can be found that the section from Wangjiaba to Sanhejian is dominated by erosion, with an amount of 1.05 million m^3. The section from Runheji to Wangji shows an equilibrium state. The other sections in the middle reach of the Huai River have a major deposition with an amount of 7.01 million m^3 of sediment, which accounts for 74% of the total deposition. In the transverse direction, all the sections have erosion in the main channel and deposition along the bank. The average erosion depth is around 30 cm, while the deposition height is 15 cm. In Scenario 2, with repetitive series, the main reach of the Huai River shows erosion in the upper reach and deposition in the downstream of Sanhejian station. Further analysis indicates that the upstream has a larger reduction in deposition, especially from Wangjiaba station to Sanhejian station. The middle reach of the Huai River shows dominated erosion behavior in the main channel under the continuous series of Scenario 1. Although the changes are not significant, the impacts of incoming sediment are very clear. Overall, the main reach of the Huai River shows increasing erosion and decreasing deposition from the simulation.

Table 5. Distribution of river erosion and deposition at the end of the simulated scenarios (unit: 10,000 m^3).

River Section		Continuous Series			Repetitive Series		
		Main Channel	Beach	Full Cross-Section	Main Channel	Beach	Full Cross-Section
The mainstream of Huai River	Wangjiaba–Sanhejian	−301.2	196	−105.2	−390.7	194.4	−196.3
	Sanhejian–Nanzhaoji	14	124.9	138.8	2.6	120.9	123.5
	Nanzhaoji–Runheji	86.3	108.4	194.7	48.4	86.5	134.9
	Runheji–Wangji	−36.3	54.3	18	−53.5	51.8	−1.7
	Wangji–Linhuagang	64	291.1	355.1	36.1	272.8	308.9
Floodway	Meng River Floodway	7.4	338.1	345.5	4.8	244.9	249.7
Total		−165.9	1112.7	946.8	−352.2	971.2	619

4.3. Recommendations

The recommendations for flooding mitigation in the middle reach of the Huai River are presented. The Wanglin section is used as the test site. For flooding mitigation, it is important to maintain stabilization of the bank. From the simulation, it is found that the overall trend of the Huai River is erosion. Some recommendations are provided, as follows:

(1) Main channel dredging

The simulation indicates that the Wanglin section will be dominated by erosion. However, the main channel in this section is around 250 m in width. The flow capacity in the main channel is around 1500 m^3/s, which is less than 1/5 of the bankfull value. The main channel is relatively small, which presents a larger threat to overland flooding. Dredging the main channel is thus recommended. The present study indicates that dredging the channel is preferable to widening the main channel.

(2) Floodplain protection

The beach bank along this section is close to 1500 m in length and has many functions, including water drainage, water storage, and sediment deposition. From the simulation, less incoming sediment results in a reduction in sediment deposition, which may impact the stability of the banks. The recommendation is to construct bank protection to mitigate the threat of bank failure. Additionally, sediment mining should be well-managed to avoid the potential impacts on bank stability.

5. Conclusions

The present study was conducted to investigate the flow rate and sediment relationship in the middle reach of the Huai River. The Wanglin section was used as an example for numerical simulation. The model is well validated and calibrated. In total, 30 years of data were used for modeling sediment behavior. Due to the special geophysical characteristics of the Wanglin section, the overall trend in this section is erosion in the main channel. Two scenarios were used to investigate the flow–sediment behavior at the Wanglin section. It was found that in the flow–sediment type of rich to medium, an equilibrium sediment transport could be reached. Based on the simulation results, recommendations were provided for flood mitigation in this section, including main channel dredging and bank protection.

Author Contributions: Conceptualization, J.N., B.Y., P.W.; methodology, J.N.; writing—original draft preparation, J.N.; writing—review and editing, P.W. and J.N.; supervision, B.Y. All authors have read and agreed to the published version of the manuscript.

Funding: This research was funded by the National Key Research and Development Program of China, grant number 2017YFC0405602 and the Hydraulic Science and Technology Program of Anhui Province, grant number SLKJ202001-07, and NSERC Discovery Grant.

Institutional Review Board Statement: Not applicable.

Data Availability Statement: Not applicable.

Acknowledgments: The authors are thankful for the financial support provided by the National Key Research and Development Program of China (2017YFC0405602), the Hydraulic Science and Technology Program of Anhui Province (SLKJ202001-07), and NSERC Discovery Grant.

Conflicts of Interest: The authors declare no conflict of interest.

References

1. Nelson, J.M.; James; Bennett, P.; Stephen, W.M. Flow and sediment–transport modeling. *Tools Fluv. Geomorphol.* **2003**, 539–576. [CrossRef]
2. Chao, W.; Zheng, S.; Wang, P.; Hou, J. Interactions between vegetation, water flow and sediment transport: A review. *J. Hydrodyn.* **2015**, *27*, 24–37.
3. Nowacki, D.J.; Andrea, S.O. Water and sediment transport of channel-flat systems in a mesotidal mudflat: Willapa Bay, Washington. *Cont. Shelf Res.* **2013**, *60*, S111–S124. [CrossRef]
4. Restrepo, J.D.; Kjerfve, B. The Pacific and Caribbean rivers of Colombia: Water discharge, sediment transport and dissolved loads. In *Environmental Geochemistry in Tropical and Subtropical Environments*; Springer: Berlin/Heidelberg, Germany, 2004; pp. 169–187.
5. Deletic, A. Modelling of water and sediment transport over grassed areas. *J. Hydrol.* **2001**, *248*, 168–182. [CrossRef]
6. Papanicolaou, N.A.T.; Elhakeem, M.; Krallis, G.; Prakash, S.; Edinger, J. Sediment transport modeling review—current and future developments. *J. Hydraul. Eng.* **2008**, *134*, 1–14. [CrossRef]
7. Simpson, G.; Sébastien, C. Coupled model of surface water flow, sediment transport and morphological evolution. *Comput. Geosci.* **2006**, *32*, 1600–1614. [CrossRef]
8. Hongling, S.; Chunhong, H.; Yangui, W.; Qingqi, T. Variation trend and cause of runoff and sediment load variation in Huaihe River. *J. Hydraul. Eng.* **2012**, *43*, 71–579. (In Chinese)
9. Bangyi, Y.U.; Yusuo, Y.U.; Jin, N.I. Analysis on sediment erosion and deposition of Huaihe River main Stream from WuJiadu to XiaoLiuxiang. *J. Sediment Res.* **2009**, *4*, 12–16. (In Chinese)
10. Bangyi, Y.U.; Yusuo, Y.U. Analysis on sediment transport balance of Huaihe River main stream from Lutaizi to Wujiadu. *J. Nanchang Univ. (Eng. Technol. Ed.)* **2009**, *31*, 162–166. (In Chinese)
11. Yusuo, Y.U.; Bangyi, Y.U. Analysis on sediment transport balance of Huaihe River main stream from Wangjiaba to Lutaiji. *Jianghuai Water Resour. Sci. Technol.* **2009**, *2*, 41–42. (In Chinese)
12. Yunian, L.I.U.; Huasong, H.E.; Bangyi, Y.U. *Study on Characteristics and Regulation of Middle Reaches of Huaihe River*; Water & Power Press: Beijing, China, 2012. (In Chinese)
13. Xingju, Y.; Bangyi, Y.; Jin, N. Analysis on recent evolution of main stream of reach from Bengbu to Fushan of Huaihe River. *Water Resour. Hydropower Eng.* **2010**, *41*, 70–72+86. (In Chinese)

14. Donglai, L.; Jingming, H.; Xinhong, W.; Jin, Q. Influence of morphological variation on flood propagation evaluated by a hydro-morphological model. *J. Sediment Res.* **2018**, *43*, 13–20.
15. Ping, Z.; Jun, X.; Lei, Z.; Xieyi, M. An Analysis of Temporal and Spatial Variation of Rainfall in Huaihe Watershed above Bengbu in Recent 50 Years. *Hydropower Water Resour. Rural. China* **2017**, *3*, 1–8. (In Chinese)
16. Nelson, J.M.; Bennett, J.P.; Wiele, S.M. Flow and sediment-transport modeling. In *Tools in Fluvial Geomorphology*; Wiley: Hoboken, NJ, USA, 2003; pp. 539–576.
17. Jing, Y.; Quanxi, X.; Bingjiang, D. Differece Between sediment-transport balance and topographic change methods and applicability in the Three Gorges reservior. *J. China Hydrol.* **2011**, *31*, 87–91. (In Chinese)
18. Yanqun, Y.; Xiaomei, Y. River regulation from Wangjiaba to Linhuaigang in Huaihe River. *Water Resour. Planning Des.* **2013**, *11*, 62–65. (In Chinese)
19. James, G.; Witten, D.; Hastie, T.; Tibshirani, R. *An Introduction to Statistical Learning: With Application in R*; Springer: Berlin/Heidelberg, Germany, 2013.
20. Feldman, A.D. Hec Models for Water Resources System Simulation: Theory and Experience. *Adv. Hydrosci.* **1981**, *12*, 297–423.
21. Zhilin, W.; Lianggui, Z.; Xiaoping, F. Research on mathematical model for sediment in Yellow River. *J. Wuhan Univ. Hydraul. Electr. Eng.* **1997**, *5*, 22–26. (In Chinese)
22. Zhou, J.; Lin, B. One-Dimensional Mathematical Model for Suspended Sediment by Lateral Integration. *J. Hydraul. Eng.* **1998**, *124*, 712–717. [CrossRef]
23. Qiwei, H. *Non-Equilibrium Transport of Non-Uniform Sediment Load*; Science Press: Beijing, China, 2014. (In Chinese)
24. Hui, Z.; Bangyi, Y.; Jin, N. Susceptibility and periodicity of water and sediment characteristics of the Huaihe River in recently 66 Years. *J. Yangtze River Sci. Res. Inst.* **2020**, *37*, 6–11. (In Chinese)

MDPI
St. Alban-Anlage 66
4052 Basel
Switzerland
Tel. +41 61 683 77 34
Fax +41 61 302 89 18
www.mdpi.com

Water Editorial Office
E-mail: water@mdpi.com
www.mdpi.com/journal/water

www.ingramcontent.com/pod-product-compliance
Lightning Source LLC
Chambersburg PA
CBHW040836250926
51850CB00007BA/1334
9783036573113